Integral Geometry and Radon Transforms

Sigurdur Helgason

Integral Geometry
and Radon Transforms

 Springer

Sigurdur Helgason
Department of Mathematics
Massachusetts Institute of Technology
Cambridge, MA 02139
USA
helgason@mit.edu

ISBN 978-1-4899-9420-2   ISBN 978-1-4419-6055-9 (eBook)
DOI 10.1007/978-1-4419-6055-9
Springer New York Dordrecht Heidelberg London

Mathematics Subject Classification (2010): 53C65, 44A12

Springer is part of Springer Science+Business Media (www.springer.com)

TO ARTIE

Skalat maðr rúnir rista
nema ráða vel kunni
þat verðr mörgum manni
es of myrkvan staf villisk.

Egils Saga Ch. 73, (ca. 1230).

Preface

This book deals with a special subject in the wide field of Geometric Analysis. The subject has its origins in results by Funk [1913] and Radon [1917] determining, respectively, a symmetric function on the two-sphere S^2 from its great circle integrals and an integrable function on R^2 from its straight line integrals. (See References.) The first of these is related to a geometric theorem of Minkowski [1911] (see Ch. III, §1).

While the above work of Funk and Radon lay dormant for a while, Fritz John revived the subject in important papers during the thirties and found significant applications to differential equations. More recent applications to X-ray technology and tomography have widened interest in the subject.

This book originated with lectures given at MIT in the Fall of 1966, based mostly on my papers during 1959–1965 on the Radon transform and its generalizations. The viewpoint of these generalizations is the following.

The set of points on S^2 and the set of great circles on S^2 are both acted on transitively by the group $O(3)$. Similarly, the set of points in R^2 and the set P^2 of lines in R^2 are both homogeneous spaces of the group $M(2)$ of rigid motions of R^2. This motivates our general Radon transform definition from [1965a] and [1966a], which forms the framework of Chapter II:

Given two homogeneous spaces $X = G/K$ and $\Xi = G/H$ of the same group G, two elements $x = gK$ and $\xi = \gamma H$ are said to be *incident* (denoted $x\#\xi$) if $gK \cap \gamma H \neq \emptyset$ (as subsets of G). We then define the *abstract Radon transform* $f \to \widehat{f}$ from $C_c(X)$ to $C(\Xi)$ and the *dual transform* $\varphi \to \check{\varphi}$ from $C_c(\Xi)$ to $C(X)$ by

$$\widehat{f}(\xi) = \int\limits_{x\#\xi} f(x)\, dm(x), \qquad \check{\varphi}(x) = \int\limits_{\xi\#x} \varphi(\xi)\, d\mu(\xi)$$

with canonical measures dm and $d\mu$. These geometrically dual operators $f \to \widehat{f}$ and $\varphi \to \check{\varphi}$ are also adjoint operators relative to the G-invariant measures dg_K, dg_H on G/K and G/H.

In the example R^2, one takes $G = M(2)$ and K the subgroup $O(2)$ fixing the origin x_o and H the subgroup mapping a line ξ_o into itself. Thus we have

$$X = G/K = R^2, \qquad \Xi = G/H = P^2$$

and here it turns out $x \in X$ is incident to $\xi \in \Xi$ if and only if their distance equals the distance p between x_o and ξ_o. It is important not just to consider the case $p = 0$. Also the abstract definition does not require the members of Ξ to be subsets of X. Some natural questions arise for the operators $f \to \widehat{f}$, $\varphi \to \check{\varphi}$, namely:

(i)   Injectivity

(ii)   Inversion formulas

(iii)   Ranges and kernels for specific function spaces on X and on Ξ

(iv)   Support problems (does $\widehat{f}$ of compact support imply f of compact support?)

We investigate these problems for a variety of examples, mainly in Chapter II. Interesting analogies and differences appear. One such instance is when the classical Poisson integral for the unit disk turns out to be a certain Radon transform and offers wide ranging analogies with the X-ray transform in $\mathbf{R}^3$. See Table II.1 in Chapter II, §4.

In the abstract framework indicated above, a specific result for a single example automatically raises a host of conjectures.

The problems above are to a large extent solved for the X-ray transform and for the horocycle transform on Riemannian symmetric spaces. When G/K is a Euclidean space (respectively, a Riemannian symmetric space) and G/H the space of hyperplanes (respectively, the space of horocycles) the transform $f \to \widehat{f}$ has applications to certain differential equations. If L is a natural differential operator on G/K, the map $f \to \widehat{f}$ transfers it into a more manageable operator $\widehat{L}$ on G/H by the relation

$$(Lf)\widehat{} = \widehat{L}\widehat{f}.$$

Then the support theorem

$$\widehat{f} \text{ compact support} \quad \Rightarrow f \text{ compact support}$$

implies the existence theorem $L\mathcal{C}^\infty(G/K) = \mathcal{C}^\infty(G/K)$ for G-invariant differential operators L on G/K.

On the other hand, the applications of the original Radon transform on $\mathbf{R}^2$ to X-ray technology and tomography are based on the fact that for an unknown density f, X-ray attenuation measurements give $\widehat{f}$ directly and thus yield f itself via Radon's inversion formula. More precisely, let B be a planar convex body, $f(x)$ its density at the point x, and suppose a thin beam of X-rays is directed at B along a line ξ. Then, as observed by Cormack, the line integral $\widehat{f}(\xi)$ of f along ξ equals $\log(I_0/I)$ where I_0 and I, respectively, are the intensities of the beam before hitting B and after leaving B. Thus while f is at first unknown, the function $\widehat{f}$ (and thus f) is determined by the X-ray data. See Ch. I, §7,B. This work, initiated by Cormack and Hounsfield and earning them a Nobel Prize, has greatly increased interest in Radon transform theory. The support theorem brings in a certain refinement that the density $f(x)$ outside a convex set C can be determined by only using X-rays that do not enter C. See Ch. I, §7, B.

This book includes and recasts some material from my earlier book, "The Radon Transform", Birkhäuser (1999). It has a large number of new examples of Radon transforms, has an extended treatment of the Radon transform on constant curvature spaces, and contains full proofs for the antipodal Radon transform on compact two-point homogeneous spaces. The X-ray transform on symmetric spaces is treated in detail with explicit inversion formulas.

In order to make the book self-contained we have added three chapters at the end of the book. Chapter VII treats Fourier transforms and distributions, relying heavily on the concise treatment in Hörmander's books. We call particular attention to his profound Theorem 4.9, which in spite of its importance does not seem to have generally entered distribution theory books. We have found this result essential in our study [1994b] of the Radon transform on a symmetric space. Chapter VIII contains a short treatment of basic Lie group theory assuming only minimal familiarity with the concept of a manifold. Chapter IX is a short exposition of the basics of the theory of Cartan's symmetric spaces. Most chapters end with some Exercises and Further Results with explicit references.

Although the Bibliography is fairly extensive no completeness is attempted. In view of the rapid development of the subject the Bibliographical Notes can not be up to date. In these notes and in the text my books [1978] and [1984] and [1994b] are abbreviated to **DS** and **GGA** and **GSS**.

I am indebted to T.H. Danielsen, S. Jensen and J. Orloff for critical reading of parts of the manuscript, to R. Melrose and R. Seeley for suggestions, to F. Gonzalez, J. Hilgert, A. Kurusa, F. Rouvière and H. Schlichtkrull for concrete contributions mentioned at specific places in the text, and for various textual suggestions. Once more my sincere thanks to Jan Wetzel for skillful preparation of the manuscript and to Kaitlin Leach at Springer for her patient cooperation.

MIT   Sigurdur Helgason
May 2009

Contents

CHAPTER III
The Radon Transform on Two-Point Homogeneous Spaces

CHAPTER IV
The X-Ray Transform on a Symmetric Space

CHAPTER V
Orbital Integrals and the Wave Operator for Isotropic Lorentz Spaces

CHAPTER VI
The Mean-Value Operator

CHAPTER VII
Fourier Transforms and Distributions. A Rapid Course

CHAPTER VIII
Lie Transformation Groups and Differential Operators

CHAPTER IX
Symmetric Spaces

CHAPTER I

THE RADON TRANSFORM ON $\mathbf{R}^n$

§1 Introduction

It was proved by J. Radon in 1917 that a differentiable function on $\mathbf{R}^3$ can be determined explicitly by means of its integrals over the planes in $\mathbf{R}^3$. Let $J(\omega, p)$ denote the integral of f over the hyperplane $\langle x, \omega \rangle = p$, ω denoting a unit vector and $\langle \, , \, \rangle$ the inner product. Then

$$f(x) = -\frac{1}{8\pi^2} L_x \left(\int_{\mathbf{S}^2} J(\omega, \langle \omega, x \rangle) \, d\omega \right),$$

where L is the Laplacian on $\mathbf{R}^3$ and $d\omega$ the area element on the sphere $\mathbf{S}^2$ (cf. Theorem 3.1).

We now observe that the formula above has built in a remarkable duality: first one integrates over the set of points in a hyperplane, then one integrates over the set of hyperplanes passing through a given point. This suggests considering the transforms $f \to \widehat{f}, \varphi \to \check{\varphi}$ defined below.

The formula has another interesting feature. For a fixed ω the integrand $x \to J(\omega, \langle \omega, x \rangle)$ is a **plane wave**, that is a function constant on each plane perpendicular to ω. Ignoring the Laplacian the formula gives a continuous decomposition of f into plane waves. Since a plane wave amounts to a function of just one variable (along the normal to the planes) this decomposition can sometimes reduce a problem for $\mathbf{R}^3$ to a similar problem for $\mathbf{R}$. This principle has been particularly useful in the theory of partial differential equations.

The analog of the formula above for the line integrals is of importance in radiography where the objective is the description of a density function by means of certain line integrals.

In this chapter we discuss relationships between a function on $\mathbf{R}^n$ and its integrals over k-dimensional planes in $\mathbf{R}^n$. The case $k = n - 1$ will be the one of primary interest. We shall occasionally use some facts about Fourier transforms and distributions. This material will be developed in sufficient detail in Chapter VII so the treatment should be self-contained. Later chapters involve some Lie groups and symmetric spaces so the needed background is developed in Chapter VIII and Chapter IX.

Following Schwartz [1966] we denote by $\mathcal{E}(\mathbf{R}^n)$ and $\mathcal{D}(\mathbf{R}^n)$, respectively, the space of complex-valued C^∞ functions (respectively C^∞ functions of compact support) on $\mathbf{R}^n$. The space $\mathcal{S}(\mathbf{R}^n)$ of rapidly decreasing functions on $\mathbf{R}^n$ is defined in connection with (6) below. $C^m(\mathbf{R}^n)$ denotes the space of m times continuously differentiable functions. We write $C(\mathbf{R}^n)$ for $C^0(\mathbf{R}^n)$, the space of continuous functions on $\mathbf{R}^n$.

S. Helgason, *Integral Geometry and Radon Transforms*,
DOI 10.1007/978-1-4419-6055-9_1, © Springer Science+Business Media, LLC 2010

For a manifold M, $C^m(M)$ (and $C(M)$) is defined similarly and we write $\mathcal{D}(M)$ for $C_c^\infty(M)$ and $\mathcal{E}(M)$ for $C^\infty(M)$.

§2   The Radon Transform of the Spaces $\mathcal{D}(\mathbf{R}^n)$ and $\mathcal{S}(\mathbf{R}^n)$. The Support Theorem

Let f be a function on $\mathbf{R}^n$, integrable on each hyperplane in $\mathbf{R}^n$. Let $\mathbf{P}^n$ denote the space of all hyperplanes in $\mathbf{R}^n$, $\mathbf{P}^n$ being furnished with the obvious topology. The **Radon transform** of f is defined as the function $\widehat{f}$ on $\mathbf{P}^n$ given by

$$\widehat{f}(\xi) = \int_\xi f(x)dm(x)\,,$$

where dm is the Euclidean measure on the hyperplane ξ. Along with the transformation $f \to \widehat{f}$ we consider also the **dual transform** $\varphi \to \check{\varphi}$ which to a continuous function φ on $\mathbf{P}^n$ associates the function $\check{\varphi}$ on $\mathbf{R}^n$ given by

$$\check{\varphi}(x) = \int_{x \in \xi} \varphi(\xi)\,d\mu(\xi)$$

where $d\mu$ is the measure on the compact set $\{\xi \in \mathbf{P}^n : x \in \xi\}$ which is invariant under the group of rotations around x and for which the measure of the whole set is 1 (see Fig. I.1). We shall relate certain function spaces on $\mathbf{R}^n$ and on $\mathbf{P}^n$ by means of the transforms $f \to \widehat{f}$, $\varphi \to \check{\varphi}$; later we obtain explicit inversion formulas.

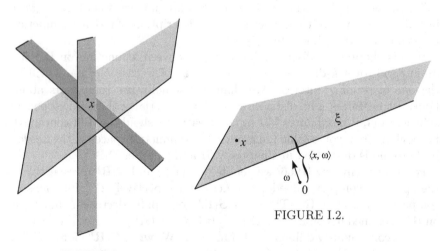

FIGURE I.2.

FIGURE I.1.

Each hyperplane $\xi \in \mathbf{P}^n$ can be written $\xi = \{x \in \mathbf{R}^n : \langle x, \omega \rangle = p\}$, where $\langle \ , \ \rangle$ is the usual inner product, $\omega = (\omega_1, \ldots, \omega_n)$ a unit vector and $p \in \mathbf{R}$ (Fig. I.2). Note that the pairs (ω, p) and $(-\omega, -p)$ give the same ξ; the mapping $(\omega, p) \rightarrow \xi$ is a double covering of $\mathbf{S}^{n-1} \times \mathbf{R}$ onto $\mathbf{P}^n$. Thus $\mathbf{P}^n$ has a canonical manifold structure with respect to which this covering map is differentiable and regular. We thus identify continuous (differentiable) function on $\mathbf{P}^n$ with continuous (differentiable) functions φ on $\mathbf{S}^{n-1} \times \mathbf{R}$ satisfying the symmetry condition $\varphi(\omega, p) = \varphi(-\omega, -p)$. Writing $\widehat{f}(\omega, p)$ instead of $\widehat{f}(\xi)$ and f_t (with $t \in \mathbf{R}^n$) for the translated function $x \rightarrow f(t + x)$ we have

$$\widehat{f_t}(\omega, p) = \int_{\langle x, \omega \rangle = p} f(x + t)\, dm(x) = \int_{\langle y, \omega \rangle = p + \langle t, \omega \rangle} f(y)\, dm(y)$$

so

$$(1) \qquad \widehat{f_t}(\omega, p) = \widehat{f}(\omega, p + \langle t, \omega \rangle).$$

Taking limits we see that if $\partial_i = \partial / \partial x_i$

$$(2) \qquad (\partial_i f)\widehat{\ }(\omega, p) = \omega_i \frac{\partial \widehat{f}}{\partial p}(\omega, p).$$

Let L denote the Laplacian $\Sigma_i \partial_i^2$ on $\mathbf{R}^n$ and let $\square$ denote the operator

$$\varphi(\omega, p) \rightarrow \frac{\partial^2}{\partial p^2} \varphi(\omega, p),$$

which is a well-defined operator on $\mathcal{E}(\mathbf{P}^n) = C^\infty(\mathbf{P}^n)$. It can be proved that if $\mathbf{M}(n)$ is the group of isometries of $\mathbf{R}^n$, then L (respectively $\square$) generates the algebra of $\mathbf{M}(n)$-invariant differential operators on $\mathbf{R}^n$ (respectively $\mathbf{P}^n$).

Lemma 2.1. *The transforms* $f \rightarrow \widehat{f}, \varphi \rightarrow \check{\varphi}$ *intertwine* L *and* $\square$, *i.e.*,

$$(Lf)\widehat{\ } = \square(\widehat{f}), \qquad (\square\varphi)^\vee = L\check{\varphi}.$$

Proof. The first relation follows from (2) by iteration. For the second we just note that for a certain constant c,

$$(3) \qquad \check{\varphi}(x) = c \int_{\mathbf{S}^{n-1}} \varphi(\omega, \langle x, \omega \rangle)\, d\omega,$$

where $d\omega$ is the usual measure on $\mathbf{S}^{n-1}$.

The Radon transform is closely connected with the Fourier transform

$$\widetilde{f}(u) = \int_{\mathbf{R}^n} f(x) e^{-i\langle x, \omega \rangle} \, dx \quad u \in \mathbf{R}^n.$$

In fact, if $s \in \mathbf{R}$, ω a unit vector,

$$\widetilde{f}(s\omega) = \int_{-\infty}^{\infty} dr \int_{\langle x, \omega \rangle = r} f(x) e^{-is\langle x, \omega \rangle} \, dm(x)$$

so

(4)
$$\widetilde{f}(s\omega) = \int_{-\infty}^{\infty} \widehat{f}(\omega, r) e^{-isr} \, dr.$$

This means that the n-dimensional Fourier transform is the 1-dimensional Fourier transform of the Radon transform. From (4) it follows that the Radon transform of the convolution

$$f(x) = \int_{\mathbf{R}^n} f_1(x - y) f_2(y) \, dy$$

is the convolution

(5)
$$\widehat{f}(\omega, p) = \int_{\mathbf{R}} \widehat{f}_1(\omega, p - q) \widehat{f}_2(\omega, q) \, dq.$$

Formula (5) can also be proved directly:

$$\widehat{f}(\omega, p) = \int_{\mathbf{R}^n} \left(\int_{\langle x, \omega \rangle = p} f_1(x - y) \, dm(x) \right) f_2(y) \, dy$$

$$= \int_{\mathbf{R}^n} \left(\int_{\langle z, \omega \rangle = p - \langle y, \omega \rangle} f_1(z) \, dm(z) \right) f_2(y) \, dy$$

$$= \int_{\mathbf{R}^n} \widehat{f}_1(\omega, p - \langle y, \omega \rangle) f_2(y) \, dy = \int_{\mathbf{R}} \widehat{f}_1(\omega, p - q) \widehat{f}_2(\omega, q) \, dq.$$

We consider now the space $\mathcal{S}(\mathbf{R}^n)$ of complex-valued rapidly decreasing functions on $\mathbf{R}^n$. We recall that $f \in \mathcal{S}(\mathbf{R}^n)$ if and only if for each polynomial P and each integer $m \geq 0$,

(6)
$$\sup_x |\,|x|^m P(\partial_1, \ldots, \partial_n) f(x)| < \infty,$$

$|x|$ denoting the norm of x. We now formulate this in a more invariant fashion.

Lemma 2.2. *A function $f \in \mathcal{E}(\mathbf{R}^n)$ belongs to $\mathcal{S}(\mathbf{R}^n)$ if and only if for each pair $k, \ell \in \mathbf{Z}^+$*

$$\sup_{x \in \mathbf{R}^n} |(1 + |x|)^k (L^\ell f)(x)| < \infty.$$

This is easily proved just by using the Fourier transforms.

In analogy with $\mathcal{S}(\mathbf{R}^n)$ we define $\mathcal{S}(\mathbf{S}^{n-1} \times \mathbf{R})$ as the space of C^∞ functions φ on $\mathbf{S}^{n-1} \times \mathbf{R}$ which for any integers $k, \ell \geq 0$ and any differential operator D on $\mathbf{S}^{n-1}$ satisfy

$$(7) \qquad \sup_{\omega \in \mathbf{S}^{n-1}, r \in \mathbf{R}} \left| (1 + |r|^k) \frac{d^\ell}{dr^\ell} (D\varphi)(\omega, r) \right| < \infty.$$

The space $\mathcal{S}(\mathbf{P}^n)$ is then defined as the set of $\varphi \in \mathcal{S}(\mathbf{S}^{n-1} \times \mathbf{R})$ satisfying $\varphi(\omega, p) = \varphi(-\omega, -p)$.

Lemma 2.3. *For each $f \in \mathcal{S}(\mathbf{R}^n)$ the Radon transform $\widehat{f}(\omega, p)$ satisfies the following condition: For $k \in \mathbf{Z}^+$ the integral*

$$\int_{\mathbf{R}} \widehat{f}(\omega, p) p^k \, dp$$

can be written as a kth degree homogeneous polynomial in $\omega_1, \ldots, \omega_n$.

Proof. This is immediate from the relation

$$(8) \qquad \int_{\mathbf{R}} \widehat{f}(\omega, p) p^k \, dp = \int_{\mathbf{R}} p^k \, dp \int_{\langle x, \omega \rangle = p} f(x) \, dm(x) = \int_{\mathbf{R}^n} f(x) \langle x, \omega \rangle^k \, dx.$$

In accordance with this lemma we define the space

$$\mathcal{S}_H(\mathbf{P}^n) = \left\{ F \in \mathcal{S}(\mathbf{P}^n) : \begin{array}{l} \text{For each } k \in \mathbf{Z}^+, \int_{\mathbf{R}} F(\omega, p) p^k \, dp \\ \text{is a homogeneous polynomial} \\ \text{in } \omega_1, \ldots, \omega_n \text{ of degree } k \end{array} \right\}.$$

With the notation $\mathcal{D}(\mathbf{P}^n) = C_c^\infty(\mathbf{P}^n)$ we write

$$\mathcal{D}_H(\mathbf{P}^n) = \mathcal{S}_H(\mathbf{P}^n) \cap \mathcal{D}(\mathbf{P}^n).$$

According to Schwartz [1966], p. 249, the Fourier transform $f \to \widetilde{f}$ maps the space $\mathcal{S}(\mathbf{R}^n)$ onto itself. See Ch. VII, Theorem 4.1. We shall now settle the analogous question for the Radon transform.

Theorem 2.4. (The Schwartz theorem) *The Radon transform $f \to \widehat{f}$ is a linear one-to-one mapping of $\mathcal{S}(\mathbf{R}^n)$ onto $\mathcal{S}_H(\mathbf{P}^n)$.*

Proof. Since

$$\frac{d}{ds}\widetilde{f}(s\omega) = \sum_{i=1}^{n} \omega_i(\partial_i \widetilde{f})$$

it is clear from (4) that for each fixed ω the function $r \to \widehat{f}(\omega, r)$ lies in $\mathcal{S}(\mathbf{R})$. For each $\omega_0 \in \mathbf{S}^{n-1}$ a subset of $(\omega_1, \ldots, \omega_n)$ will serve as local coordinates on a neighborhood of ω_0 in $\mathbf{S}^{n-1}$. To see that $\widehat{f} \in \mathcal{S}(\mathbf{P}^n)$, it therefore suffices to verify (7) for $\varphi = \widehat{f}$ on an open subset $N \subset \mathbf{S}^{n-1}$ where ω_n is bounded away from 0 and $\omega_1, \ldots, \omega_{n-1}$ serve as coordinates, in terms of which D is expressed. Since

$$(9) \qquad u_1 = s\omega_1, \ldots, u_{n-1} = s\omega_{n-1}, \quad u_n = s(1 - \omega_1^2 - \cdots - \omega_{n-1}^2)^{1/2},$$

we have

$$\frac{\partial}{\partial \omega_i}(\widetilde{f}(s\omega)) = s\frac{\partial \widetilde{f}}{\partial u_i} - s\omega_i(1 - \omega_1^2 - \cdots - \omega_{n-1}^2)^{-1/2}\frac{\partial \widetilde{f}}{\partial u_n}.$$

It follows that if D is any differential operator on $\mathbf{S}^{n-1}$ and if $k, \ell \in \mathbf{Z}^+$ then

$$(10) \qquad \sup_{\omega \in N, s \in \mathbf{R}} \left|(1 + s^{2k})\frac{d^\ell}{ds^\ell}(D\widetilde{f})(\omega, s)\right| < \infty.$$

We can therefore apply D under the integral sign in the inversion formula to (4),

$$\widehat{f}(\omega, r) = \frac{1}{2\pi}\int_{\mathbf{R}} \widetilde{f}(s\omega)e^{isr}\, ds$$

and obtain

$$(1 + r^{2k})\frac{d^\ell}{dr^\ell}\left(D_\omega(\widehat{f}(\omega, r))\right) = \frac{1}{2\pi}\int\left(1 + (-1)^k\frac{d^{2k}}{ds^{2k}}\right)\left((is)^\ell D_\omega(\widetilde{f}(s\omega))\right)e^{isr}\, ds.$$

Now (10) shows that $\widehat{f} \in \mathcal{S}(\mathbf{P}^n)$ so by Lemma 2.3, $\widehat{f} \in \mathcal{S}_H(\mathbf{P}^n)$.

Because of (4) and the fact that the Fourier transform is one-to-one it only remains to prove the surjectivity in Theorem 2.4. Let $\varphi \in \mathcal{S}_H(\mathbf{P}^n)$. In order to prove $\varphi = \widehat{f}$ for some $f \in \mathcal{S}(\mathbf{R}^n)$ we put

$$\psi(s, \omega) = \int_{-\infty}^{\infty} \varphi(\omega, r)e^{-irs}\, dr.$$

Then $\psi(s, \omega) = \psi(-s, -\omega)$ and $\psi(0, \omega)$ is a homogeneous polynomial of degree 0 in $\omega_1, \ldots, \omega_n$, hence constant. Thus there exists a function F on $\mathbf{R}^n$ such that

$$F(s\omega) = \int_{\mathbf{R}} \varphi(\omega, r)e^{-irs}\, dr.$$

While F is clearly smooth away from the origin we shall now prove it to be smooth at the origin too; this is where the homogeneity condition in the definition of $\mathcal{S}_H(\mathbf{P}^n)$ enters decisively. Consider the coordinate neighborhood $N \subset \mathbf{S}^{n-1}$ above and if $h \in C^\infty(\mathbf{R}^n - 0)$ let $h^*(\omega_1, \ldots, \omega_{n-1}, s)$ be the function obtained from h by means of the substitution (9). Then

$$\frac{\partial h}{\partial u_i} = \sum_{j=1}^{n-1} \frac{\partial h^*}{\partial \omega_j} \frac{\partial \omega_j}{\partial u_i} + \frac{\partial h^*}{\partial s} \cdot \frac{\partial s}{\partial u_i} \quad (1 \le i \le n)$$

and

$$\frac{\partial \omega_j}{\partial u_i} = \frac{1}{s}(\delta_{ij} - \frac{u_i u_j}{s^2}) \quad (1 \le i \le n, \quad 1 \le j \le n-1),$$

$$\frac{\partial s}{\partial u_i} = \omega_i \quad (1 \le i \le n-1), \frac{\partial s}{\partial u_n} = (1 - \omega_1^2 - \cdots - \omega_{n-1}^2)^{1/2}.$$

Hence

$$\frac{\partial h}{\partial u_i} = \frac{1}{s}\frac{\partial h^*}{\partial \omega_i} + \omega_i \left(\frac{\partial h^*}{\partial s} - \frac{1}{s}\sum_{j=1}^{n-1} \omega_j \frac{\partial h^*}{\partial \omega_j} \right) \quad (1 \le i \le n-1)$$

$$\frac{\partial h}{\partial u_n} = (1 - \omega_1^2 - \cdots - \omega_{n-1}^2)^{1/2} \left(\frac{\partial h^*}{\partial s} - \frac{1}{s}\sum_{j=1}^{n-1} \omega_j \frac{\partial h^*}{\partial \omega_j} \right).$$

In order to use this for $h = F$ we write

$$F(s\omega) = \int_{-\infty}^{\infty} \varphi(\omega, r)\, dr + \int_{-\infty}^{\infty} \varphi(\omega, r)(e^{-irs} - 1)\, dr.$$

By assumption the first integral is independent of ω. Thus using (7) we have for constant $K > 0$

$$\left| \frac{1}{s}\frac{\partial}{\partial \omega_i}(F(s\omega)) \right| \le K \int (1 + r^4)^{-1} s^{-1} |e^{-isr} - 1|\, dr \le K \int \frac{|r|}{1 + r^4}\, dr$$

and a similar estimate is obvious for $\partial F(s\omega)/\partial s$. The formulas above therefore imply that all the derivatives $\partial F/\partial u_i$ are bounded in a punctured ball $0 < |u| < \epsilon$ so F is certainly continuous at $u = 0$.

More generally, we prove by induction that

(11)
$$\frac{\partial^q h}{\partial u_{i_1} \cdots \partial u_{i_q}} = \sum_{1 \le i+j \le q, 1 \le k_1, \cdots, k_i \le n-1} A_{j,k_1 \ldots k_i}(\omega, s) \frac{\partial^{i+j} h^*}{\partial \omega_{k_1} \cdots \partial \omega_{k_i} \partial s^j},$$

where the coefficients A have the form

(12)
$$A_{j,k_1 \ldots k_i}(\omega, s) = a_{j,k_1 \ldots k_i}(\omega) s^{j-q}.$$

For $q = 1$ this is in fact proved above. Assuming (11) for q we calculate

$$\frac{\partial^{q+1} h}{\partial u_{i_1} \dots \partial u_{i_{q+1}}}$$

using the above formulas for $\partial / \partial u_i$. If $A_{j,k_1 \dots k_i}(\omega, s)$ is differentiated with respect to $u_{i_{q+1}}$ we get a formula like (12) with q replaced by $q + 1$. If on the other hand the $(i + j)^{\text{th}}$ derivative of h^* in (11) is differentiated with respect to $u_{i_{q+1}}$ we get a combination of terms

$$s^{-1} \frac{\partial^{i+j+1} h^*}{\partial \omega_{k_1} \dots \partial \omega_{k_{i+1}} \partial s^j} \,, \quad \frac{\partial^{i+j+1} h^*}{\partial \omega_{k_1} \dots \partial \omega_{k_i} \partial s^{j+1}} \,,$$

and in both cases we get coefficients satisfying (12) with q replaced by $q+1$, noting $s^{j-q} = s^{j+1-(q+1)}$. This proves (11)–(12) in general. Now

$$(13) \qquad F(s\omega) = \int\limits_{-\infty}^{\infty} \varphi(\omega, r) \sum_{0}^{q-1} \frac{(-isr)^k}{k!} \, dr + \int\limits_{-\infty}^{\infty} \varphi(\omega, r) e_q(-irs) \, dr \,,$$

where

$$e_q(t) = \frac{t^q}{q!} + \frac{t^{q+1}}{(q+1)!} + \cdots$$

Our assumption on φ implies that the first integral in (13) is a polynomial in $u_1, \dots, u_n$ of degree $\leq q-1$ and is therefore annihilated by the differential operator (11). If $0 \leq j \leq q$, we have

$$(14) \qquad \left| s^{j-q} \frac{\partial^j}{\partial s^j} (e_q(-irs)) \right| = |(-ir)^q (-irs)^{j-q} e_{q-j}(-irs)| \leq k_j r^q \,,$$

where k_j is a constant because the function $t \to (it)^{-p} e_p(it)$ is obviously bounded on $\mathbf{R}\,(p \geq 0)$. Since $\varphi \in \mathcal{S}(\mathbf{P}^n)$ it follows from (11)–(14) that each qth order derivative of F with respect to $u_1, \dots, u_n$ is bounded in a punctured ball $0 < |u| < \epsilon$. Thus we have proved $F \in \mathcal{E}(\mathbf{R}^n)$. That F is rapidly decreasing is now clear from (7), Lemma 2.2 and (11). Finally, if f is the function in $\mathcal{S}(\mathbf{R}^n)$ whose Fourier transform is F then

$$\tilde{f}(s\omega) = F(s\omega) = \int\limits_{-\infty}^{\infty} \varphi(\omega, r) e^{-irs} \, dr \,;$$

hence by (4), $\hat{f} = \varphi$ and the theorem is proved.

To make further progress we introduce some useful notation. Let $S_r(x)$ denote the sphere $\{y : |y - x| = r\}$ in $\mathbf{R}^n$ and $A(r)$ its area. Let $B_r(x)$ denote the open ball $\{y : |y - x| < r\}$. For a continuous function f on $S_r(x)$ let $(M^r f)(x)$ denote the mean value

$$(M^r f)(x) = \frac{1}{A(r)} \int\limits_{S_r(x)} f(\omega) \, d\omega \,,$$

where $d\omega$ is the Euclidean measure. Let K denote the orthogonal group $\mathbf{O}(n)$, dk its Haar measure, normalized by $\int dk = 1$. If $y \in \mathbf{R}^n, r = |y|$ then

$$(15) \quad (M^r f)(x) = \int_K f(x + k \cdot y)\, dk \,.$$

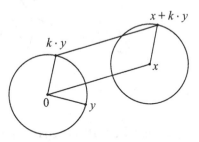

(Fig. I.3) In fact, for x, y fixed both sides represent rotation-invariant functionals on $C(S_r(x))$, having the same value for the function $f \equiv 1$. The rotations being transitive on $S_r(x)$, (15) follows from the uniqueness of such invariant functionals. Formula (3) can similarly be written

FIGURE I.3.

$$(16) \quad \check{\varphi}(x) = \int_K \varphi(x + k \cdot \xi_0)\, dk$$

if ξ_0 is some fixed hyperplane through the origin. We see then that if $f(x) = 0(|x|^{-n})$, Ω_k the area of the unit sphere in $\mathbf{R}^k$, i.e., $\Omega_k = 2\frac{\pi^{k/2}}{\Gamma(k/2)}$,

$$(\widehat{f})^\vee(x) = \int_K \widehat{f}(x + k \cdot \xi_0)\, dk = \int_K \left(\int_{\xi_0} f(x + k \cdot y)\, dm(y) \right) dk$$

$$= \int_{\xi_0} (M^{|y|} f)(x)\, dm(y) = \Omega_{n-1} \int_0^\infty r^{n-2} \left(\frac{1}{\Omega_n} \int_{S^{n-1}} f(x + r\omega)\, d\omega \right) dr,$$

so

$$(17) \quad (\widehat{f})^\vee(x) = \frac{\Omega_{n-1}}{\Omega_n} \int_{\mathbf{R}^n} |x - y|^{-1} f(y)\, dy \,.$$

We consider now the analog of Theorem 2.4 for the transform $\varphi \to \check{\varphi}$. But $\varphi \in \mathcal{S}_H(\mathbf{P}^n)$ does not imply $\check{\varphi} \in \mathcal{S}(\mathbf{R}^n)$. (If this were so and we by Theorem 2.4 write $\varphi = \widehat{f}, f \in \mathcal{S}(\mathbf{R}^n)$ then the inversion formula in Theorem 3.1 for $n = 3$ would imply $\int f(x)\, dx = 0$.) On a smaller space we shall obtain a more satisfactory result.

Let $\mathcal{S}^*(\mathbf{R}^n)$ denote the space of all functions $f \in \mathcal{S}(\mathbf{R}^n)$ which are orthogonal to all polynomials, i.e.,

$$\int_{\mathbf{R}^n} f(x)P(x)\, dx = 0 \quad \text{for all polynomials } P \,.$$

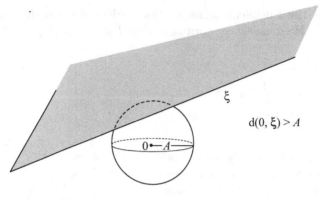

FIGURE I.4.

Similarly, let $\mathcal{S}^*(\mathbf{P}^n) \subset \mathcal{S}(\mathbf{P}^n)$ be the space of φ satisfying

$$\int_{\mathbf{R}} \varphi(\omega, r) p(r)\, dr = 0 \quad \text{for all polynomials } p.$$

Note that under the Fourier transform the space $\mathcal{S}^*(\mathbf{R}^n)$ corresponds to the subspace $\mathcal{S}_0(\mathbf{R}^n) \subset \mathcal{S}(\mathbf{R}^n)$ of functions all of whose derivatives vanish at 0.

Corollary 2.5. *The transforms* $f \to \widehat{f}, \varphi \to \widecheck{\varphi}$ *are bijections of* $\mathcal{S}^*(\mathbf{R}^n)$ *onto* $\mathcal{S}^*(\mathbf{P}^n)$ *and of* $\mathcal{S}^*(\mathbf{P}^n)$ *onto* $\mathcal{S}^*(\mathbf{R}^n)$, *respectively.*

The first statement is clear from (8) if we take into account the elementary fact that the polynomials $x \to \langle x, \omega \rangle^k$ span the space of homogeneous polynomials of degree k. To see that $\varphi \to \widecheck{\varphi}$ is a bijection of $\mathcal{S}^*(\mathbf{P}^n)$ onto $\mathcal{S}^*(\mathbf{R}^n)$ we use (17), knowing that $\varphi = \widehat{f}$ for some $f \in \mathcal{S}^*(\mathbf{R}^n)$. The right hand side of (17) is the convolution of f with the tempered distribution $|x|^{-1}$ whose Fourier transform is by Chapter VII, §6 a constant multiple of $|u|^{1-n}$. (Here we leave out the trivial case $n = 1$.) By Chapter VII, §4 this convolution is a tempered distribution whose Fourier transform is a constant multiple of $|u|^{1-n}\widehat{f}(u)$. But, by Lemma 6.6, Chapter VII this lies in the space $\mathcal{S}_0(\mathbf{R}^n)$ since $\widetilde{f}$ does. Now (17) implies that $\widecheck{\varphi} = (\widehat{f})^{\vee} \in \mathcal{S}^*(\mathbf{R}^n)$ and that $\widecheck{\varphi} \not\equiv 0$ if $\varphi \not\equiv 0$. Finally we see that the mapping $\varphi \to \widecheck{\varphi}$ is surjective because the function

$$((\widehat{f})^{\vee}\widetilde{)}(u) = c|u|^{1-n}\widetilde{f}(u)$$

(where c is a constant) runs through $\mathcal{S}_0(\mathbf{R}^n)$ as f runs through $\mathcal{S}^*(\mathbf{R}^n)$.

We now turn to the space $\mathcal{D}(\mathbf{R}^n)$ and its image under the Radon transform. We begin with a preliminary result. (See Fig. I.4.)

Theorem 2.6. (The support theorem.) *Let* $f \in C(\mathbf{R}^n)$ *satisfy the following conditions:*

(i)   *For each integer $k > 0$, $|x|^k f(x)$ is bounded.*

(ii)   *There exists a constant $A > 0$ such that*

$$\widehat{f}(\xi) = 0 \ \ for \ d(0, \xi) > A,$$

d denoting distance.

   Then

$$f(x) = 0 \ \ for \ |x| > A.$$

Proof. Replacing f by the convolution $\varphi * f$, where φ is a radial C^∞ function with support in a small ball $B_\epsilon(0)$, we see that it suffices to prove the theorem for $f \in \mathcal{E}(\mathbf{R}^n)$. In fact, $\varphi * f$ is smooth, it satisfies (i) and by (5) it satisfies (ii) with A replaced by $A + \epsilon$. Assuming the theorem for the smooth case we deduce that support $(\varphi * f) \subset B_{A+\epsilon}(0)$ so letting $\epsilon \to 0$ we obtain support $(f) \subset$ Closure $B_A(0)$.

   To begin with we assume f is a radial function. Then $f(x) = F(|x|)$ where $F \in \mathcal{E}(\mathbf{R})$ and even. Then $\widehat{f}$ has the form $\widehat{f}(\xi) = \widehat{F}(d(0, \xi))$ where $\widehat{F}$ is given by

$$\widehat{F}(p) = \int_{\mathbf{R}^{n-1}} F((p^2 + |y|^2)^{1/2}) \, dm(y), \quad (p \geq 0)$$

because of the definition of the Radon transform. Using polar coordinates in $\mathbf{R}^{n-1}$ we obtain

(18) $$\widehat{F}(p) = \Omega_{n-1} \int_0^\infty F((p^2 + t^2)^{1/2}) t^{n-2} \, dt.$$

Here we substitute $s = (p^2 + t^2)^{-1/2}$ and then put $u = p^{-1}$. Then (18) becomes

$$u^{n-3}\widehat{F}(u^{-1}) = \Omega_{n-1} \int_0^u (F(s^{-1})s^{-n})(u^2 - s^2)^{(n-3)/2} \, ds.$$

We write this equation for simplicity

(19) $$h(u) = \int_0^u g(s)(u^2 - s^2)^{(n-3)/2} \, ds.$$

This integral equation is very close to Abel's integral equation (Whittaker-Watson [1927], Ch. XI) and can be inverted as follows. Multiplying both sides by $u(t^2 - u^2)^{(n-3)/2}$ and integrating over $0 \le u \le t$, we obtain

$$\int_0^t h(u)(t^2 - u^2)^{(n-3)/2} u \, du$$

$$= \int_0^t \left[\int_0^u g(s)[(u^2 - s^2)(t^2 - u^2)]^{(n-3)/2} \, ds \right] u \, du$$

$$= \int_0^t g(s) \left[\int_{u=s}^t u[(t^2 - u^2)(u^2 - s^2)]^{(n-3)/2} \, du \right] ds .$$

The substitution $(t^2 - s^2)V = (t^2 + s^2) - 2u^2$ gives an explicit evaluation of the inner integral and we obtain

$$\int_0^t h(u)(t^2 - u^2)^{(n-3)/2} u \, du = C \int_0^t g(s)(t^2 - s^2)^{n-2} \, ds ,$$

where $C = 2^{1-n} \pi^{\frac{1}{2}} \Gamma((n-1)/2)/\Gamma(n/2)$. Here we apply the operator $\frac{d}{d(t^2)} = \frac{1}{2t} \frac{d}{dt}$ $(n-1)$ times whereby the right hand side gives a constant multiple of $t^{-1} g(t)$. Hence we obtain

$$(20) \qquad F(t^{-1}) t^{-n} = ct \left[\frac{d}{d(t^2)} \right]^{n-1} \int_0^t (t^2 - u^2)^{(n-3)/2} u^{n-2} \widehat{F}(u^{-1}) \, du$$

where $c^{-1} = (n-2)! \Omega_n / 2^n$. By assumption (ii) we have $\widehat{F}(u^{-1}) = 0$ if $u^{-1} \ge A$, that is if $u \le A^{-1}$. But then (20) implies $F(t^{-1}) = 0$ if $t \le A^{-1}$, that is if $t^{-1} \ge A$. This proves the theorem for the case when f is radial.

We consider next the case of a general f. Fix $x \in \mathbf{R}^n$ and consider the function

$$g_x(y) = \int_K f(x + k \cdot y) \, dk$$

as in (15). Then g_x satisfies (i) and

$$(21) \qquad \widehat{g}_x(\xi) = \int_K \widehat{f}(x + k \cdot \xi) \, dk ,$$

$x + k \cdot \xi$ denoting the translate of the hyperplane $k \cdot \xi$ by x. The triangle inequality shows that

$$d(0, x + k \cdot \xi) \ge d(0, \xi) - |x| , \qquad x \in \mathbf{R}^n, k \in K .$$

Hence we conclude from assumption (ii) and (21) that

(22) $\widehat{g}_x(\xi) = 0$ if $d(0, \xi) > A + |x|$.

But g_x is a radial function so (22) implies by the first part of the proof that

(23) $\displaystyle\int_K f(x + k \cdot y) \, dk = 0$ if $|y| > A + |x|$.

Geometrically, this formula reads: The surface integral of f over $S_{|y|}(x)$ is 0 if the ball $B_{|y|}(x)$ contains the ball $B_A(0)$. The theorem is therefore a consequence of the following lemma.

Lemma 2.7. *Let $f \in C(\mathbf{R}^n)$ be such that for each integer $k > 0$,*

$$\sup_{x \in \mathbf{R}^n} |x|^k |f(x)| < \infty .$$

Suppose f has surface integral 0 over every sphere S which encloses the unit ball. Then $f(x) \equiv 0$ for $|x| > 1$.

Proof. The idea is to perturb S in the relation

(24) $\displaystyle\int_S f(s) \, d\omega(s) = 0$

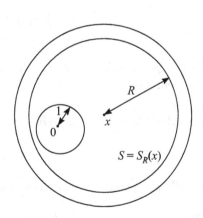

FIGURE I.5.

slightly, and differentiate with respect to the parameter of the perturbations, thereby obtaining additional relations. (See Fig. I.5.) Replacing, as above, f with a suitable convolution $\varphi * f$ we see that it suffices to prove the lemma for f in $\mathcal{E}(\mathbf{R}^n)$. Writing $S = S_R(x)$ and viewing the exterior of the ball $B_R(x)$ as a union of spheres with center x we have by the assumptions,

$$\int_{B_R(x)} f(y) \, dy = \int_{\mathbf{R}^n} f(y) \, dy ,$$

which is a constant. Differentiating with respect to x_i we obtain

(25) $\displaystyle\int_{B_R(0)} (\partial_i f)(x + y) \, dy = 0 .$

We use now the divergence theorem

(26)
$$\int_{B_R(0)} (\mathrm{div} F)(y)\, dy = \int_{S_R(0)} \langle F, \mathbf{n} \rangle (s)\, d\omega(s)$$

for a vector field F on $\mathbf{R}^n$, $\mathbf{n}$ denoting the outgoing unit normal and $d\omega$ the surface element on $S_R(0)$. For the vector field $F(y) = f(x+y)\frac{\partial}{\partial y_i}$ we obtain from (25) and (26), since $\mathbf{n} = R^{-1}(s_1, \ldots, s_n)$,

(27)
$$\int_{S_R(0)} f(x+s)s_i\, d\omega(s) = 0.$$

But by (24)

$$\int_{S_R(0)} f(x+s)x_i\, d\omega(s) = 0$$

so by adding

$$\int_S f(s)s_i\, d\omega(s) = 0.$$

This means that the hypotheses of the lemma hold for $f(x)$ replaced by the function $x_i f(x)$. By iteration

$$\int_S f(s)P(s)\, d\omega(s) = 0$$

for any polynomial P, so $f \equiv 0$ on S. This proves the lemma as well as Theorem 2.6.

Corollary 2.8. *Let $f \in C(\mathbf{R}^n)$ satisfy (i) in Theorem 2.6 and assume*

$$\widehat{f}(\xi) = 0$$

for all hyperplanes ξ disjoint from a certain compact convex set C. Then

(28)
$$f(x) = 0 \quad \text{for} \quad x \notin C.$$

In fact, if B is a closed ball containing C we have by Theorem 2.6, $f(x) = 0$ for $x \notin B$. But C is the intersection of such balls. In fact, if $x \notin C$, x and C are separated by a hyperplane ξ. Let $z \in \xi$. The half space bounded by ξ, containing C is the union of an increasing sequence of balls B tangential to ξ at z. Finitely many cover C and the largest of those contains C. Thus, $x \notin \bigcap_{B \supset C} B$, so (28) follows.

Remark 2.9. While condition (i) of rapid decrease entered in the proof of Lemma 2.7 (we used $|x|^k f(x) \in L^1(\mathbf{R}^n)$ for each $k > 0$) one may wonder whether it could not be weakened in Theorem 2.6 and perhaps even dropped in Lemma 2.7.

As an example, showing that the condition of rapid decrease can not be dropped in either result consider for $n = 2$ the function

$$f(x, y) = (x + iy)^{-5}$$

made smooth in $\mathbf{R}^2$ by changing it in a small disk around 0. Using Cauchy's theorem for a large semicircle we have $\int_\ell f(x) \, dm(x) = 0$ for every line ℓ outside the unit circle. Thus (ii) is satisfied in Theorem 2.6. Hence (i) cannot be dropped or weakened substantially.

This same example works for Lemma 2.7. In fact, let S be a circle $|z - z_0| = r$ enclosing the unit disk. Then $d\omega(s) = -ir \frac{dz}{z - z_0}$, so, by expanding the contour or by residue calculus,

$$\int_S z^{-5}(z - z_0)^{-1} \, dz = 0 \,,$$

(the residue at $z = 0$ and $z = z_0$ cancel) so we have in fact

$$\int_S f(s) \, d\omega(s) = 0 \,.$$

We recall now that $\mathcal{D}_H(\mathbf{P}^n)$ is the space of symmetric C^∞ functions $\varphi(\xi) = \varphi(\omega, p)$ on $\mathbf{P}^n$ of compact support such that for each $k \in \mathbf{Z}^+$, $\int_\mathbf{R} \varphi(\omega, p) p^k \, dp$ is a homogeneous kth degree polynomial in $\omega_1, \ldots, \omega_n$. Combining Theorems 2.4, 2.6 we obtain the following characterization of the Radon transform of the space $\mathcal{D}(\mathbf{R}^n)$. This can be regarded as the analog for the Radon transform of the Paley-Wiener theorem for the Fourier transform (see Chapter VII).

Theorem 2.10. (The Paley-Wiener theorem.) *The Radon transform is a bijection of $\mathcal{D}(\mathbf{R}^n)$ onto $\mathcal{D}_H(\mathbf{P}^n)$.*

We conclude this section with a variation and a consequence of Theorem 2.6.

Lemma 2.11. *Let $f \in C_c(\mathbf{R}^n)$, $A > 0$, ω_0 a fixed unit vector and $N \subset S$ a neighborhood of ω_0 in the unit sphere $S \subset \mathbf{R}^n$. Assume*

$$\widehat{f}(\omega, p) = 0 \quad \text{for} \quad \omega \in N, p > A \,.$$

Then

(29) $$f(x) = 0 \text{ in the half-space } \langle x, \omega_0 \rangle > A \,.$$

Proof. Let B be a closed ball around the origin containing the support of f. Let $\epsilon > 0$ and let H_ϵ be the union of the half spaces $\langle x, \omega \rangle > A + \epsilon$ as ω runs through N. Then by our assumption,

$$(30) \qquad \widehat{f}(\xi) = 0 \quad \text{if} \quad \xi \subset H_\epsilon.$$

Now choose a ball B_ϵ with a center on the ray from 0 through $-\omega_0$, with the point $(A + 2\epsilon)\omega_0$ on the boundary, and with radius so large that any hyperplane ξ intersecting B but not B_ϵ must be in H_ϵ. Then by (30),

$$\widehat{f}(\xi) = 0 \quad \text{whenever} \quad \xi \in \mathbf{P}^n, \xi \cap B_\epsilon = \emptyset.$$

Hence by Theorem 2.6, $f(x) = 0$ for $x \notin B_\epsilon$. In particular, $f(x) = 0$ for $\langle x, \omega_0 \rangle > A + 2\epsilon$; since $\epsilon > 0$ is arbitrary, the lemma follows.

Corollary 2.12. *Let N be any open subset of the unit sphere $\mathbf{S}^{n-1}$. If $f \in C_c(\mathbf{R}^n)$ and*

$$\widehat{f}(\omega, p) = 0 \quad \text{for } p \in \mathbf{R}, \ \omega \in N,$$

then

$$f \equiv 0.$$

Since $\widehat{f}(-\omega, -p) = \widehat{f}(\omega, p)$ this is obvious from Lemma 2.11.

§3   The Inversion Formula. Injectivity Questions

We shall now establish explicit inversion formulas for the Radon transform $f \to \widehat{f}$ and its dual $\varphi \to \check{\varphi}$.

Theorem 3.1. *The function f can be recovered from the Radon transform by means of the following inversion formula*

$$(31) \qquad cf = (-L)^{(n-1)/2}((\widehat{f})^\vee), \quad f \in \mathcal{E}(\mathbf{R}^n),$$

provided $f(x) = 0(|x|^{-N})$ for some $N > n - 1$. Here c is the constant

$$c = (4\pi)^{(n-1)/2}\Gamma(n/2)/\Gamma(1/2),$$

and the power of the Laplacian L is given in Chapter VII, §6.

Proof. We first give a geometric proof of (31) for n odd. We start with some general useful facts about the mean value operator M^r. It is a familiar fact that if $f \in C^2(\mathbf{R}^n)$ is a radial function, i.e., $f(x) = F(r)$, $r = |x|$, then

$$(32) \qquad (Lf)(x) = \frac{d^2 F}{dr^2} + \frac{n-1}{r}\frac{dF}{dr}.$$

This is immediate from the relations

$$\frac{\partial^2 f}{\partial x_i^2} = \frac{\partial^2 f}{\partial r^2}\left(\frac{\partial r}{\partial x_i}\right)^2 + \frac{\partial f}{\partial r}\frac{\partial^2 r}{\partial x_i^2}.$$

Lemma 3.2. *(i)* $LM^r = M^r L$ *for each* $r > 0$.

(ii)   For $f \in C^2(\mathbf{R}^n)$ *the mean value* $(M^r f)(x)$ *satisfies the "Darboux equation"*

$$L_x\left((M^r f)(x)\right) = \left(\frac{\partial^2}{\partial r^2} + \frac{n-1}{r}\frac{\partial}{\partial r}\right)(M^r f(x)),$$

that is, the function $F(x, y) = (M^{|y|} f)(x)$ *satisfies*

$$L_x(F(x, y)) = L_y(F(x, y)).$$

Proof. We prove this group theoretically, using expression (15) for the mean value. For $z \in \mathbf{R}^n$, $k \in K$, let T_z denote the translation $x \to x + z$ and R_k the rotation $x \to k \cdot x$. Since L is invariant under these transformations, we have if $r = |y|$,

$$(LM^r f)(x) = \int_K L_x(f(x + k \cdot y)) \, dk = \int_K (Lf)(x + k \cdot y) \, dk = (M^r Lf)(x)$$

$$= \int_K [(Lf) \circ T_x \circ R_k](y) \, dk = \int_K [L(f \circ T_x \circ R_k)](y) \, dk$$

$$= L_y\left(\int_K f(x + k \cdot y)\right) dk,$$

which proves the lemma.

Now suppose $f \in \mathcal{S}(\mathbf{R}^n)$. Fix a hyperplane ξ_0 through 0, and an isometry $g \in \mathbf{M}(n)$. As k runs through $\mathbf{O}(n)$, $gk \cdot \xi_0$ runs through the set of hyperplanes through $g \cdot 0$, and we have

$$\check{\varphi}(g \cdot 0) = \int_K \varphi(gk \cdot \xi_0) \, dk$$

and therefore

$$(\hat{f})^\vee(g \cdot 0) = \int_K \left(\int_{\xi_0} f(gk \cdot y) \, dm(y)\right) dk$$

$$= \int_{\xi_0} dm(y) \int_K f(gk \cdot y) \, dk = \int_{\xi_0} (M^{|y|} f)(g \cdot 0) \, dm(y).$$

Hence

(33) $$((\hat{f}))^\vee(x) = \Omega_{n-1} \int_0^\infty (M^r f)(x) r^{n-2} \, dr,$$

where Ω_{n-1} is the area of the unit sphere in $\mathbf{R}^{n-1}$. Applying L to (33), using (32) and Lemma 3.2, we obtain

$$(34) \qquad L((\widehat{f})^{\vee}) = \Omega_{n-1} \int\limits_0^\infty \left(\frac{d^2 F}{dr^2} + \frac{n-1}{r} \frac{dF}{dr} \right) r^{n-2} \, dr,$$

where $F(r) = (M^r f)(x)$. Integrating by parts and using

$$F(0) = f(x), \qquad \lim_{r \to \infty} r^k F(r) = 0,$$

we get

$$L((\widehat{f})^{\vee}) = \begin{cases} -\Omega_{n-1} f(x) & \text{if} \quad n = 3, \\ -\Omega_{n-1}(n-3) \int_0^\infty F(r) r^{n-4} \, dr & (n > 3). \end{cases}$$

More generally,

$$L_x \left(\int\limits_0^\infty (M^r f)(x) r^k \, dr \right) = \begin{cases} -(n-2) f(x) & \text{if } k = 1, \\ -(n-1-k)(k-1) \int_0^\infty F(r) r^{k-2} dr, & (k > 1). \end{cases}$$

If n is odd, the formula in Theorem 3.1 follows by iteration. Although we assumed $f \in \mathcal{S}(\mathbf{R}^n)$ the proof is valid under much weaker assumptions.

Another proof (for all n) uses the Riesz potential in Ch. VII. In fact, (17) implies

$$(35) \qquad (\widehat{f})^{\vee} = 2^{n-1} \pi^{\frac{n}{2}-1} \Gamma(n/2) I^{n-1} f,$$

so the inversion formula follows from Theorem 6.11 in Ch. VII.

However, since the fractional power $(-L)^{(n-1)/2}$ is only defined by means of holomorphic continuation of the Riesz potential, the geometric proof above for n odd and the alternative proof of Theorem 1.4 in Ch. III have to be considered much more explicit and direct inversion formulas.

Remark 3.3. It is interesting to observe that while the inversion formula requires $f(x) = 0(|x|^{-N})$ for *one* $N > n - 1$ the support theorem requires $f(x) = 0(|x|^{-N})$ for *all* N as mentioned in Remark 2.9.

In connection with the inversion formula we shall now discuss the weaker question of injectivity.

Proposition 3.4. *Let* $f \in L^1(\mathbf{R}^n)$. *Then* $\widehat{f}(\omega, p)$ *exists for almost all* $(\omega, p) \in \mathbf{S}^{n-1} \times \mathbf{R}$. *Also the map* $f \to \widehat{f}$ *is injective on* $L^1(\mathbf{R}^n)$.

Proof. For a fixed $\omega \in \mathbf{S}^{n-1}$, $x = x' + p\omega$ where $\langle x', \omega \rangle = 0$, $p \in \mathbf{R}$. This gives a product representation $\mathbf{R}^n = \mathbf{R} \times H$ with $H \simeq \mathbf{R}^{n-1}$. By the Fubini theorem $\widehat{f}(\omega, p)$ exist for almost all p, and

$$\int\limits_{\mathbf{R}^n} f(x) \, dx = \int\limits_{\mathbf{R}} \widehat{f}(\omega, p) \, dp.$$

Then

$$\int_{\mathbf{S}^{n-1}\times\mathbf{R}} |\widehat{f}(\omega,p)|\, d\omega\, dp \leq \int_{\mathbf{S}^{n-1}\times\mathbf{R}} |f\widehat{\;}(\omega,p)\, d\omega\, dp$$

$$= \int_{\mathbf{S}^{n-1}} d\omega \int_{\mathbf{R}^n} |f(x)|\, dx < \infty,$$

proving the first statement.

Using this on the function $f(x)e^{-i\langle x,\omega\rangle}$, we derive (4) for $f \in L^1(\mathbf{R}^n)$. Thus $\widehat{f}(\omega,p) = 0$ for almost all ω,p implies $\widetilde{f} = 0$ a.e. so $f = 0$ by the injectivity of the Fourier transform.

An example of non-injectivity

We shall now give an example of a smooth function $f \not\equiv 0$ on $\mathbf{R}^2$ which is integrable on each line $\xi \subset \mathbf{R}^2$, yet $\widehat{f}(\xi) = 0$ for all ξ. Such an example was first constructed by Zalcman [1982], using a delicate approximation theorem by Arakelyan [1964]. A more elementary construction, which we follow below, was given by Armitage [1994].

Lemma 3.5. *Let $z_1, z_2 \in \mathbf{C}$ and $|z_1 - z_2| < 1$. If f_1 is holomorphic on $\mathbf{C} - \{z_1\}$ and $\epsilon > 0$ there exists a function f_2 holomorphic on $\mathbf{C} - \{z_2\}$ such that*

$$(36) \qquad |f_1(z) - f_2(z)| < \frac{\epsilon}{(1 + |z|)^2} \quad \text{for } |z - z_2| > 1.$$

Proof. We have a Laurent expansion of f_1 centered at z_2:

$$f_1(z) = f_0(z) + \sum_{j=1}^{\infty} a_j(z - z_2)^{-j}, \quad |z - z_2| > |z_1 - z_2|$$

f_0 being an entire function. For $m \in \mathbf{Z}^+$ define the function

$$f_2(z) = f_0(z) + \sum_{j=1}^{m} a_j(z - z_2)^{-j} \quad z \neq z_2$$

so

$$(37) \qquad f_1(z) - f_2(z) = \sum_{m+1}^{\infty} a_j(z - z_2)^{-j}.$$

We shall choose m such that (36) holds. The coefficients a_j are given by

$$a_j = \frac{1}{2\pi i} \int_{|\zeta - z_2| = r} \frac{f_1(\zeta)}{(\zeta - z_2)^{j+1}}\, d\zeta$$

with any $r > |z_1 - z_2|$. In order to estimate (37) it is convenient to take r such that $r|z_1 - z_2| = 2$. Let M be the maximum of $f_1(\zeta)$ on the circle $|\zeta - z_2| = r$. Then

$$|a_j| \leq \frac{M}{r^j}.$$

We can then estimate the right hand side of (37) by a geometric series, noting that $r|z - z_2| - 1 > r|z_1 - z_2| - 1$. Then

(38)
$$|f_1(z) - f_2(z)| \leq \frac{M}{(r|z - z_2|)^m}.$$

Given $K > 1$ we can choose $m = m_1$ such that the right hand side is $\leq \epsilon(1 + |z|)^{-2}$ for $|z - z_2| > K$. We can also find $m = m_2$ such that it is bounded by $\epsilon(1 + |z|)^{-2}$ for $1 < |z - z_2| \leq K$. With $m = m_1 + m_2$, (36) holds.

Theorem 3.6. *There exists a holomorphic function $f \not\equiv 0$ in $\mathbf{C}$ such that each derivative $f^{(n)}$ is integrable on every line ξ and*

(39)
$$\int_\xi f^{(n)}(z)\, dm(z) = 0, \quad n = 0, 1, 2, \ldots.$$

Proof. Choose a sequence (ζ_k) on the parabolic arc $P = \{t + it^2 | t \geq 0\}$ such that

$$\zeta_0 = 0, \ |\zeta_k - \zeta_{k-1}| < 1 \quad k \geq 1, \ \zeta_k \to \infty.$$

Let $g_0(z) = 1/z^2$. By iteration of Lemma 3.5 we obtain g_k holomorphic on $\mathbf{C} - \{\zeta_k\}$ and

(40)
$$|g_k(z) - g_{k-1}(z)| \leq \frac{1}{2^k(1 + |z|)^2}, \quad k \geq 1, |z - \zeta_k| > 1.$$

For each z_0 there is an integer N and a neighborhood $|z - z_0| < \delta$ on which (40) holds for $k > N$. Thus

$$|g_{k+p}(z) - g_k(z)| \leq \frac{1}{(1 + |z|)^2} \frac{1}{2^k}, \quad |z - z_0| < \delta, p > 0.$$

Thus $g_k(z)$ converges uniformly to an entire function $g(z)$. Let

$$P_a = \{z : \inf_{w \in P} |z - w| > a\}.$$

If $z \in P_1$ then by (40)

$$|g(z) - g_0(z)| \leq \sum_{k=1}^\infty |g_k(z) - g_{k-1}(z)| \leq \frac{1}{1 + |z|^2} < |g_0(z))|,$$

so $g(z) \not\equiv 0$ and

$$|g(\zeta)| \leq 2\frac{1}{|\zeta|^2} \qquad (\zeta \in P_1).$$

For $z \in P_2$ consider a circle $S_1(z)$ and express $g^{(n)}(z)$ by Cauchy's formula

$$g^{(n)}(z) = \frac{n!}{2\pi i} \int\limits_{S_1(z)} \frac{g(\zeta)}{(\zeta - z)^{n+1}} \, d\zeta.$$

Since $S_1(z) \subset P_1$ we have

$$|g(\zeta)| \leq 2\frac{1}{|\zeta|^2} \leq 2\frac{1}{(|z| - 1)^2},$$

so

$$|g^{(n)}(z)| \leq \frac{2n!}{(|z| - 1)^2}, \qquad z \in P_2.$$

Since $\xi \setminus P_2$ is bounded for each line ξ we have

$$\int\limits_\xi |g^{(n+1)}(z)| \, dm(z) < \infty$$

and $g^{(n)}(z) \to 0$ as $z \to \infty$ on ξ. Then the function $f(z) = g'(z)$ satisfies (39).

We shall now prove an inversion formula for the dual transform $\varphi \to \check{\varphi}$ on the subspace $\mathcal{S}^*(\mathbf{P}^n)$ similar to Theorem 3.1.

Theorem 3.7. *We have*

$$c\varphi = (-\Box)^{(n-1)/2}(\check{\varphi})\hat{\ }, \qquad \varphi \in \mathcal{S}^*(\mathbf{P}^n),$$

where c is the constant $(4\pi)^{(n-1)/2}\Gamma(n/2)/\Gamma(1/2)$.

Here $\Box$ denotes as before the operator $\frac{d^2}{dp^2}$ and its fractional powers are again defined in terms of the Riesz' potentials on the 1-dimensional p-space.

If n is odd our inversion formula follows from the odd-dimensional case in Theorem 3.1 if we put $f = \check{\varphi}$ and take Lemma 2.1 and Corollary 2.5 into account. Suppose now n is even. We claim that

(41) $$((-L)^{\frac{n-1}{2}}f)\hat{\ } = (-\Box)^{\frac{n-1}{2}}\hat{f} \qquad f \in \mathcal{S}^*(\mathbf{R}^n).$$

By Lemma 6.6 in Chapter VII, $(-L)^{(n-1)/2}f$ belongs to $\mathcal{S}^*(\mathbf{R}^n)$. Taking the 1-dimensional Fourier transform of $((-L)^{(n-1)/2}f)\hat{\ }$, we obtain

$$((-L)^{(n-1)/2}f)\widetilde{\hat{\ }}(s\omega) = |s|^{n-1}\widetilde{f}(s\omega).$$

On the other hand, for a fixed ω, $p \to \widehat{f}(\omega, p)$ is in $\mathcal{S}^*(\mathbf{R})$. By the lemma quoted, the function $p \to ((-\Box)^{(n-1)/2}\widehat{f})(\omega, p)$ also belongs to $\mathcal{S}^*(\mathbf{R})$ and its Fourier transform equals $|s|^{n-1}\widetilde{f}(s\omega)$. This proves (41). Now Theorem 3.7 follows from (41) if we put in (41)

$$\varphi = \widehat{g}, \quad f = (\widehat{g})^{\vee}, \quad g \in \mathcal{S}^*(\mathbf{R}^n),$$

because, by Corollary 2.5, $\widehat{g}$ belongs to $\mathcal{S}^*(\mathbf{P}^n)$.

Because of its theoretical importance we now prove the inversion theorem (3.1) in a different form. The proof is less geometric and involves just the one variable Fourier transform.

Let $\mathcal{H}$ denote the **Hilbert transform**

$$(\mathcal{H}F)(t) = \frac{i}{\pi} \int\limits_{-\infty}^{\infty} \frac{F(p)}{t - p}\, dp, \quad F \in \mathcal{S}(\mathbf{R}),$$

the integral being considered as the **Cauchy principal value** (see Lemma 3.9 below). For $\varphi \in \mathcal{S}(\mathbf{P}^n)$ let $\Lambda\varphi$ be defined by

$$(42) \qquad (\Lambda\varphi)(\omega, p) = \begin{cases} \frac{d^{n-1}}{dp^{n-1}}\varphi(\omega, p) & n \text{ odd}, \\ \mathcal{H}_p \frac{d^{n-1}}{dp^{n-1}}\varphi(\omega, p) & n \text{ even}. \end{cases}$$

Note that in both cases $(\Lambda\varphi)(-\omega, -p) = (\Lambda\varphi)(\omega, p)$ so $\Lambda\varphi$ is a function on $\mathbf{P}^n$.

Theorem 3.8. Let Λ be as defined by (42). Then

$$cf = (\Lambda\widehat{f})^{\vee}, \quad f \in \mathcal{S}(\mathbf{R}^n),$$

where as before

$$c = (-4\pi)^{(n-1)/2}\Gamma(n/2)/\Gamma(1/2).$$

Proof. By the inversion formula for the Fourier transform and by (4),

$$f(x) = (2\pi)^{-n} \int\limits_{\mathbf{S}^{n-1}} d\omega \int\limits_{0}^{\infty}\left(\int\limits_{-\infty}^{\infty} e^{-isp}\widehat{f}(\omega, p)\, dp\right) e^{is\langle x,\omega\rangle} s^{n-1}\, ds,$$

which we write as

$$f(x) = (2\pi)^{-n} \int\limits_{\mathbf{S}^{n-1}} F(\omega, x)\, d\omega = (2\pi)^{-n} \int\limits_{\mathbf{S}^{n-1}} \tfrac{1}{2}(F(\omega, x) + F(-\omega, x))\, d\omega.$$

Using $\widehat{f}(-\omega, p) = \widehat{f}(\omega, -p)$ this gives the formula

$$(43) \quad f(x) = \tfrac{1}{2}(2\pi)^{-n} \int\limits_{\mathbf{S}^{n-1}} d\omega \int\limits_{-\infty}^{\infty} |s|^{n-1}e^{is\langle x,\omega\rangle}\, ds \int\limits_{-\infty}^{\infty} e^{-isp}\widehat{f}(\omega, p)\, dp.$$

If n is odd the absolute value on s can be dropped. The factor s^{n-1} can be removed by replacing $\widehat{f}(\omega, p)$ by $(-i)^{n-1} \frac{d^{n-1}}{dp^{n-1}} \widehat{f}(\omega, p)$. The inversion formula for the Fourier transform on $\mathbf{R}$ then gives

$$f(x) = \tfrac{1}{2}(2\pi)^{-n}(2\pi)^{+1}(-i)^{n-1} \int_{\mathbf{S}^{n-1}} \left\{ \frac{d^{n-1}}{dp^{n-1}} \widehat{f}(\omega, p) \right\}_{p=\langle x, \omega \rangle} d\omega$$

as desired.

In order to deal with the case n even we recall some general facts.

Lemma 3.9. *Let S denote the Cauchy principal value*

$$S : \psi \rightarrow \lim_{\epsilon \to 0} \int_{|x| \geq \epsilon} \frac{\psi(x)}{x} dx.$$

Then S is a tempered distribution and $\widetilde{S}$ is the function

$$\widetilde{S}(s) = -\pi i \operatorname{sgn}(s) = \begin{cases} -\pi i & s \geq 0 \\ \pi i & s < 0 \end{cases}.$$

Proof. If $\operatorname{supp}(\psi) \subset (-K, K)$ we can write

$$\int_{|x| \geq \epsilon} \frac{\psi(x)}{x} dx = \int_{K \geq |x| \geq \epsilon} \frac{\psi(x) - \psi(0)}{x} dx + \int_{K \geq |x| \geq \epsilon} \frac{\psi(0)}{x} dx$$

and the last term is 0. Thus S is indeed a distribution. To see that S is tempered let $\varphi \in \mathcal{E}(\mathbf{R})$ be such that $\varphi(x) = 0$ for $|x| \leq 1$ and $\varphi(x) = 1$ for $|x| \geq 2$. For $\psi \in \mathcal{S}(\mathbf{R})$ we put

$$\psi = \varphi\psi + (1 - \varphi)\psi = \xi + \eta.$$

Then $\xi \in \mathcal{S}(\mathbf{R})$ but $\xi = 0$ near 0 and η has compact support. If $\psi_i \to 0$ in $\mathcal{S}(\mathbf{R})$, then $\xi_i \to 0$ in $\mathcal{S}$ and $S(\xi_i) \to 0$. Also $S(\eta_i) \to 0$ is obvious. Thus S is tempered. Also $xS = 1$ so

$$2\pi\delta = \widetilde{1} = (\widetilde{xS}) = i(\widetilde{S})'.$$

But $\operatorname{sgn}' = 2\delta$, so $\widetilde{S} = -\pi i \operatorname{sgn} + C$. But $\widetilde{S}$ and sgn are odd, so $C = 0$.

By Ch. VII, Proposition 4.4, $\mathcal{H}F$ is a tempered distribution and by Theorem 4.6,

$$(44) \qquad\qquad (\widetilde{\mathcal{H}F})(s) = \operatorname{sgn}(s)\widetilde{F}(s).$$

This in turn implies $(\mathcal{H}F')(s) = \frac{d}{ds}\mathcal{H}F(s)$ since both sides have the same Fourier transform. For n even we write in (43), $|s|^{n-1} = \operatorname{sgn}(s)s^{n-1}$ and then (43) implies

$$(45) \quad f(x) = c_0 \int_{\mathbf{S}^{n-1}} d\omega \int_{\mathbf{R}} \operatorname{sgn}(s)e^{is\langle x, \omega \rangle} ds \int_{\mathbf{R}} \frac{d^{n-1}}{dp^{n-1}} \widehat{f}(\omega, p)e^{-isp} dp,$$

where $c_0 = \frac{1}{2}(-i)^{n-1}(2\pi)^{-n}$. Now we have for each $F \in \mathcal{S}(\mathbf{R})$ the identity

$$\int_{\mathbf{R}} \operatorname{sgn}(s)e^{ist}\left(\int_{\mathbf{R}} F(p)e^{-ips}\,dp\right)ds = 2\pi(\mathcal{H}F)(t)\,.$$

In fact, if we apply both sides to $\widetilde{\psi}$ with $\psi \in \mathcal{S}(\mathbf{R})$, the left hand side is by (44)

$$\int_{\mathbf{R}}\left(\int_{\mathbf{R}} \operatorname{sgn}(s)e^{ist}\widetilde{F}(s)\,ds\right)\widetilde{\psi}(t)\,dt$$

$$= \int_{\mathbf{R}} \operatorname{sgn}(s)\widetilde{F}(s)2\pi\psi(s)\,ds = 2\pi(\mathcal{H}F)\widetilde{}(\psi) = 2\pi(\mathcal{H}F)(\widetilde{\psi})\,.$$

Putting $F(p) = \frac{d^{n-1}}{dp^{n-1}}\widehat{f}(\omega,p)$ and using (45), Theorem 3.8 follows also for n even.

For later use we add here a few remarks concerning $\mathcal{H}$. Let $F \in \mathcal{D}$ have support contained in $(-R, R)$. Then

$$-i\pi(\mathcal{H}F)(t) = \lim_{\epsilon \to 0} \int_{\epsilon < |t-p|} \frac{F(p)}{t-p}\,dp = \lim_{\epsilon \to 0} \int_I \frac{F(p)}{t-p}\,dp\,,$$

where $I = \{p : |p| < R, \epsilon < |t-p|\}$. We decompose this last integral

$$\int_I \frac{F(p)}{t-p}\,dp = \int_I \frac{F(p)-F(t)}{t-p}\,dp + F(t)\int_I \frac{dp}{t-p}\,.$$

The last term vanishes for $|t| > R$ and all $\epsilon > 0$. The first term on the right is majorized by

$$\int_{|p|<R} \left|\frac{F(t)-F(p)}{t-p}\right|\,dp \le 2R \sup|F'|\,.$$

Thus by the dominated convergence theorem

$$\lim_{|t|\to\infty}(\mathcal{H}F)(t) = 0\,.$$

Also if $J \subset (-R, R)$ is a compact subset the mapping $F \to \mathcal{H}F$ is continuous from $\mathcal{D}_J$ into $\mathcal{E}(\mathbf{R})$ (with the topologies in Chapter VII, §1).

Later we prove one more version of the inversion formula from the point of view of double fibrations. See Theorem 1.4 in Chapter III.

§4   The Plancherel Formula

We recall that the functions on $\mathbf{P}^n$ have been identified with the functions φ on $\mathbf{S}^{n-1} \times \mathbf{R}$ which are even: $\varphi(-\omega, -p) = \varphi(\omega, p)$. The functional

$$(46) \qquad \varphi \rightarrow \int_{\mathbf{S}^{n-1}} \int_{\mathbf{R}} \varphi(\omega, p)\, d\omega\, dp \qquad \varphi \in C_c(\mathbf{P}^n),$$

is therefore a well defined measure on $\mathbf{P}^n$, denoted $d\omega\, dp$. The group $\mathbf{M}(n)$ of rigid motions of $\mathbf{R}^n$ acts transitively on $\mathbf{P}^n$: it also leaves the measure $d\omega\, dp$ invariant. It suffices to verify this latter statement for the translations T in $\mathbf{M}(n)$ because $\mathbf{M}(n)$ is generated by them together with the rotations around 0, and these rotations clearly leave $d\omega\, dp$ invariant. But

$$(\varphi \circ T)(\omega, p) = \varphi(\omega, p + q(\omega, T))$$

where $q(\omega, T) \in \mathbf{R}$ is independent of p so

$$\iint (\varphi \circ T)(\omega, p)\, d\omega\, dp = \iint \varphi(\omega, p + q(\omega, T))\, d\omega\, dp = \iint \varphi(\omega, p)\, dp\, d\omega,$$

proving the invariance.

In accordance with the definition of $(-L)^p$ in Ch. VII the fractional power $\Box^k$ is defined on $\mathcal{S}(\mathbf{P}^n)$ by

$$(47) \qquad (-\Box^k)\varphi(\omega, p) = \frac{1}{H_1(-2k)} \int_{\mathbf{R}} \varphi(\omega, q)|p - q|^{-2k-1}\, dq$$

and then the 1-dimensional Fourier transform satisfies

$$(48) \qquad ((-\Box)^k \varphi)\widetilde{\;}(\omega, s) = |s|^{2k} \widetilde{\varphi}(\omega, s).$$

Now, if $f \in \mathcal{S}(\mathbf{R}^n)$ we have by (4)

$$\widehat{f}(\omega, p) = (2\pi)^{-1} \int \widetilde{f}(s\omega)e^{isp}\, ds$$

and

$$(49) \qquad (-\Box)^{\frac{n-1}{4}} \widehat{f}(\omega, p) = (2\pi)^{-1} \int_{\mathbf{R}} |s|^{\frac{n-1}{2}} \widetilde{f}(s\omega)e^{isp}\, ds.$$

Theorem 4.1. *The mapping $f \rightarrow \Box^{\frac{n-1}{4}} \widehat{f}$ extends to an isometry of $L^2(\mathbf{R}^n)$ onto the space $L_e^2(\mathbf{S}^{n-1} \times \mathbf{R})$ of even functions in $L^2(\mathbf{S}^{n-1} \times \mathbf{R})$, the measure on $\mathbf{S}^{n-1} \times \mathbf{R}$ being*

$$\tfrac{1}{2}(2\pi)^{1-n}\, d\omega\, dp.$$

Proof. By (49) we have from the Plancherel formula on $\mathbf{R}$

$$(2\pi) \int\limits_{\mathbf{R}} |(-\Box)^{\frac{n-1}{4}} \widehat{f}(\omega, p)|^2 \, dp = \int\limits_{\mathbf{R}} |s|^{n-1} |\widetilde{f}(s\omega)|^2 \, ds \,,$$

so by integration over $\mathbf{S}^{n-1}$ and using the Plancherel formula for $f(x) \to \widetilde{f}(s\omega)$ we obtain

$$\int\limits_{\mathbf{R}^n} |f(x)|^2 \, dx = \tfrac{1}{2}(2\pi)^{1-n} \left(\int\limits_{\mathbf{S}^{n-1} \times \mathbf{R}} |\Box^{\frac{n-1}{4}} \widehat{f}(\omega, p)|^2 \, d\omega \, dp \right).$$

It remains to prove that the mapping is surjective. For this it would suffice to prove that if $\varphi \in L^2(\mathbf{S}^{n-1} \times \mathbf{R})$ is even and satisfies

$$\int\limits_{\mathbf{S}^{n-1}} \int\limits_{\mathbf{R}} \varphi(\omega, p)(-\Box)^{\frac{n-1}{4}} \widehat{f}(\omega, p) \, d\omega \, dp = 0$$

for all $f \in \mathcal{S}(\mathbf{R}^n)$, then $\varphi = 0$. Taking Fourier transforms we must prove that if $\psi \in L^2(\mathbf{S}^{n-1} \times \mathbf{R})$ is even and satisfies

$$(50) \qquad \int\limits_{\mathbf{S}^{n-1}} \int\limits_{\mathbf{R}} \psi(\omega, s)|s|^{\frac{n-1}{2}} \widetilde{f}(s\omega) \, ds \, d\omega = 0$$

for all $f \in \mathcal{S}(\mathbf{R}^n)$, then $\psi = 0$. Using the condition $\psi(-\omega, -s) = \psi(\omega, s)$, we see that

$$\int\limits_{\mathbf{S}^{n-1}} \int\limits_{-\infty}^{0} \psi(\omega, s)|s|^{\frac{1}{2}(n-1)} \widetilde{f}(s\omega) \, ds \, d\omega$$

$$= \int\limits_{\mathbf{S}^{n-1}} \int\limits_{0}^{\infty} \psi(\omega, t)|t|^{\frac{1}{2}(n-1)} \widetilde{f}(t\omega) \, dt \, d\omega$$

so (50) holds with $\mathbf{R}$ replaced with the positive axis $\mathbf{R}^+$. But then the function

$$\Psi(u) = \psi\left(\frac{u}{|u|}, |u| \right) |u|^{-\frac{1}{2}(n-1)}, \quad u \in \mathbf{R}^n - \{0\}$$

satisfies

$$\int\limits_{\mathbf{R}^n} \Psi(u) \widetilde{f}(u) \, du = 0, \quad f \in \mathcal{S}(\mathbf{R}^n)$$

so $\Psi = 0$ almost everywhere, whence $\psi = 0$.

If we combine the inversion formula in Theorem 3.8 with (51) below we obtain the following version of the Plancherel formula

$$c \int\limits_{\mathbf{R}^n} f(x)g(x) \, dx = \int\limits_{\mathbf{P}^n} (\Lambda \widehat{f})(\xi) \widehat{g}(\xi) \, d\xi \,.$$

§5   Radon Transform of Distributions

It will be proved in a general context in Chapter II (Proposition 2.2) that

$$(51) \qquad \int_{\mathbf{P}^n} \widehat{f}(\xi)\varphi(\xi) \, d\xi = \int_{\mathbf{R}^n} f(x)\check{\varphi}(x) \, dx$$

for $f \in C_c(\mathbf{R}^n), \varphi \in C(\mathbf{P}^n)$ if $d\xi$ is a suitable fixed $\mathbf{M}(n)$-invariant measure on $\mathbf{P}^n$. Thus $d\xi = \gamma \, d\omega \, dp$ where γ is a constant, independent of f and φ. With applications to distributions in mind we shall prove (51) in a somewhat stronger form.

Lemma 5.1. *Formula (51) holds (with $\widehat{f}$ and $\check{\varphi}$ existing almost anywhere) in the following two situations:*

(a) $f \in L^1(\mathbf{R}^n)$ vanishing outside a compact set; $\varphi \in C(\mathbf{P}^n)$.

(b) $f \in C_c(\mathbf{R}^n)$, φ locally integrable.

  Also $d\xi = \Omega_n^{-1} \, d\omega \, dp$.

Proof. We shall use the Fubini theorem repeatedly both on the product $\mathbf{R}^n \times \mathbf{S}^{n-1}$ and on the product $\mathbf{R}^n = \mathbf{R} \times \mathbf{R}^{n-1}$. Since $f \in L^1(\mathbf{R}^n)$ we have (as noted before) that for each $\omega \in \mathbf{S}^{n-1}$, $\widehat{f}(\omega, p)$ exists for almost all p and

$$\int_{\mathbf{R}^n} f(x) \, dx = \int_{\mathbf{R}} \widehat{f}(\omega, p) \, dp \, .$$

We also proved that $\widehat{f}(\omega, p)$ exists for almost all $(\omega, p) \in \mathbf{S}^{n-1} \times \mathbf{R}$. Next we consider the measurable function

$$(x, \omega) \rightarrow f(x)\varphi(\omega, \langle \omega, x \rangle) \text{ on } \mathbf{R}^n \times \mathbf{S}^{n-1} \, .$$

We have

$$\int_{\mathbf{S}^{n-1} \times \mathbf{R}^n} |f(x)\varphi(\omega, \langle \omega, x \rangle)| \, d\omega \, dx$$

$$= \int_{\mathbf{S}^{n-1}} \left(\int_{\mathbf{R}^n} |f(x)\varphi(\omega, \langle \omega, x \rangle)| \, dx \right) d\omega$$

$$= \int_{\mathbf{S}^{n-1}} \left(\int_{\mathbf{R}} |f\widehat{}(\omega, p)| |\varphi(\omega, p)| \, dp \right) d\omega \, ,$$

which in both cases is finite. Thus $f(x) \cdot \varphi(\omega, \langle \omega, x \rangle)$ is integrable on $\mathbf{R}^n \times \mathbf{S}^{n-1}$ and its integral can be calculated by removing the absolute values above. This gives the left hand side of (51). Reversing the integrations we conclude that $\check{\varphi}(x)$ exists for almost all x and that the double integral reduces to the right hand side of (51).

The formula (51) dictates how to define the Radon transform and its dual for distributions (see Chapter VII). In order to make the definitions formally consistent with those for functions we would require $\widehat{S}(\varphi) = S(\check{\varphi})$, $\check{\Sigma}(f) = \Sigma(\widehat{f})$ if S and Σ are distributions on $\mathbf{R}^n$ and $\mathbf{P}^n$, respectively. But while $f \in \mathcal{D}(\mathbf{R}^n)$ implies $\widehat{f} \in \mathcal{D}(\mathbf{P}^n)$ a similar implication does not hold for φ; we do not even have $\check{\varphi} \in \mathcal{S}(\mathbf{R}^n)$ for $\varphi \in \mathcal{D}(\mathbf{P}^n)$ so $\widehat{S}$ cannot be defined as above even if S is assumed to be tempered. Using the notation $\mathcal{E}$ (resp. $\mathcal{D}$) for the space of C^∞ functions (resp. of compact support) and $\mathcal{D}'$ (resp. $\mathcal{E}'$) for the space of distributions (resp. of compact support) we make the following definition.

Definition. For $S \in \mathcal{E}'(\mathbf{R}^n)$ we define the functional $\widehat{S}$ by

$$\widehat{S}(\varphi) = S(\check{\varphi}) \quad \text{for } \varphi \in \mathcal{E}(\mathbf{P}^n);$$

for $\Sigma \in \mathcal{D}'(\mathbf{P}^n)$ we define the functional $\check{\Sigma}$ by

$$\check{\Sigma}(f) = \Sigma(\widehat{f}) \quad \text{for } f \in \mathcal{D}(\mathbf{R}^n).$$

Lemma 5.2. *(i)   For each $\Sigma \in \mathcal{D}'(\mathbf{P}^n)$ we have $\check{\Sigma} \in \mathcal{D}'(\mathbf{R}^n)$.*

(ii)   For each $S \in \mathcal{E}'(\mathbf{R}^n)$ we have $\widehat{S} \in \mathcal{E}'(\mathbf{P}^n)$.

Proof. For $A > 0$ let $\mathcal{D}_A(\mathbf{R}^n)$ denote the set of functions $f \in \mathcal{D}(\mathbf{R}^n)$ with support in the closure of $B_A(0)$. Similarly let $\mathcal{D}_A(\mathbf{P}^n)$ denote the set of functions $\varphi \in \mathcal{D}(\mathbf{P}^n)$ with support in the closure of the "ball"

$$\beta_A(0) = \{\xi \in \mathbf{P}^n : d(0, \xi) < A\}.$$

The mapping of $f \to \widehat{f}$ from $\mathcal{D}_A(\mathbf{R}^n)$ to $\mathcal{D}_A(\mathbf{P}^n)$ being continuous (with the topologies defined in Chapter VII, §1) the restriction of $\check{\Sigma}$ to each $\mathcal{D}_A(\mathbf{R}^n)$ is continuous so (i) follows. That $\widehat{S}$ is a distribution is clear from (3). Concerning its support select $R > 0$ such that S has support inside $B_R(0)$. Then if $\varphi(\omega, p) = 0$ for $|p| \le R$ we have $\check{\varphi}(x) = 0$ for $|x| \le R$ whence $\widehat{S}(\varphi) = S(\check{\varphi}) = 0$.

Lemma 5.3. *For $S \in \mathcal{E}'(\mathbf{R}^n), \Sigma \in \mathcal{D}'(\mathbf{P}^n)$ we have*

$$(LS)\widehat{} = \square\widehat{S}, \quad (\square\Sigma)^\vee = L\check{\Sigma}.$$

Proof. In fact by Lemma 2.1,

$$(LS)\widehat{}(\varphi) = (LS)(\check{\varphi}) = S(L\check{\varphi}) = S((\square\varphi)^\vee) = \widehat{S}(\square\varphi) = (\square\widehat{S})(\varphi).$$

The other relation is proved in the same manner.

We shall now prove an analog of the support theorem (Theorem 2.6) for distributions. For $A > 0$ let $\beta_A(0)$ be defined as above and let supp denote support.

Theorem 5.4 (Support theorem for distributions). *Let $T \in \mathcal{E}'(\mathbf{R}^n)$ satisfy the condition*

$$\operatorname{supp} \widehat{T} \subset C\ell(\beta_A(0)), \quad (C\ell = \text{ closure}).$$

Then

$$\operatorname{supp}(T) \subset C\ell(B_A(0)).$$

Proof. For $f \in \mathcal{D}(\mathbf{R}^n)$, $\varphi \in \mathcal{D}(\mathbf{P}^n)$ we can consider the "convolution"

$$(f \times \varphi)(\xi) = \int_{\mathbf{R}^n} f(y)\varphi(\xi - y)\, dy,$$

where for $\xi \in \mathbf{P}^n$, $\xi - y$ denotes the translate of the hyperplane ξ by $-y$. Then

$$(f \times \varphi)^{\vee} = f * \check{\varphi}.$$

In fact, if ξ_0 is any hyperplane through 0,

$$
\begin{aligned}
(f \times \varphi)^{\vee}(x) &= \int_K dk \int_{\mathbf{R}^n} f(y)\varphi(x + k \cdot \xi_0 - y)\, dy \\
&= \int_K dk \int_{\mathbf{R}^n} f(x - y)\varphi(y + k \cdot \xi_0)\, dy = (f * \check{\varphi})(x).
\end{aligned}
$$

By the definition of $\widehat{T}$, the support assumption on $\widehat{T}$ is equivalent to

$$T(\check{\varphi}) = 0$$

for all $\varphi \in \mathcal{D}(\mathbf{P}^n)$ with support in $\mathbf{P}^n - C\ell(\beta_A(0))$. Let $\epsilon > 0$, let $f \in \mathcal{D}(\mathbf{R}^n)$ be a symmetric function with support in $C\ell(B_\epsilon(0))$ and let $\varphi = \mathcal{D}(\mathbf{P}^n)$ have support contained in $\mathbf{P}^n - C\ell(\beta_{A+\epsilon}(0))$. Since $d(0, \xi - y) \le d(0, \xi) + |y|$ it follows that $f \times \varphi$ has support in $\mathbf{P}^n - C\ell(\beta_A(0))$; thus by the formulas above, and the symmetry of f,

$$(f * T)(\check{\varphi}) = T(f * \check{\varphi}) = T((f \times \varphi)^{\vee}) = 0.$$

But then

$$(f * T)\widehat{\;}(\varphi) = (f * T)(\check{\varphi}) = 0,$$

which means that $(f*T)\widehat{\;}$ has support in $C\ell(\beta_{A+\epsilon}(0))$. But now Theorem 2.6 implies that $f * T$ has support in $C\ell(B_{A+\epsilon})(0)$. Letting $\epsilon \to 0$ we obtain by Prop. 3.4, Ch. VII the desired conclusion, $\operatorname{supp}(T) \subset C\ell(B_A(0))$.

We can now extend the inversion formulas for the Radon transform to distributions. First we observe that the Hilbert transform $\mathcal{H}$ can be extended to distributions T on $\mathbf{R}$ of compact support. It suffices to put

$$\mathcal{H}(T)(F) = T(-\mathcal{H}F), \quad F \in \mathcal{D}(\mathbf{R}).$$

In fact, as remarked at the end of §3, the mapping $F \longrightarrow \mathcal{H}F$ is a continuous mapping of $\mathcal{D}(\mathbf{R})$ into $\mathcal{E}(\mathbf{R})$. In particular $\mathcal{H}(T) \in \mathcal{D}'(\mathbf{R})$.

Theorem 5.5. *The Radon transform $S \longrightarrow \widehat{S}$ $(S \in \mathcal{E}'(\mathbf{R}^n))$ is inverted by the following formula*

$$cS = (\Lambda\widehat{S})^{\vee}, \quad S \in \mathcal{E}'(\mathbf{R}^n),$$

where the constant $c = (-4\pi)^{(n-1)/2}\Gamma(n/2)/\Gamma(1/2)$.
In the case when n is odd we have also

$$cS = L^{(n-1)/2}((\widehat{S})^{\vee}).$$

Remark 5.6. Since $\widehat{S}$ has compact support and since Λ is defined by means of the Hilbert transform, the remarks above show that $\Lambda\widehat{S} \in \mathcal{D}'(\mathbf{P}^n)$, so the right hand side is well defined.

Proof. Using Theorem 3.8 we have

$$(\Lambda\widehat{S})^{\vee}(f) = (\Lambda\widehat{S})(\widehat{f}) = \widehat{S}(\Lambda\widehat{f}) = S((\Lambda\widehat{f})^{\vee}) = cS(f).$$

The other inversion formula then follows, using the lemma.

In analogy with β_A we define the "sphere" σ_A in $\mathbf{P}^n$ as

$$\sigma_A = \{\xi \in \mathbf{P}^n : d(0, \xi) = A\}.$$

From Theorem 5.5 we can then deduce the following complement to Theorem 5.4.

Corollary 5.7. *Suppose n is odd. Then if $S \in \mathcal{E}'(\mathbf{R}^n)$,*

$$\operatorname{supp}(\widehat{S}) \subset \sigma_R \Rightarrow \operatorname{supp}(S) \subset S_R(0).$$

To see this let $\epsilon > 0$ and let $f \in \mathcal{D}(\mathbf{R}^n)$ have $\operatorname{supp}(f) \subset B_{R-\epsilon}(0)$. Then $\operatorname{supp} \widehat{f} \in \beta_{R-\epsilon}$ and since Λ is a differential operator, $\operatorname{supp}(\Lambda\widehat{f}) \subset \beta_{R-\epsilon}$. Hence

$$cS(f) = S((\Lambda\widehat{f})^{\vee}) = \widehat{S}(\Lambda\widehat{f}) = 0,$$

so $\operatorname{supp}(S) \cap B_{R-\epsilon}(0) = \emptyset$. Since $\epsilon > 0$ is arbitrary,

$$\operatorname{supp}(S) \cap B_R(0) = \emptyset.$$

On the other hand, by Theorem 5.4, $\operatorname{supp}(S) \subset C\ell(B_R(0))$. This proves the corollary.

Let M be a manifold and $d\mu$ a measure such that on each local coordinate patch with coordinates $(t_1, \ldots, t_n)$ the Lebesque measure $dt_1, \ldots, dt_n$ and $d\mu$ are absolutely continuous with respect to each other. If h is a function on M locally integrable with respect to $d\mu$ the distribution $\varphi \to \int \varphi h \, d\mu$ will be denoted T_h.

Proposition 5.8. *(a)   Let $f \in L^1(\mathbf{R}^n)$ vanish outside a compact set. Then the distribution T_f has Radon transform given by*

$$\widehat{T_f} = T_{\widehat{f}} \, . \tag{52}$$

(b)   Let φ be a locally integrable function on $\mathbf{P}^n$. Then

$$(T_\varphi)^\vee = T_{\check{\varphi}} \, . \tag{53}$$

Proof. The existence and local integrability of $\widehat{f}$ and $\check{\varphi}$ was established during the proof of Lemma 5.1. The two formulas now follow directly from Lemma 5.1.

As a result of this proposition the smoothness assumption can be dropped in the inversion formula. In particular, we can state the following result.

Corollary 5.9. *(n odd.) The inversion formula*

$$cf = L^{(n-1)/2}((\widehat{f})^\vee) \, ,$$

$c = (-4\pi)^{(n-1)/2}\Gamma(n/2)/\Gamma(1/2)$, holds for all $f \in L^1(\mathbf{R}^n)$ vanishing outside a compact set, the derivative interpreted in the sense of distributions.

Examples. If μ is a measure (or a distribution) on a closed submanifold S of a manifold M, the distribution on M given by $\varphi \to \mu(\varphi|S)$ will also be denoted by μ.

(a)   Let δ_0 be the delta distribution $f \to f(0)$ on $\mathbf{R}^n$. Then

$$\widehat{\delta_0}(\varphi) = \delta_0(\check{\varphi}) = \Omega_n^{-1} \int\limits_{S^{n-1}} \varphi(w, 0) \, dw \, ,$$

so

$$\widehat{\delta_0} = \Omega_n^{-1} m_{\mathbf{S}^{n-1}} \, , \tag{54}$$

the normalized measure on $\mathbf{S}^{n-1}$ considered as a distribution on $\mathbf{S}^{n-1} \times \mathbf{R}$.

(b)   Let ξ_0 denote the hyperplane $x_n = 0$ in $\mathbf{R}^n$, and δ_{ξ_0} the delta distribution $\varphi \to \varphi(\xi_0)$ on $\mathbf{P}^n$. Then

$$(\delta_{\xi_0})^\vee (f) = \int\limits_{\xi_0} f(x) \, dm(x) \, ,$$

so

(55) $(\delta_{\xi_0})^\vee = m_{\xi_0}$,

the Euclidean measure of ξ_0.

(c)   Let χ_B be the characteristic function of the unit ball $B \subset \mathbf{R}^n$. Then by (52),

$$\widehat{\chi_B}(\omega,p) = \begin{cases} \frac{\Omega_{n-1}}{n-1}(1-p^2)^{(n-1)/2} & , \ |p| \le 1 \\ 0 & , \ |p| > 1 \end{cases}.$$

(d)   Let Ω be a bounded convex region in $\mathbf{R}^n$ whose boundary is a smooth surface. We shall obtain a formula for the characteristic funtion of Ω in terms of the areas of its hyperplane sections. For simplicity we assume n odd. The characteristic function χ_Ω is a distribution of compact support and $(\chi_\Omega)\widehat{}$ is thus well defined. Approximating χ_Ω in the L^2-norm by a sequence $(\psi_n) \subset \mathcal{D}(\Omega)$ we see from Theorem 4.1 that $\partial_p^{(n-1)/2}\widehat{\psi}_n(\omega,p)$ converges in the L^2-norm on $\mathbf{P}^n$. Since

$$\int \widehat{\psi}(\xi)\varphi(\xi)\,d\xi = \int \psi(x)\breve{\varphi}(x)\,dx$$

it follows from Schwarz' inequality that $\widehat{\psi}_n \longrightarrow (\chi_\Omega)\widehat{}$ in the sense of distributions and accordingly $\partial^{(n-1)/2}\widehat{\psi}_n$ converges as a distribution to $\partial^{(n-1)/2}((\chi_\Omega)\widehat{})$. Since the L^2 limit is also a limit in the sense of distributions this last function equals the L^2 limit of the sequence $\partial^{(n-1)/2}\widehat{\psi}_n$. From Theorem 4.1 we can thus conclude the following result:

Theorem 5.10. *Let $\Omega \subset \mathbf{R}^n$ (n odd) be a convex region as above. Let $A(\omega,p)$ denote the $(n-1)$-dimensional area of the intersection of Ω with the hyperplane $\langle x, \omega \rangle = p$. Then*

$$(56) \qquad c_0\chi_\Omega(x) = L_x^{(\frac{1}{2})(n-1)}\left(\int_{\mathbf{S}^{n-1}} A(\omega,\langle x,\omega\rangle)\,d\omega \right), \quad c_0 = 2(2\pi i)^{n-1}$$

in the sense of distributions.

§6   Integration over d-planes. X-ray Transforms. The Range of the d-plane Transform

Let d be a fixed integer in the range $0 < d < n$. We define the **d-dimensional Radon transform $f \to \widehat{f}$** by

$$(57) \qquad \widehat{f}(\xi) = \int_\xi f(x)\,dm(x), \quad \text{where } \xi \text{ is a } d\text{-plane}.$$

Because of the applications to radiology indicated in § 7,B) the 1-dimensional Radon transform is often called the **X-ray transform**. Since a hyperplane can be viewed as a disjoint union of parallel d-planes parameterized by $\mathbf{R}^{n-1-d}$ it is obvious from (4) that the transform $f \to \widehat{f}$ is injective. Similarly we deduce the following consequence of Theorem 2.6.

Corollary 6.1. *Let $f, g \in C(\mathbf{R}^n)$ satisfy the rapid decrease condition: For each $m > 0$, $|x|^m f(x)$ and $|x|^m g(x)$ are bounded on $\mathbf{R}^n$. Assume for the d-dimensional Radon transforms*

$$\widehat{f}(\xi) = \widehat{g}(\xi)$$

whenever the d-plane ξ lies outside the unit ball. Then

$$f(x) = g(x) \quad \text{for} \quad |x| > 1 \,.$$

We shall now generalize the inversion formula in Theorem 3.1. If φ is a continuous function on the space of d-planes in $\mathbf{R}^n$ we denote by $\check{\varphi}$ the point function

$$\check{\varphi}(x) = \int_{x \in \xi} \varphi(\xi) \, d\mu(\xi) \,,$$

where μ is the unique measure on the (compact) space of d-planes passing through x, invariant under all rotations around x and with total measure 1. If σ is a fixed d-plane through the origin we have in analogy with (16),

$$(58) \qquad \check{\varphi}(x) = \int_K \varphi(x + k \cdot \sigma) \, dk \,,$$

where again $K = \mathbf{O}(n)$.

Theorem 6.2. *The d-dimensional Radon transform in $\mathbf{R}^n$ is inverted by the formula*

$$(59) \qquad cf = (-L)^{d/2}((\widehat{f})^\vee) \,,$$

where $c = (4\pi)^{d/2} \Gamma(n/2) / \Gamma((n-d)/2)$. Here it is assumed that $f(x) = 0(|x|^{-N})$ for some $N > d$.

Proof. We have, in analogy with (34),

$$
\begin{aligned}
(\widehat{f})^\vee(x) &= \int_K \left(\int_\sigma f(x + k \cdot y) \, dm(y) \right) dk \\
&= \int_\sigma dm(y) \int_K f(x + k \cdot y) \, dk = \int_\sigma (M^{|y|} f)(x) \, dm(y) \,.
\end{aligned}
$$

Hence

$$(\widehat{f})^\vee(x) = \Omega_d \int_0^\infty (M^r f)(x) r^{d-1}\, dr\,.$$

Using polar coordinates around x, in the integral below, we obtain

(60) $$(\widehat{f})^\vee(x) = \frac{\Omega_d}{\Omega_n} \int_{\mathbf{R}^n} |x - y|^{d-n} f(y)\, dy\,.$$

The theorem now follows from Theorem 6.11 in Chapter VII.

As a consequence of Theorem 2.10 we now obtain a generalization, characterizing the image of the space $\mathcal{D}(\mathbf{R}^n)$ under the d-dimensional Radon transform.

The set $\mathbf{G}(d, n)$ of d-planes in $\mathbf{R}^n$ is a manifold, in fact a homogeneous space of the group $\mathbf{M}(n)$ of all isometries of $\mathbf{R}^n$. Let $\mathbf{G}_{d,n}$ denote the manifold of all d-dimensional subspaces (d-planes through 0) of $\mathbf{R}^n$. The parallel translation of a d-plane to one through 0 gives a mapping π of $\mathbf{G}(d, n)$ onto $\mathbf{G}_{d,n}$. The inverse image $\pi^{-1}(\sigma)$ of a member $\sigma \in \mathbf{G}_{d,n}$ is naturally identified with the orthogonal complement $\sigma^\perp$. Let us write

$$\xi = (\sigma, x'') = x'' + \sigma \text{ if } \sigma = \pi(\xi) \text{ and } x'' = \sigma^\perp \cap \xi\,.$$

(See Fig. I.6.) Then (57) can be written

(61)

$$\widehat{f}(x'' + \sigma) = \int_\sigma f(x' + x'')\, dx'\,.$$

For $k \in \mathbf{Z}^+$ we consider the polynomial

(62)

$$P_k(u) = \int_{\mathbf{R}^n} f(x)\langle x, u\rangle^k\, dx\,.$$

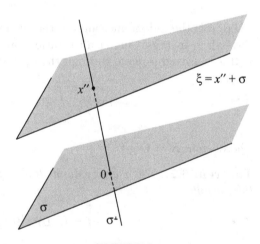

FIGURE I.6.

If $u = u'' \in \sigma^\perp$ this can be written

$$\int_{\mathbf{R}^n} f(x)\langle x, u''\rangle^k\, dx = \int_{\sigma^\perp} \int_\sigma f(x' + x'')\langle x'', u''\rangle^k\, dx'\, dx''\,,$$

so the polynomial

$$P_{\sigma,k}(u'') = \int_{\sigma^\perp} \widehat{f}(x'' + \sigma)\langle x'', u''\rangle^k\, dx''$$

is the restriction to $\sigma^\perp$ of the polynomial P_k.

In analogy with the space $\mathcal{D}_H(\mathbf{P}^n)$ in §2 we define the space $\mathcal{D}_H(\mathbf{G}(d,n))$ as the set of C^∞ functions

$$\varphi(\xi) = \varphi_\sigma(x'') = \varphi(x'' + \sigma) \quad (\text{if } \xi = (\sigma, x''))$$

on $\mathbf{G}(d,n)$ of compact support satisfying the following condition.

(**H**) : *For each $k \in \mathbf{Z}^+$ there exists a homogeneous k^{th} degree polynomial P_k on $\mathbf{R}^n$ such that for each $\sigma \in \mathbf{G}_{d,n}$ the polynomial*

$$P_{\sigma,k}(u'') = \int_{\sigma^\perp} \varphi(x'' + \sigma)\langle x'', u''\rangle^k \, dx'', \quad u'' \in \sigma^\perp,$$

coincides with the restriction $P_k|\sigma^\perp$.

Theorem 6.3 (The Range Theorem). *The d-dimensional Radon transform is a bijection of*

$$\mathcal{D}(\mathbf{R}^n) \text{ onto } \mathcal{D}_H(\mathbf{G}(d,n)).$$

Proof. For $d = n - 1$ this is Theorem 2.10. We shall now reduce the case of general $d \leq n - 2$ to the case $d = n - 1$. It remains just to prove the surjectivity in Theorem 6.3.

Let $\varphi \in \mathcal{D}(\mathbf{G}(d,n))$. Let $\omega \in \mathbf{R}^n$ be a unit vector. Choose a d-dimensional subspace σ perpendicular to ω and consider the $(n-d-1)$-dimensional integral

$$(63) \qquad \Psi_\sigma(\omega, p) = \int_{\langle \omega, x''\rangle = p, x'' \in \sigma^\perp} \varphi(x'' + \sigma) d_{n-d-1}(x''), \qquad p \in \mathbf{R}.$$

We claim that this is independent of the choice of σ. In fact

$$\int_{\mathbf{R}} \Psi_\sigma(\omega, p)p^k \, dp = \int_{\mathbf{R}} p^k \left(\int \varphi(x'' + \sigma) d_{n-d-1}(x'') \right) dp$$

$$= \int_{\sigma^\perp} \varphi(x'' + \sigma)\langle x'', \omega\rangle^k \, dx'' = P_k(\omega).$$

If we had chosen another σ, say σ_1, perpendicular to ω, then, by the above, $\Psi_\sigma(\omega, p) - \Psi_{\sigma_1}(\omega, p)$ would have been orthogonal to all polynomials in p; having compact support it would have been identically 0. Thus we have a well-defined function $\Psi(\omega, p) = \Psi_\sigma(\omega, p)$ to which Theorem 2.10 applies. In fact, the smoothness of $\Psi(\omega, p)$ is clear from (63) since for ω in a neighborhood of a fixed ω_0 we can let σ depend smoothly on ω. From this theorem we get a function $f \in \mathcal{D}(\mathbf{R}^n)$ such that

$$(64) \qquad \Psi(\omega, p) = \int_{\langle x, \omega\rangle = p} f(x) \, dm(x).$$

It remains to prove that

(65)
$$\varphi(x'' + \sigma) = \int_\sigma f(x' + x'') \, dx' \, .$$

But as x'' runs through an arbitrary hyperplane in $\sigma^\perp$ it follows from (63) and (64) that both sides of (65) have the same integral. By the injectivity of the $(n-d-1)$-dimensional Radon transform on $\sigma^\perp$, equation (65) follows. This proves Theorem 6.3.

Modifying the argument we shall now prove a stronger statement.

Theorem 6.4. *Let $\varphi \in \mathcal{D}(\mathbf{G}(d,n))$ have the property: For each pair $\sigma, \tau \in \mathbf{G}_{d,n}$ and each $k \in \mathbf{Z}^+$ the polynomials*

$$P_{\sigma,k}(u) = \int_{\sigma^\perp} \varphi(x'' + \sigma)\langle x'', u\rangle^k \, dx'' \qquad u \in \mathbf{R}^n$$

$$P_{\tau,k}(u) = \int_{\tau^\perp} \varphi(y'' + \tau)\langle y'', u\rangle^k \, dy'' \qquad u \in \mathbf{R}^n$$

agree for $u \in \sigma^\perp \cap \tau^\perp$. Then $\varphi = \hat{f}$ for some $f \in \mathcal{D}(\mathbf{R}^n)$.

Proof. Let $\varphi = \mathcal{D}(\mathbf{G}(d,n))$ have the property above. Let $\omega \in \mathbf{R}^n$ be a unit vector. Let $\sigma, \tau \in \mathbf{G}_{d,n}$ be perpendicular to ω. Consider again the $(n-d-1)$-dimensional integral

(66)
$$\Psi_\sigma(\omega, p) = \int_{\langle \omega, x''\rangle = p, \, x'' \in \sigma^\perp} \varphi(x'' + \sigma) d_{n-d-1}(x''), \quad p \in \mathbf{R} \, .$$

We claim that
$$\Psi_\sigma(\omega, p) = \Psi_\tau(\omega, p) \, .$$

To see this consider the moment

$$\int_{\mathbf{R}} \Psi_\sigma(\omega, p) p^k \, dp$$

$$= \int_{\mathbf{R}} p^k \left(\int \varphi(x'' + \sigma) d_{n-d-1}(x'') \right) dp = \int_{\sigma^\perp} \varphi(x'' + \sigma)\langle x'', \omega\rangle^k \, dx''$$

$$= \int_{\tau^\perp} \varphi(y'' + \tau)\langle y'', \omega\rangle^k \, dy'' = \int_{\mathbf{R}} \Psi_\tau(\omega, p) p^k \, dp \, .$$

Thus $\Psi_\sigma(\omega, p) - \Psi_\tau(\omega, p)$ is perpendicular to all polynomials in p; having compact support it would be identically 0. We therefore put $\Psi(\omega, p) = \Psi_\sigma(\omega, p)$. Observe that Ψ is smooth; in fact for ω in a neighborhood of a fixed ω_0 we can let σ depend smoothly on ω, so by (66), $\Psi_\sigma(\omega, p)$ is smooth.

Writing

$$\langle x'', \omega \rangle^k = \sum_{|\alpha|=k} p_\alpha(x'')\omega^\alpha, \qquad \omega^\alpha = \omega_1^{\alpha_1} \ldots \omega_n^{\alpha_n}$$

we have

$$\int_{\mathbf{R}} \Psi(\omega, p) p^k \, dp = \sum_{|\alpha|=k} A_\alpha \omega^\alpha,$$

where

$$A_\alpha = \int_{\sigma^\perp} \varphi(x'' + \sigma) p_\alpha(x'') \, dx''.$$

Here A_α is independent of σ if $\omega \in \sigma^\perp$; in other words, viewed as a function of ω, A_α has for each σ a constant value as ω varies in $\sigma^\perp \cap S_1(0)$. To see that this value is the same as the value on $\tau^\perp \cap S_1(0)$ we observe that there exists a $\rho \in \mathbf{G}_{d,n}$ such that $\rho^\perp \cap \sigma^\perp \neq 0$ and $\rho^\perp \cap \tau^\perp \neq 0$. (Extend the 2-plane spanned by a vector in $\sigma^\perp$ and a vector in $\tau^\perp$ to an $(n-d)$-plane.) This shows that A_α is constant on $S_1(0)$ so $\Psi \in \mathcal{D}_H(\mathbf{P}^n)$. Thus by Theorem 2.10,

$$(67) \qquad \Psi(\omega, p) = \int_{\langle x, \omega \rangle = p} f(x) \, dm(x)$$

for some $f \in \mathcal{D}(\mathbf{R}^n)$. It remains to prove that

$$(68) \qquad \varphi(x'' + \sigma) = \int_\sigma f(x' + x'') \, dx'.$$

But as x'' runs through an arbitrary hyperplane in $\sigma^\perp$ it follows from (63) and (64) that both sides of (68) have the same integral. By the injectivity of the $(n-d-1)$-dimensional Radon transform on $\sigma^\perp$ equation (68) follows. This proves Theorem 6.4.

Theorem 6.4 raises the following elementary question: If a function f on $\mathbf{R}^n$ is a polynomial on each k-dimensional subspace, is f itself a polynomial? The answer is no for $k = 1$ but yes if $k > 1$. See Proposition 6.13 below, kindly communicated by Schlichtkrull.

We shall now prove another characterization of the range of $\mathcal{D}(\mathbf{R}^n)$ under the d-plane transform (for $d \leq n - 2$). The proof will be based on Theorem 6.4.

Given any $d + 1$ points $(x_0, \ldots, x_d)$ in general position let $\xi(x_0, \ldots, x_d)$ denote the d-plane passing through them. If $\varphi \in \mathcal{E}(\mathbf{G}(d, n))$ we shall write $\varphi(x_0, \ldots, x_d)$ for the value $\varphi(\xi(x_0, \ldots, x_d))$. We also write $V(\{x_i - x_0\}_{i=1,d})$ for the volume of the parallelepiped spanned by vectors $(x_i - x_0), (1 \leq i \leq d)$. The mapping

$$(\lambda_1, \ldots, \lambda_d) \rightarrow x_0 + \Sigma_{i=1}^d \lambda_i(x_i - x_0)$$

is a bijection of $\mathbf{R}^d$ onto $\xi(x_0, \dots, x_d)$ and

$$(69) \quad \widehat{f}(x_0, \dots, x_d) = V(\{x_i - x_0\}_{i=1,d}) \int_{\mathbf{R}^d} f(x_0 + \Sigma_i \lambda_i(x_i - x_0)) \, d\lambda \,.$$

The range $\mathcal{D}(\mathbf{R}^n)$ can now be described by the following alternative to Theorem 6.4. Let x_i^k denote the kth coordinate of x_i.

Theorem 6.5. *If $f \in \mathcal{D}(\mathbf{R}^n)$ then $\varphi = \widehat{f}$ satisfies the system*

$$(70) \quad (\partial_{i,k}\partial_{j,\ell} - \partial_{j,k}\partial_{i,\ell})(\varphi(x_0, \dots, x_d)/V(\{x_i - x_0\}_{i=1,d})) = 0 \,,$$

where

$$0 \le i, j \le d, \ \ 1 \le k, \ell \le n, \ \ \partial_{i,k} = \partial/\partial x_i^k \,.$$

Conversely, if $\varphi \in \mathcal{D}(\mathbf{G}(d,n))$ satisfies (70) then $\varphi = \widehat{f}$ for some $f \in \mathcal{D}(\mathbf{R}^n)$.

The validity of (70) for $\varphi = \widehat{f}$ is obvious from (69) just by differentiation under the integral sign. For the converse we first prove a simple lemma.

Lemma 6.6. *Let $\varphi \in \mathcal{E}(\mathbf{G}(d,n))$ and $A \in \mathbf{O}(n)$. Let $\psi = \varphi \circ A$. Then if $\varphi(x_0, \dots, x_d)$ satisfies (70) so does the function*

$$\psi(x_0, \dots, x_d) = \varphi(Ax_0, \dots, Ax_d) \,.$$

Proof. Let $y_i = Ax_i$ so $y_i^\ell = \Sigma_p a_{\ell p} x_i^p$. Then, if $D_{i,k} = \partial/\partial y_i^k$,

$$(71) \quad (\partial_{i,k}\partial_{j,\ell} - \partial_{j,k}\partial_{i,\ell}) = \Sigma_{p,q=1}^n a_{pk} a_{q\ell}(D_{i,p}D_{j,q} - D_{i,q}D_{j,p}) \,.$$

Since A preserves volumes, the lemma follows.

Suppose now φ satisfies (70). We write $\sigma = (\sigma_1, \dots, \sigma_d)$ if (σ_j) is an orthonormal basis of σ. If $x'' \in \sigma^\perp$, the $(d+1)$-tuple

$$(x'', x'' + \sigma_1, \dots, x'' + \sigma_d)$$

represents the d-plane $x'' + \sigma$ and the polynomial

$$(72) \ P_{\sigma,k}(u'') = \int_{\sigma^\perp} \varphi(x'' + \sigma)\langle x'', u''\rangle^k \, dx''$$

$$= \int_{\sigma^\perp} \varphi(x'', x'' + \sigma_1, \dots, x'' + \sigma_d)\langle x'', u''\rangle^k \, dx'' \,, \quad u'' \in \sigma^\perp \,,$$

depends only on σ. In particular, it is invariant under orthogonal transformations of $(\sigma_1, \dots, \sigma_d)$. In order to use Theorem 6.4 we must show that for any $\sigma, \tau \in \mathbf{G}_{d,n}$ and any $k \in \mathbf{Z}^+$,

$$(73) \quad P_{\sigma,k}(u) = P_{\tau,k}(u) \text{ for } u \in \sigma^\perp \cap \tau^\perp \,, \quad |u| = 1 \,.$$

The following lemma is a basic step towards (73).

Lemma 6.7. *Assume* $\varphi \in \mathbf{G}(d, n)$ *satisfies* (70). *Let*

$$\sigma = (\sigma_1, \ldots, \sigma_d), \tau = (\tau_1, \ldots, \tau_d)$$

be two members of $\mathbf{G}_{d,n}$. *Assume*

$$\sigma_j = \tau_j \quad for\ 2 \le j \le d.$$

Then

$$P_{\sigma,k}(u) = P_{\tau,k}(u) \quad for\ u \in \sigma^{\perp} \cap \tau^{\perp}, \quad |u| = 1.$$

Proof. Let e_i $(1 \le i \le n)$ be the natural basis of $\mathbf{R}^n$ and $\epsilon = (e_1, \ldots, e_d)$. Select $A \in \mathbf{O}(n)$ such that

$$\sigma = A\epsilon, \quad u = Ae_n.$$

Let

$$\eta = A^{-1}\tau = (A^{-1}\tau_1, \ldots, A^{-1}\tau_d) = (A^{-1}\tau_1, e_2, \ldots, e_d).$$

The vector $E = A^{-1}\tau_1$ is perpendicular to e_j $(2 \le j \le d)$ and to e_n (since $u \in \tau^{\perp}$). Thus

$$E = a_1 e_1 + \sum_{d+1}^{n-1} a_i e_i \quad (a_1^2 + \sum_i a_i^2 = 1).$$

In (72) we write $P_{\sigma,k}^{\varphi}$ for $P_{\sigma,k}$. Putting $x'' = Ay$ and $\psi = \varphi \circ A$ we have

$$P_{\sigma,k}^{\varphi}(u) = \int_{\epsilon^{\perp}} \varphi(Ay, A(y + e_1), \ldots, A(y + e_d))\langle y, e_n \rangle^k \, dy = P_{\epsilon,k}^{\psi}(e_n)$$

and similarly

$$P_{\tau,k}^{\varphi}(u) = P_{\eta,k}^{\psi}(e_n).$$

Thus, taking Lemma 6.6 into account, we have to prove the statement:

(74) $$P_{\epsilon,k}(e_n) = P_{\eta,k}(e_n),$$

where $\epsilon = (e_1, \ldots, e_d), \eta = (E, e_2, \ldots, e_d)$, E being any unit vector perpendicular to e_j $(2 \le j \le d)$ and to e_n. First we take

$$E = E_t = \sin t\, e_1 + \cos t\, e_i \quad (d < i < n)$$

and put $\epsilon_t = (E_t, e_2, \ldots, e_d)$. We shall prove

(75) $$P_{\epsilon_t,k}(e_n) = P_{\epsilon,k}(e_n).$$

With no loss of generality we can take $i = d + 1$. The space $\epsilon_t^{\perp}$ consists of the vectors

(76) $$x_t = (-\cos t\, e_1 + \sin t\, e_{d+1})\lambda_{d+1} + \sum_{i=d+2}^{n} \lambda_i e_i, \quad \lambda_i \in \mathbf{R}.$$

Putting $P(t) = P_{\epsilon_t, k}(e_n)$ we have

(77) $$P(t) = \int\limits_{\mathbf{R}^{n-d}} \varphi(x_t, x_t + E_t, x_t + e_2, \ldots, x_t + e_d) \lambda_n^k \, d\lambda_n \ldots d\lambda_{d+1} \, .$$

In order to use (70) we replace φ by the function

$$\psi(x_0, \ldots, x_d) = \varphi(x_0, \ldots, x_d) / V(\{x_i - x_0\}_{i=1,d}) \, .$$

Since the vectors in (77) span volume 1, replacing φ by ψ in (77) does not change $P(t)$. Applying $\partial/\partial t$ we get (with $d\lambda = d\lambda_n \ldots d\lambda_{d+1}$),

(78) $$P'(t) = \int\limits_{\mathbf{R}^{n-d}} \left[\sum_{j=0}^{d} \lambda_{d+1}(\sin t\, \partial_{j,1}\psi + \cos t\, \partial_{j,d+1}\psi) \right.$$
$$\left. + \quad \cos t\, \partial_{1,1}\psi - \sin t\, \partial_{1,d+1}\psi \right] \lambda_n^k \, d\lambda \, .$$

Now φ is a function on $\mathbf{G}(d, n)$. Thus for each $i \neq j$ it is invariant under the substitution

$$y_k = x_k \ (k \neq i), \ y_i = sx_i + (1-s)x_j = x_j + s(x_i - x_j), \quad s > 0 \, ,$$

whereas the volume changes by the factor s. Thus

$$\psi(y_0, \ldots, y_d) = s^{-1}\psi(x_0, \ldots, x_d) \, .$$

Taking $\partial/\partial s$ at $s = 1$ we obtain

(79) $$\psi(x_0, \ldots, x_d) + \sum_{k=1}^{n}(x_i^k - x_j^k)(\partial_{i,k}\psi)(x_0, \ldots, x_d) = 0 \, .$$

Note that in (78) the derivatives are evaluated at

(80) $$(x_0, \ldots, x_d) = (x_t, x_t + E_t, x_t + e_2, \ldots, x_t + e_d) \, .$$

Using (79) for $(i, j) = (1, 0)$ and $(i, j) = (0, 1)$ and adding we obtain

(81) $$\sin t\, (\partial_{0,1}\psi + \partial_{1,1}\psi) + \cos t\, (\partial_{0,d+1}\psi + \partial_{1,d+1}\psi) = 0 \, .$$

For $i \geq 2$ we have

$$x_i - x_0 = e_i, \ x_i - x_1 = -\sin t\, e_1 - \cos t\, e_{d+1} + e_i \, ,$$

and this gives the relations (for $j = 0$ and $j = 1$)

(82) $$\psi(x_0, \ldots, x_d) + (\partial_{i,i}\psi)(x_0, \ldots, x_d) = 0 \, ,$$
(83) $$\psi - \sin t\, (\partial_{i,1}\psi) - \cos t\, (\partial_{i,d+1}\psi) + \partial_{i,i}\psi = 0 \, .$$

Thus by (81)–(83) formula (78) simplifies to

$$P'(t) = \int\limits_{\mathbf{R}^{n-d}} [\cos t \, (\partial_{1,1}\psi) - \sin t \, (\partial_{1,d+1}\psi)] \, \lambda_n^k \, d\lambda \, .$$

In order to bring in 2nd derivatives of ψ we integrate by parts in λ_n,

$$(84) \quad (k+1)P'(t) = \int\limits_{\mathbf{R}^{n-d}} -\frac{\partial}{\partial\lambda_n} [\cos t \, (\partial_{1,1}\psi) - \sin t \, (\partial_{1,d+1}\psi)] \, \lambda_n^{k+1} \, d\lambda \, .$$

Since the derivatives $\partial_{j,k}\psi$ are evaluated at the point (80) we have in (84)

$$(85) \qquad\qquad \frac{\partial}{\partial\lambda_n}(\partial_{j,k}\psi) = \sum_{i=0}^{d} \partial_{i,n}(\partial_{j,k}\psi)$$

and also, by (76) and (80),

$$(86) \quad \frac{\partial}{\partial\lambda_{d+1}}(\partial_{j,k}\psi) = -\cos t \sum_{0}^{d} \partial_{i,1}(\partial_{j,k}\psi) + \sin t \sum_{0}^{d} \partial_{i,d+1}(\partial_{j,k}\psi) \, .$$

We now plug (85) into (84) and then invoke equations (70) for ψ which give

$$(87) \quad \sum_{0}^{d} \partial_{i,n}\partial_{1,1}\psi = \partial_{1,n} \sum_{0}^{d} \partial_{i,1}\psi \, , \quad \sum_{0}^{d} \partial_{i,n}\partial_{1,d+1}\psi = \partial_{1,n} \sum_{0}^{d} \partial_{i,d+1}\psi \, .$$

Using (85) and (87) we see that (84) becomes

$$-(k+1)P'(t) =$$

$$\int\limits_{\mathbf{R}^{n-d}} [\partial_{1,n}(\cos t \, \Sigma_i\partial_{i,1}\psi - \sin t \, \Sigma_i\partial_{i,d+1}\psi)] \, (x_t, x_t+E_t, \dots, x_t+e_d)\lambda_n^{k+1} \, d\lambda,$$

so by (86)

$$(k+1)P'(t) = \int\limits_{\mathbf{R}^{n-d}} \frac{\partial}{\partial\lambda_{d+1}}(\partial_{1,n}\psi)\lambda_n^{k+1} \, d\lambda \, .$$

Since $d+1 < n$, the integration in λ_{d+1} shows that $P'(t) = 0$, proving (75).

This shows that without changing $P_{\epsilon,k}(e_n)$ we can pass from $\epsilon = (e_1, \dots, e_d)$ to

$$\epsilon_t = (\sin t \, e_1 + \cos t \, e_{d+1}, e_2, \dots, e_d) \, .$$

By iteration we can replace e_1 by

$$\sin t_{n-d-1} \dots \sin t_1 e_1 + \sin t_{n-d-1} \dots \sin t_2 \cos t_1 e_{d+1} + \dots + \cos t_{n-d-1}e_{n-1},$$

but keeping $e_2, \dots, e_d$ unchanged. This will reach an arbitrary E, so (74) is proved.

We shall now prove (73) in general. We write σ and τ in orthonormal bases, $\sigma = (\sigma_1, \ldots, \sigma_d), \tau = (\tau_1, \ldots, \tau_d)$. Using Lemma 6.7 we shall pass from σ to τ without changing $P_{\sigma,k}(u)$, u being fixed.

Consider τ_1. If two members of σ, say σ_j and σ_k, are both not orthogonal to τ_1 that is ($\langle \sigma_j, \tau_1 \rangle \neq 0, \langle \sigma_k, \tau_1 \rangle \neq 0$) we rotate them in the (σ_j, σ_k)-plane so that one of them becomes orthogonal to τ_1. As remarked after (72) this has no effect on $P_{\sigma,k}(u)$. We iterate this process (with the same τ_1) and end up with an orthogonal frame ($\sigma_1^*, \ldots, \sigma_d^*$) of σ in which at most one entry σ_i^* is not orthogonal to τ_1. In this frame we replace this σ_i^* by τ_1. By Lemma 6.7 this change of σ does not alter $P_{\sigma,k}(u)$.

We now repeat this process with $\tau_2, \tau_3 \ldots$, etc. Each step leaves $P_{\sigma,k}(u)$ unchanged (and u remains fixed) so this proves (73) and the theorem.

We consider now the case $d = 1, n = 3$ in more detail. Here $f \to \widehat{f}$ is the X-ray transform in $\mathbf{R}^3$. We also change the notation and write ξ for x_0, η for x_1 so $V(\{x_1 - x_0\})$ equals $|\xi - \eta|$. Then Theorem 6.5 reads as follows.

Theorem 6.8. *The X-ray transform $f \to \widehat{f}$ in $\mathbf{R}^3$ is a bijection of $\mathcal{D}(\mathbf{R}^3)$ onto the space of $\varphi \in \mathcal{D}(\mathbf{G}(1,3))$ satisfying*

$$(88) \qquad \left(\frac{\partial}{\partial \xi_k} \frac{\partial}{\partial \eta_\ell} - \frac{\partial}{\partial \xi_\ell} \frac{\partial}{\partial \eta_k} \right) \left(\frac{\varphi(\xi, \eta)}{|\xi - \eta|} \right) = 0, \quad 1 \le k, \ell \le 3.$$

Now let $\mathbf{G}'(1,3) \subset \mathbf{G}(1,3)$ denote the open subset consisting of the *non-horizontal* lines. We shall now show that for $\varphi \in \mathcal{D}(\mathbf{G}(1,n))$ (and even for $\varphi \in \mathcal{E}(\mathbf{G}'(1,n))$) the validity of (88) for $(k, \ell) = (1, 2)$ implies (87) for general (k, ℓ). Note that (79) (which is also valid for $\varphi \in \mathcal{E}(\mathbf{G}'(1,n))$) implies

$$\frac{\varphi(\xi, \eta)}{|\xi - \eta|} + \sum_1^3 (\xi_i - \eta_i) \frac{\partial}{\partial \xi_i} \left(\frac{\varphi(\xi, \eta)}{|\xi - \eta|} \right) = 0.$$

Here we apply $\partial / \partial \eta_k$ and obtain

$$\left(\sum_{i=1}^3 (\xi_i - \eta_i) \frac{\partial^2}{\partial \xi_i \partial \eta_k} - \frac{\partial}{\partial \xi_k} + \frac{\partial}{\partial \eta_k} \right) \left(\frac{\varphi(\xi, \eta)}{|\xi - \eta|} \right) = 0.$$

Exchanging ξ and η and adding we derive

$$(89) \qquad \sum_{i=1}^3 (\xi_i - \eta_i) \left(\frac{\partial^2}{\partial \xi_i \partial \eta_k} - \frac{\partial^2}{\partial \xi_k \partial \eta_i} \right) \left(\frac{\varphi(\xi, \eta)}{|\xi - \eta|} \right) = 0$$

for $k = 1, 2, 3$. Now assume (88) for $(k, \ell) = (1, 2)$. Taking $k = 1$ in (89) we derive (88) for $(k, \ell) = (1, 3)$. Then taking $k = 3$ in (89) we deduce (88) for $(k, \ell) = (3, 2)$. This verifies the claim above.

We can now put this in a simpler form. Let $\ell(\xi, \eta)$ denote the line through the points $\xi \neq \eta$. Then the mapping

$$(\xi_1, \xi_2, \eta_1, \eta_2) \to \ell((\xi_1, \xi_2, 0), (\eta_1, \eta_2, -1))$$

is a bijection of $\mathbf{R}^4$ onto $\mathbf{G}'(1,3)$. The operator

$$(90) \qquad \Lambda = \frac{\partial^2}{\partial \xi_1 \partial \eta_2} - \frac{\partial^2}{\partial \xi_2 \partial \eta_1}$$

is a well defined differential operator on the dense open set $\mathbf{G}'(1,3)$. If $\varphi \in \mathcal{E}(\mathbf{G}(1,3))$ we denote by ψ the restriction of the function $(\xi, \eta) \to \varphi(\xi, \eta)/|\xi - \eta|$ to $\mathbf{G}'(1,3)$. Then we have proved the following result.

Theorem 6.9. *The X-ray transform $f \to \widehat{f}$ is a bijection of $\mathcal{D}(\mathbf{R}^3)$ onto the space*

$$(91) \qquad \{\varphi \in \mathcal{D}(\mathbf{G}(1,3)) : \Lambda\psi = 0\}.$$

We shall now rewrite the differential equation (91) in Plücker coordinates. The line joining ξ and η has Plücker coordinates $(p_1, p_2, p_3, q_1, q_2, q_3)$ given by

$$\begin{vmatrix} \mathbf{i} & \mathbf{j} & \mathbf{k} \\ \xi_1 & \xi_2 & \xi_3 \\ \eta_1 & \eta_2 & \eta_3 \end{vmatrix} = p_1 \mathbf{i} + p_2 \mathbf{j} + p_3 \mathbf{k}, \qquad q_i = \begin{vmatrix} \xi_i & 1 \\ \eta_i & 1 \end{vmatrix}$$

which satisfy

$$(92) \qquad p_1 q_1 + p_2 q_2 + p_3 q_3 = 0.$$

Conversely, each ratio $(p_1 : p_2 : p_3 : q_1 : q_2 : q_3)$ determines uniquely a line provided (92) is satisfied. The set $\mathbf{G}'(1,3)$ is determined by $q_3 \neq 0$. Since the common factor can be chosen freely we fix q_3 as 1. Then we have a bijection $\tau : \mathbf{G}'(1,3) \to \mathbf{R}^4$ given by

$$(93) \qquad x_1 = p_2 + q_2, \ x_2 = -p_1 - q_1, \ x_3 = p_2 - q_2, \ x_4 = -p_1 + q_1$$

with inverse

$$(p_1, p_2, p_3, q_1, q_2) =$$
$$\left(\tfrac{1}{2}(-x_2 - x_4), \tfrac{1}{2}(x_1 + x_3), \tfrac{1}{4}(-x_1^2 - x_2^2 + x_3^2 + x_4^2), \tfrac{1}{2}(-x_2 + x_4), \tfrac{1}{2}(x_1 - x_3) \right).$$

Theorem 6.10. *If $\varphi \in \mathcal{D}(\mathbf{G}(1,3))$ satisfies (91) then the restriction $\varphi | \mathbf{G}'(1,3)$ (with $q_3 = 1$) has the form*

$$(94) \qquad \varphi(\xi, \eta) = |\xi - \eta| \, u(p_2 + q_2, -p_1 - q_1, p_2 - q_2, -p_1 + q_1)$$

where u satisfies

$$(95) \qquad \frac{\partial^2 u}{\partial x_1^2} + \frac{\partial^2 u}{\partial x_2^2} - \frac{\partial^2 u}{\partial x_3^2} - \frac{\partial^2 u}{\partial x_4^2} = 0.$$

On the other hand, if u satisfies (95) then (94) defines a function φ on $\mathbf{G}'(1,3)$ which satisfies (91).

Proof. First assume $\varphi \in \mathcal{D}(\mathbf{G}(1,3))$ satisfies (91) and define $u \in \mathcal{E}(\mathbf{R}^4)$ by

$$(96) \qquad u(\tau(\ell)) = \varphi(\ell)(1 + q_1^2 + q_2^2)^{-\frac{1}{2}},$$

where $\ell \in \mathbf{G}'(1,3)$ has Plücker coordinates $(p_1, p_2, p_3, q_1, q_2, 1)$. On the line ℓ consider the points ξ, η for which $\xi_3 = 0, \eta_3 = -1$ (so $q_3 = 1$). Then since

$$p_1 = -\xi_2, \; p_2 = \xi_1, \; q_1 = \xi_1 - \eta_1, \; q_2 = \xi_2 - \eta_2$$

we have

$$(97) \qquad \frac{\varphi(\xi, \eta)}{|\xi - \eta|} = u(\xi_1 + \xi_2 - \eta_2, -\xi_1 + \xi_2 + \eta_1, \xi_1 - \xi_2 + \eta_2, \xi_1 + \xi_2 - \eta_1).$$

Now (91) implies (95) by use of the chain rule.

On the other hand, suppose $u \in \mathcal{E}(\mathbf{R}^4)$ satisfies (95). Define φ by (96). Then $\varphi \in \mathcal{E}(\mathbf{G}'(1,3))$ and by (97),

$$\Lambda\left(\frac{\varphi(\xi, \eta)}{|\xi - \eta|}\right) = 0.$$

As shown before the proof of Theorem 6.9 this implies that the whole system (88) is verified.

We shall now see what implications Ásgeirsson's mean-value theorem Chapter VI, §2 has for the range of the X-ray transform. We have from (95),

$$(98) \qquad \int_0^{2\pi} u(r\cos\varphi, r\sin\varphi, 0, 0)\, d\varphi = \int_0^{2\pi} u(0, 0, r\cos\varphi, r\sin\varphi)\, d\varphi.$$

The first points $(r\cos\varphi, r\sin\varphi, 0, 0)$ correspond via (93) to the lines with

$$(p_1, p_2, p_3, q_1, q_2, q_3) = (-\tfrac{r}{2}\sin\varphi, \tfrac{r}{2}\cos\varphi, -\tfrac{r^2}{4}, -\tfrac{r}{2}\sin\varphi, \tfrac{r}{2}\cos\varphi, 1)$$

containing the points

$$(\xi_1, \xi_2, \xi_3) = (\tfrac{r}{2}\cos\varphi, \tfrac{r}{2}\sin\varphi, 0)$$
$$(\eta_1, \eta_2, \eta_3) = (\tfrac{r}{2}(\sin\varphi + \cos\varphi), +\tfrac{r}{2}(\sin\varphi - \cos\varphi), -1)$$

with $|\xi - \eta|^2 = 1 + \tfrac{r^2}{4}$. The points $(0, 0, r\cos\varphi, r\sin\varphi)$ correspond via (93) to the lines with

$$(p_1, p_2, p_3, q_1, q_2, q_3) = (-\tfrac{r}{2}\sin\varphi, \tfrac{r}{2}\cos\varphi, \tfrac{r^2}{4}, \tfrac{r}{2}\sin\varphi, -\tfrac{r}{2}\cos\varphi, 1)$$

containing the points

$$(\xi_1, \xi_2, \xi_3) = (\tfrac{r}{2}\cos\varphi, \tfrac{r}{2}\sin\varphi, 0),$$
$$(\eta_1, \eta_2, \eta_3) = (\tfrac{r}{2}(\cos\varphi - \sin\varphi), \tfrac{r}{2}(\cos\varphi + \sin\varphi), -1),$$

with $|\xi - \eta|^2 = 1 + \frac{r^2}{4}$. Thus (98) takes the form

$$(99) \quad \int_0^{2\pi} \varphi(\tfrac{r}{2}\cos\theta, \tfrac{r}{2}\sin\theta, 0, \tfrac{r}{2}(\sin\theta + \cos\theta), \tfrac{r}{2}(\sin\theta - \cos\theta), -1)\, d\theta$$

$$= \int_0^{2\pi} \varphi(\tfrac{r}{2}\cos\theta, \tfrac{r}{2}\sin\theta, 0, \tfrac{r}{2}(\cos\theta - \sin\theta), \tfrac{r}{2}(\cos\theta + \sin\theta), -1)\, d\theta\,.$$

The lines forming the arguments of φ in these integrals are the two families of generating lines for the hyperboloid (see Fig. I.7)

$$x^2 + y^2 = \tfrac{r^2}{4}(z^2 + 1)\,.$$

Definition. A function $\varphi \in \mathcal{E}(\mathbf{G}'(1,3))$ is said to be a **harmonic line function** if

$$\Lambda\left(\frac{\varphi(\xi,\eta)}{|\xi - \eta|}\right) = 0\,.$$

Theorem 6.11. *A function $\varphi \in \mathcal{E}(\mathbf{G}'(1,3))$ is a harmonic line function if and only if for each hyperboloid of revolution H of one sheet and vertical axis the mean values of φ over the two families of generating lines of H are equal. (The variable of integration is the polar angle in the equatorial plane of H.).*

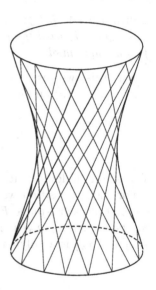

FIGURE I.7.

The proof of (99) shows that φ harmonic implies the mean value property for φ. The converse is also true since Ásgeirsson's theorem has a converse; for $\mathbf{R}^n$ this is obvious from the relation between L and M^r in Ch. VII, §6.

Corollary 6.12. *Let $\varphi \in \mathcal{D}(\mathbf{G}(1,3))$. Then φ is in the range of the X-ray transform if and only if φ has the mean value property for arbitrary hyperboloid of revolution of one sheet (and arbitrary axis).*

We conclude this section with the following result due to Schlichtkrull mentioned in connection with Theorem 6.4.

Proposition 6.13. *Let f be a function on $\mathbf{R}^n$ and $k \in \mathbf{Z}^+$, $1 < k < n$. Assume that for each k-dimensional subspace $E_k \subset \mathbf{R}^n$ the restriction $f|E_k$ is a polynomial on E_k. Then f is a polynomial on $\mathbf{R}^n$.*

For $k = 1$ the result is false as the example $f(x,y) = xy^2/(x^2 + y^2)$, $f(0,0) = 0$ shows. We recall now the Lagrange interpolation formula. Let

$a_0, \ldots, a_m$ be distinct numbers in $\mathbf{C}$. Then each polynomial $P(x)$ $(x \in \mathbf{R})$ of degree $\leq m$ can be written

$$P(x) = P(a_0)Q_0(x) + \cdots + P(a_m)Q_m(x) \,,$$

where

$$Q_i(x) = \prod_{j=0}^{m} (x - a_j)/(x - a_i) \prod_{j \neq i} (a_i - a_j) \,.$$

In fact, the two sides agree at $m + 1$ distinct points. This implies the following result.

Lemma 6.14. *Let $f(x_1, \ldots, x_n)$ be a function on $\mathbf{R}^n$ such that for each i with $x_j(j \neq i)$ fixed the function $x_i \to f(x_1, \ldots, x_n)$ is a polynomial. Then f is a polynomial.*

For this we use Lagrange's formula on the polynomial $x_1 \longrightarrow f(x_1, x_2, \ldots, x_n)$ and get

$$f(x_1, \ldots, x_n) = \sum_{j=0}^{m} f(a_j, x_2, \ldots, x_m)Q_j(x_1) \,.$$

The lemma follows by iteration.

For the proposition we observe that the assumption implies that f restricted to each 2-plane E_2 is a polynomial on E_2. For a fixed $(x_2, \ldots, x_n)$ the point $(x_1, \ldots, x_n)$ is in the span of $(1, 0, \ldots, 0)$ and $(0, x_2, \ldots, x_n)$ so $f(x_1, \ldots, x_n)$ is a polynomial in x_1. Now the lemma implies the result.

§7   Applications

A.   Partial Differential Equations. The Wave Equation

The inversion formula in Theorem 3.1 is very well suited for applications to partial differential equations. To explain the underlying principle we write the inversion formula in the form

$$(100) \qquad f(x) = \gamma L_x^{\frac{n-1}{2}} \left(\int_{\mathbf{S}^{n-1}} \widehat{f}(\omega, \langle x, \omega \rangle) \, d\omega \right) ,$$

where the constant γ equals $\frac{1}{2}(2\pi i)^{1-n}$. Note that the function $f_\omega(x) = \widehat{f}(\omega, \langle x, \omega \rangle)$ is a **plane wave with normal** ω, that is, it is constant on each hyperplane perpendicular to ω.

Consider now a differential operator

$$D = \sum_{(k)} a_{k_1 \ldots k_n} \partial_1^{k_1} \ldots \partial_n^{k_n}$$

with constant coefficients $a_{k_1,\ldots,k_n}$, and suppose we want to solve the differential equation

(101) $$Du = f,$$

where f is a given function in $\mathcal{S}(\mathbf{R}^n)$. To simplify the use of (100) we assume n to be odd. We begin by considering the differential equation

(102) $$Dv = f_\omega,$$

where f_ω is the plane wave defined above and we look for a solution v which is also a plane wave with normal ω. But a plane wave with normal ω is just a function of one variable; also if v is a plane wave with normal ω so is the function Dv. The differential equation (102) (with v a plane wave) is therefore an **ordinary** differential equation with constant coefficients. Suppose $v = u_\omega$ is a solution and assume that this choice can be made smoothly in ω. Then the function

(103) $$u = \gamma L^{\frac{n-1}{2}} \int_{\mathbf{S}^{n-1}} u_\omega \, d\omega$$

is a solution to the differential equation (101). In fact, since D and $L^{\frac{n-1}{2}}$ commute we have

$$Du = \gamma L^{\frac{n-1}{2}} \int_{\mathbf{S}^{n-1}} Du_\omega \, d\omega = \gamma L^{\frac{n-1}{2}} \int_{\mathbf{S}^{n-1}} f_\omega \, d\omega = f.$$

This method only assumes that the plane wave solution u_ω to the ordinary differential equation $Dv = f_\omega$ exists and can be chosen so as to depend smoothly on ω. This cannot always be done because D might annihilate all plane waves with normal ω. (For example, take $D = \partial^2/\partial x_1 \partial x_2$ and $\omega = (1,0)$.) However, if this restriction to plane waves is never 0 it follows from a theorem of Trèves [1963] that the solution u_ω can be chosen depending smoothly on ω. Thus we can state the following result.

Theorem 7.1. *If the restriction D_ω of D to the space of plane waves with normal ω is $\neq 0$ for each ω, then formula (103) gives a solution to the differential equation $Du = f$ $(f \in \mathcal{S}(\mathbf{R}^n))$.*

The method of plane waves can also be used to solve the Cauchy problem for hyperbolic differential equations with constant coefficients. We illustrate this method by means of the wave equation $\mathbf{R}^n$,

(104) $$Lu = \frac{\partial^2 u}{\partial t^2}, \quad u(x,0) = f_0(x), \quad u_t(x,0) = f_1(x),$$

f_0, f_1 being given functions in $\mathcal{D}(\mathbf{R}^n)$.

Lemma 7.2. *Let $h \in C^2(\mathbf{R})$ and $\omega \in \mathbf{S}^{n-1}$. Then the function*

$$v(x, t) = h(\langle x, \omega \rangle + t)$$

satisfies $Lv = (\partial^2/\partial t^2)v$.

The proof is obvious. It is now easy, on the basis of Theorem 3.8, to write down the unique solution of the Cauchy problem (104).

Theorem 7.3. *The solution to (104) is given by*

$$(105) \qquad u(x, t) = \int_{\mathbf{S}^{n-1}} (Sf)(\omega, \langle x, \omega \rangle + t)\, d\omega,$$

where

$$Sf = \begin{cases} c(\partial^{n-1}\widehat{f_0} + \partial^{n-2}\widehat{f_1}), & n \text{ odd} \\ c\,\mathcal{H}(\partial^{n-1}\widehat{f_0} + \partial^{n-2}\widehat{f_1}), & n \text{ even}. \end{cases}$$

Here $\partial = \partial/\partial p$ and the constant c equals

$$c = \tfrac{1}{2}(2\pi i)^{1-n}.$$

Lemma 7.2 shows that (105) is annihilated by the operator $L - \partial^2/\partial t^2$, so we just have to check the initial conditions in (104).

(a)   If $n > 1$ is odd then $\omega \to (\partial^{n-1}\widehat{f_0})(\omega, \langle x, \omega \rangle)$ is an even function on $\mathbf{S}^{n-1}$ but the other term in Sf, that is the function $\omega \to (\partial^{n-2}\widehat{f_1})(\omega, \langle x, \omega \rangle)$, is odd. Thus by Theorem 3.8, $u(x, 0) = f_0(x)$. Applying $\partial/\partial t$ to (105) and putting $t = 0$ gives $u_t(x, 0) = f_1(x)$, this time because the function $\omega \to (\partial^n \widehat{f_0})(\omega, \langle x, \omega \rangle)$ is odd and the function $\omega \to (\partial^{n-1}\widehat{f_1})(\omega, \langle x, \omega \rangle)$ is even.

(b)   If n is even the same proof works if we take into account the fact that $\mathcal{H}$ interchanges odd and even functions on $\mathbf{R}$.

Definition. For the pair $f = \{f_0, f_1\}$ we refer to the function Sf in (105) as the **source**.

In the terminology of Lax-Philips [1967] the wave $u(x, t)$ is said to be

(a)   **outgoing** if $u(x, t) = 0$ in the **forward cone** $|x| < t$;

(b)   **incoming** if $u(x, t) = 0$ in the **backward cone** $|x| < -t$.

The notation is suggestive because "outgoing" means that the function $x \to u(x, t)$ vanishes in larger balls around the origin as t increases.

Corollary 7.4. *The solution $u(x, t)$ to (104) is*

(i)   *outgoing if and only if* $(Sf)(\omega, s) = 0$ *for* $s > 0$, *all* ω.

(ii)   *incoming if and only if* $(Sf)(\omega, s) = 0$ *for* $s < 0$, *all* ω.

Proof. For (i) suppose $(Sf)(\omega, s) = 0$ for $s > 0$. For $|x| < t$ we have $\langle x, \omega \rangle + t \geq -|x| + t > 0$ so by (105) $u(x, t) = 0$ so u is outgoing. Conversely, suppose $u(x, t) = 0$ for $|x| < t$. Let $t_0 > 0$ be arbitrary and let $\varphi(t)$ be a smooth function with compact support contained in (t_0, ∞).
    Then if $|x| < t_0$ we have

$$
\begin{aligned}
0 &= \int_{\mathbf{R}} u(x, t)\varphi(t)\, dt = \int_{\mathbf{S}^{n-1}} d\omega \int_{\mathbf{R}} (Sf)(\omega, \langle x, \omega \rangle + t)\varphi(t)\, dt \\
&= \int_{\mathbf{S}^{n-1}} d\omega \int_{\mathbf{R}} (Sf)(\omega, p)\varphi(p - \langle x, \omega \rangle)\, dp\,.
\end{aligned}
$$

Taking arbitrary derivative $\partial^k / \partial x_{i_1} \ldots \partial x_{i_k}$ at $x = 0$ we deduce

$$
\int_{\mathbf{R}} \left(\int_{\mathbf{S}^{n-1}} (Sf)(\omega, p)\omega_{i_1} \ldots \omega_{i_k}\, d\omega \right) (\partial^k \varphi)(p)\, dp = 0
$$

for each k and each $\varphi \in \mathcal{D}(t_0, \infty)$. Integrating by parts in the p variable we conclude that the function

$$
(106) \qquad p \longrightarrow \int_{\mathbf{S}^{n-1}} (Sf)(\omega, p)\omega_{i_1} \ldots \omega_{i_k}\, d\omega\,, \qquad p \in \mathbf{R}
$$

has its kth derivative $\equiv 0$ for $p > t_0$. Thus it equals a polynomial for $p > t_0$. However, if n is odd the function (106) has compact support so it must vanish identically for $p > t_0$.
    On the other hand, if n is even and $F \in \mathcal{D}(\mathbf{R})$ then as remarked at the end of §3, $\lim_{|t| \to \infty} (\mathcal{H}F)(t) = 0$. Thus we conclude again that expression (106) vanishes identically for $p > t_0$.
    Thus in both cases, if $p > t_0$, the function $\omega \to (Sf)(\omega, p)$ is orthogonal to all polynomials on $\mathbf{S}^{n-1}$, hence must vanish identically.

    One can also solve (104) by means of the **Fourier transform** (see Ch. VII)

$$
\widetilde{f}(\zeta) = \int_{\mathbf{R}^n} f(x)e^{-i\langle x, \zeta \rangle}\, dx\,.
$$

Assuming the function $x \to u(x, t)$ in $\mathcal{S}(\mathbf{R}^n)$, for a given t we obtain

$$
\widetilde{u}_{tt}(\zeta, t) + \langle \zeta, \zeta \rangle \widetilde{u}(\zeta, t) = 0\,.
$$

Solving this ordinary differential equation with initial data given in (104), we get

$$(107) \qquad \widetilde{u}(\zeta, t) = \widetilde{f_0}(\zeta)\cos(|\zeta|t) + \widetilde{f_1}(\zeta)\frac{\sin(|\zeta|t)}{|\zeta|}\,.$$

The function $\zeta \to \sin(|\zeta|t)/|\zeta|$ is entire of exponential type $|t|$ on $\mathbf{C}^n$ in the sense of (18), Ch. VII. In fact, if $\varphi(\lambda)$ is even, holomorphic on $\mathbf{C}$ and satisfies the exponential type estimate (18) in Theorem 4.7, Ch. VII, then the same holds for the function Φ on $\mathbf{C}^n$ given by $\Phi(\zeta) = \Phi(\zeta_1, \ldots, \zeta_n) = \varphi(\lambda)$ where $\lambda^2 = \zeta_1^2 + \cdots + \zeta_n^2$. To see this put

$$\lambda = \mu + i\nu\,, \quad \zeta = \xi + i\eta \qquad \mu, \nu \in \mathbf{R}, \quad \xi, \eta \in \mathbf{R}^n\,.$$

Then

$$\mu^2 - \nu^2 = |\xi|^2 - |\eta|^2\,, \quad \mu^2\nu^2 = (\xi \cdot \eta)^2\,,$$

so

$$|\lambda|^4 = (|\xi|^2 - |\eta|^2)^2 + 4(\xi \cdot \eta)^2$$

and

$$2|\mathrm{Im}\,\lambda|^2 = |\eta|^2 - |\xi|^2 + \left[(|\xi|^2 - |\eta|^2)^2 + 4(\xi \cdot \eta)^2\right]^{1/2}\,.$$

Since $|(\xi \cdot \eta)| \le |\xi|\,|\eta|$ this implies $|\mathrm{Im}\,\lambda| \le |\eta|$ so the estimate (18) follows for Φ. Thus by Chapter VII, §4 there exists a $T_t \in \mathcal{E}'(\mathbf{R}^n)$ with support in $\overline{B_{|t|}(0)}$ such that

$$\frac{\sin(|\zeta|t)}{|\zeta|} = \int\limits_{\mathbf{R}^n} e^{-i\langle\zeta, x\rangle}\,dT_t(x)\,.$$

Theorem 7.5. *Given $f_0, f_1 \in \mathcal{E}(\mathbf{R}^n)$ the function*

$$(108) \qquad u(x, t) = (f_0 * T_t')(x) + (f_1 * T_t)(x)$$

satisfies (104). Here T_t' stands for $\partial_t(T_t)$.

Note that (104) implies (108) if f_0 and f_1 have compact support. The converse holds without this support condition.

Corollary 7.6. *If f_0 and f_1 have support in $B_R(0)$ then u has support in the region*

$$|x| \le |t| + R\,.$$

In fact, by (108) and support property of convolutions (Ch. VII, §3), the function $x \to u(x, t)$ has support in $B_{R+|t|}(0)^-$. While Corollary 7.6 implies that for $f_0, f_1 \in \mathcal{D}(\mathbf{R}^n)$ u has support in a suitable solid cone, we shall now see that Theorem 7.3 implies that if n is odd u has support in a conical shell (see Fig. I.8).

Corollary 7.7. *Let n be odd. Assume f_0 and f_1 have support in the ball $B_R(0)$.*

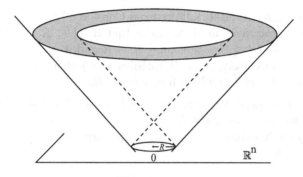

FIGURE I.8.

(i) **Huygens' Principle.** *The solution u to (104) has support in the conical shell*

(109) $$|t| - R \le |x| \le |t| + R,$$

which is the union for $|y| \le R$ of the light cones,

$$C_y = \{(x, t) : |x - y| = |t|\}.$$

(ii) The solution to (104) is outgoing if and only if

(110) $$\widehat{f_0}(\omega, p) = \int_p^\infty \widehat{f_1}(\omega, s) \, ds, \quad p > 0, \text{ all } \omega,$$

and incoming if and only if

$$\widehat{f_0}(\omega, p) = -\int_{-\infty}^p \widehat{f_1}(\omega, s) \, ds, \quad p < 0, \text{ all } \omega.$$

Note that Part (ii) can also be stated: The solution is outgoing (incoming) if and only if

$$\int_\pi f_0 = \int_{H_\pi} f_1 \quad \left(\int_\pi f_0 = -\int_{H_\pi} f_1 \right)$$

for an arbitrary hyperplane π $(0 \notin \pi)$ H_π being the halfspace with boundary π which does not contain 0.

To verify (i) note that since n is odd, Theorem 7.3 implies

(111) $$u(0, t) = 0 \quad \text{for} \quad |t| \ge R.$$

If $z \in \mathbf{R}^n$, $F \in \mathcal{E}(\mathbf{R}^n)$ we denote by F^z the translated function $y \to F(y+z)$. Then u^z satisfies (104) with initial data f_0^z, f_1^z which have support contained in $B_{R+|z|}(0)$. Hence by (111),

(112) $$u(z, t) = 0 \quad \text{for} \quad |t| > R + |z|.$$

The other inequality in (109) follows from Corollary 7.6.

For the final statement in (i) we note that if $|y| \leq R$ and $(x,t) \in C_y$ then $|x - y| = t$ so $|x| \leq |x - y| + |y| \leq |t| + R$ and $|t| = |x - y| \leq |x| + R$ proving (109). Conversely, if (x,t) satisfies (109) then $(x,t) \in C_y$ with $y = x - |t|\frac{x}{|x|} = \frac{x}{|x|}(|x| - t)$ which has norm $\leq R$.

For (ii) we just observe that since $\widehat{f}_i(\omega, p)$ has compact support in p, (110) is equivalent to (i) in Corollary 7.4.

Thus (110) implies that for $t > 0$, $u(x,t)$ has support in the thinner shell $|t| \leq |x| \leq |t| + R$.

B. X-ray Reconstruction

The classical interpretation of an X-ray picture is an attempt at reconstructing properties of a 3-dimensional body by means of the X-ray projection on a plane.

In modern X-ray technology the picture is given a more refined mathematical interpretation. Let $B \subset \mathbf{R}^3$ be a body (for example a part of a human body) and let $f(x)$ denote its density at a point x. Let ξ be a line in $\mathbf{R}^3$ and suppose a thin beam of X-rays is directed at B along ξ. Let I_0 and I respectively, denote the intensity of the beam before entering B and after leaving B (see

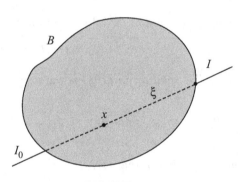

FIGURE I.9.

Fig. I.9). As the X-ray traverses the distance Δx along ξ it will undergo the relative intensity loss $\Delta I / I = f(x)\,\Delta x$. Thus $dI/I = -f(x)\,dx$ whence

$$(113) \qquad \log(I_0/I) = \int_\xi f(x)\,dx\,,$$

the integral $\widehat{f}(\xi)$ of f along ξ. Since the left hand side is determined by the X-ray picture, the **X-ray reconstruction problem** amounts to the determination of the function f by means of its line integrals $\widehat{f}(\xi)$. *The inversion formula in Theorem 3.1 gives an explicit solution of this problem.*

If $B_0 \subset B$ is a convex subset (for example the heart) it may be of interest to determine the density of f outside B_0 using only X-rays which do not intersect B_0. **The support theorem (Theorem 2.6, Cor. 2.8 and Cor. 6.1) implies that f is determined outside B_0 on the basis of the integrals $\widehat{f}(\xi)$ for which ξ does not intersect B_0. Thus the**

density outside the heart can be determined by means of X-rays which bypass the heart.

In practice one can of course only determine the integrals $\widehat{f}(\xi)$ in (113) for **finitely** many directions. A compensation for this is the fact that only an approximation to the density f is required. One then encounters the mathematical problem of selecting the directions so as to optimize the approximation.

As before we represent the line ξ as the pair $\xi = (\omega, z)$ where $\omega \in \mathbf{R}^n$ is a unit vector in the direction of ξ and $z = \xi \cap \omega^{\perp}$ ($\perp$ denoting orthogonal complement). We then write

$$(114) \qquad \widehat{f}(\xi) = \widehat{f}(\omega, z) = (P_\omega f)(z).$$

The function $P_\omega f$ is the X-ray picture or the **radiograph** in the direction ω. Here f is a function on $\mathbf{R}^n$ vanishing outside a ball B around the origin, and for the sake of Hilbert space methods to be used it is convenient to assume in addition that $f \in L^2(B)$. Then $f \in L^1(\mathbf{R}^n)$ so by the Fubini theorem we have: for each $\omega \in \mathbf{S}^{n-1}$, $P_\omega f(z)$ is defined for almost all $z \in \omega^{\perp}$. Moreover, we have in analogy with (4),

$$(115) \qquad \widetilde{f}(\zeta) = \int_{\omega^{\perp}} (P_\omega f)(z) e^{-i\langle z, \zeta\rangle}\, dz \quad (\zeta \in \omega^{\perp}).$$

Proposition 7.8. *An object is determined by any infinite set of radiographs.*

In other words, a compactly supported function f is determined by the functions $P_\omega f$ for any infinite set of ω.

Proof. Since f has compact support $\widetilde{f}$ is an analytic function on $\mathbf{R}^n$. But if $\widetilde{f}(\zeta) = 0$ for $\zeta \in \omega^{\perp}$ we have $\widetilde{f}(\eta) = \langle \omega, \eta\rangle g(\eta)$ $(\eta \in \mathbf{R}^n)$ where g is also analytic. If $P_{\omega_1} f, \ldots, P_{\omega_k} f \ldots$ all vanish identically for an infinite set $\omega_1, \ldots, \omega_k \ldots$, we see that for each k

$$\widetilde{f}(\eta) = \prod_{i=1}^{k} \langle \omega_i, \eta\rangle g_k(\eta),$$

where g_k is analytic. But this would contradict the power series expansion of $\widetilde{f}$ which shows that for a suitable $\omega \in \mathbf{S}^{n-1}$ and integer $r \geq 0$, $\lim_{t\to 0} \widetilde{f}(t\omega) t^{-r} \neq 0$.

If only finitely many radiographs are used we get the opposite result.

Proposition 7.9. *Let $\omega_1, \ldots, \omega_k \in \mathbf{S}^{n-1}$ be an arbitrary finite set. Then there exists a function $f \in \mathcal{D}(\mathbf{R}^n)$, $f \not\equiv 0$, such that*

$$P_{\omega_i} f \equiv 0 \quad \text{for all} \quad 1 \leq i \leq k.$$

Proof. We have to find $f \in \mathcal{D}(\mathbf{R}^n), f \not\equiv 0$, such that $\widetilde{f}(\zeta) = 0$ for $\zeta \in \omega_i^\perp (1 \leq i \leq k)$. For this let D be the constant coefficient differential operator such that

$$(Du\widetilde{)}(\eta) = \prod_1^k \langle \omega_i, \eta \rangle \widetilde{u}(\eta) \quad \eta \in \mathbf{R}^n .$$

If $u \not\equiv 0$ is any function in $\mathcal{D}(\mathbf{R}^n)$ then $f = Du$ has the desired property.

We next consider the problem of **approximate reconstruction** of the function f from a finite set of radiographs $P_{\omega_1} f, \ldots, P_{\omega_k} f$.

Let N_j denote the null space of P_{ω_j} and let P_j denote the orthogonal projection of $L^2(B)$ on the plane $f + N_j$; in other words,

(116) $$P_j g = Q_j(g - f) + f ,$$

where Q_j is the (linear) projection onto the subspace $N_j \subset L^2(B)$. Put $P = P_k \ldots P_1$. Let $g \in L^2(B)$ be arbitrary (the initial guess for f) and form the sequence $P^m g, m = 1, 2, \ldots$. Let $N_0 = \cap_1^k N_j$ and let P_0 (resp. Q_0) denote the orthogonal projection of $L^2(B)$ on the plane $f + N_0$ (subspace N_0). We shall prove that the sequence $P^m g$ converges to the projection $P_0 g$. This is natural since by $P_0 g - f \in N_0$, $P_0 g$ and f have the same radiographs in the directions $\omega_1, \ldots, \omega_k$.

Theorem 7.10. *With the notations above,*

$$P^m g \longrightarrow P_0 g \quad as \quad m \longrightarrow \infty$$

for each $g \in L^2(B)$.

Proof. We have, by iteration of (116)

$$(P_k \ldots P_1)g - f = (Q_k \ldots Q_1)(g - f)$$

and, putting $Q = Q_k \ldots Q_1$ we obtain

$$P^m g - f = Q^m(g - f) .$$

We shall now prove that $Q^m g \longrightarrow Q_0 g$ for each g; since

$$P_0 g = Q_0(g - f) + f$$

this would prove the result. But the statement about Q^m comes from the following general result about abstract Hilbert space.

Theorem 7.11. *Let $\mathcal{H}$ be a Hilbert space and Q_i the projection of $\mathcal{H}$ onto a subspace $N_i \subset \mathcal{H}(1 \leq i \leq k)$. Let $N_0 = \cap_1^k N_i$ and $Q_0 : \mathcal{H} \longrightarrow N_0$ the projection. Then if $Q = Q_k \ldots Q_1$,*

$$Q^m g \longrightarrow Q_0 g \quad for \ each \quad g \in \mathcal{H}.$$

Since Q is a contraction ($\|Q\| \leq 1$) we begin by proving a simple lemma about such operators.

Lemma 7.12. *Let $T : \mathcal{H} \longrightarrow \mathcal{H}$ be a linear operator of norm ≤ 1. Then*

$$\mathcal{H} = C\ell((I - T)\mathcal{H}) \oplus \text{ Null space } (I - T)$$

is an orthogonal decomposition, $C\ell$ denoting closure, and I the identity.

Proof. If $Tg = g$ then since $\|T^*\| = \|T\| \leq 1$ we have

$$\|g\|^2 = (g, g) = (Tg, g) = (g, T^*g) \leq \|g\| \|T^*g\| \leq \|g\|^2,$$

so all terms in the inequalities are equal. Hence

$$\|g - T^*g\|^2 = \|g\|^2 - (g, T^*g) - (T^*g, g) + \|T^*g\|^2 = 0,$$

so $T^*g = g$. Thus $I - T$ and $I - T^*$ have the same null space. But $(I - T^*)g = 0$ is equivalent to $(g, (I - T)\mathcal{H}) = 0$, so the lemma follows.

Definition. An operator T on a Hilbert space $\mathcal{H}$ is said to have **property S** if

$$(117) \qquad \|f_n\| \leq 1, \|Tf_n\| \longrightarrow 1 \text{ implies } \|(I - T)f_n\| \longrightarrow 0.$$

Lemma 7.13. *A projection, and more generally a finite product of projections, has property (S).*

Proof. If T is a projection then

$$\|(I - T)f_n\|^2 = \|f_n\|^2 - \|Tf_n\|^2 \leq 1 - \|Tf_n\|^2 \longrightarrow 0$$

whenever

$$\|f_n\| \leq 1 \text{ and } \|Tf_n\| \longrightarrow 1.$$

Let T_2 be a projection and suppose T_1 has property (S) and $\|T_1\| \leq 1$. Suppose $f_n \in \mathcal{H}$ and $\|f_n\| \leq 1, \|T_2T_1f_n\| \longrightarrow 1$. The inequality implies $\|T_1f_n\| \leq 1$ and since

$$\|T_1f_n\|^2 = \|T_2T_1f_n\|^2 + \|(I - T_2)(T_1f_n)\|^2$$

we also deduce $\|T_1f_n\| \longrightarrow 1$. Writing

$$(I - T_2T_1)f_n = (I - T_1)f_n + (I - T_2)T_1f_n$$

we conclude that T_2T_1 has property (S). The lemma now follows by induction.

Lemma 7.14. *Suppose T has property (S) and $\|T\| \leq 1$. Then for each $f \in \mathcal{H}$*

$$T^n f \longrightarrow \pi f \quad as \quad n \longrightarrow \infty,$$

where π is the projection onto the fixed point space of T.

Proof. Let $f \in \mathcal{H}$. Since $\|T\| \leq 1$, $\|T^n f\|$ decreases monotonically to a limit $\alpha \geq 0$. If $\alpha = 0$ we have $T^n f \longrightarrow 0$. By Lemma 7.12 $\pi T = T\pi$ so $\pi f = T^n \pi f = \pi T^n f$ so $\pi f = 0$ in this case. If $\alpha > 0$ we put $g_n = \|T^n f\|^{-1}(T^n f)$. Then $\|g_n\| = 1$ and $\|T g_n\| \to 1$. Since T has property (S) we deduce

$$T^n(I - T)f = \|T^n f\|(I - T)g_n \longrightarrow 0.$$

Thus $T^n h \longrightarrow 0$ for all h in the range of $I - T$. If g is in the closure of this range then given $\epsilon > 0$ there exist $h \in (I - T)\mathcal{H}$ such that $\|g - h\| < \epsilon$. Then

$$\|T^n g\| \leq \|T^n(g - h)\| + T^n h\| < \epsilon + \|T^n h\|,$$

whence $T^n g \longrightarrow 0$. On the other hand, if h is in the null space of $I - T$ then $Th = h$ so $T^n h \longrightarrow h$. Now the lemma follows from Lemma 7.12.

In order to deduce Theorem 7.11 from Lemmas 7.13 and 7.14 we just have to verify that N_0 is the fixed point space of Q. But if $Qg = g$, then

$$\|g\| = \|Q_k \ldots Q_1 g\| \leq \|Q_{k-1} \ldots Q_1 g\| \leq \ldots \leq \|Q_1 g\| \leq \|g\|$$

so equality signs hold everywhere. But the Q_i are projections so the norm identities imply

$$g = Q_1 g = Q_2 Q_1 g = \ldots = Q_k \ldots Q_1 g$$

which shows $g \in N_0$. This proves Theorem 7.11.

Exercises and Further Results

1.   Radon Transform on Measures. (Hertle [1979], Boman-Lindskog [2009])

Let $C_0(\mathbf{R}^n)$ denote the space of continuous complex-valued functions on $\mathbf{R}^n$ vanishing at ∞, taken with the uniform norm. The dual space is the space $\mathbf{M}(\mathbf{R}^n)$ of bounded measures on $\mathbf{R}^n$. The Radon transform can be defined on $\mathbf{M}(\mathbf{R}^n)$ in analogy with the distribution definition of §5. The spaces $C_0(\mathbf{P}^n)$ and $\mathbf{M}(\mathbf{P}^n)$ are defined in the obvious fashion.

(i)   If $\varphi \in C_0(\mathbf{P}^n)$ then $\check{\varphi} \in C_0(\mathbf{R}^n)$.

(ii)   For $\mu \in M(\mathbf{R}^n)$ define $\widehat{\mu}$ by

$$\widehat{\mu}(\varphi) = \mu(\check{\varphi}).$$

Then $\widehat{\mu} \in \mathbf{M}(\mathbf{P}^n)$.

(iii) Given $\mu \in \mathbf{M}(\mathbf{R}^n)$ its Fourier transform $\tilde{\mu}$ is defined by

$$\tilde{\mu}(\xi) = \int\limits_{\mathbf{R}^n} e^{-i\langle x,\xi \rangle}\, d\mu(x)\,, \quad \xi \in \mathbf{R}^n\,.$$

Then if $\sigma \in \mathbf{R}$, $|\omega| = 1$,

$$\tilde{\mu}(\sigma\omega) = \int\limits_{\mathbf{R}} \hat{\mu}(\omega, p)e^{-i\sigma p}\, dp\,.$$

(iv) The map $\mu \to \hat{\mu}$ is injective.

(v) The measure $\mu \in \mathbf{M}(\mathbf{R}^n)$ with norm $\|\mu\|$ is said to be rapidly decreasing at ∞ if $\|\chi_r \mu\| = O(r^{-m})$ as $r \to \infty$ for each $m > 0$. Here χ_r is a continuous approximation to the characteristic function of $\mathbf{R}^n - B_r(0)$ (i.e., $\chi_r(x) = 0$ for $|x| \leq r - 1$, $\chi_r = 1$ for $|x| \geq r$ and $0 \leq \chi_r(x) \leq 1$ for all x).

(vi) Let $K \subset \mathbf{R}^n$ be a compact and convex subset. Assume $\hat{\mu}$ rapidly decreasing and vanishing on the open set of hyperplanes not intersecting K. Then $\mu = 0$ on $\mathbf{R}^n \backslash K$. (Use the convolution method of Theorem 5.4.)

2.   The Hilbert Transform.

(i) $\mathcal{H}$ extends to an isometry of $L^2(\mathbf{R})$ into $L^2(\mathbf{R})$.

(ii)   (Titchmarsh [1948]). For $f \in L^2(\mathbf{R})$ put

$$g(x) = i\, \mathcal{H}f(x)\,.$$

Then

$$f(x) = -i\, \mathcal{H}g(x)\,.$$

The functions f and g are then called *conjugate*. Thus $\mathcal{H}^2 = I$ on $L^2(\mathbf{R})$. Moreover, the following conditions are equivalent:

(a) A function $\Phi_0 \in L^2(\mathbf{R})$ is the limit of a holomorphic function $\Phi(z)$ in $\operatorname{Im} z > 0$ satisfying

$$\int\limits_{\mathbf{R}} |\Phi(x + iy)|^2\, dx < K\,.$$

(b) $\Phi_0(x) = f(x) - ig(x)$ where $f, g \in L^2$ are conjugate.
(c) The Fourier transform of Φ_0,

$$\int \Phi_0(x)e^{-ix\xi}\, dx\,,$$

vanishes for $\xi > 0$.

(iii) Let φ_a be the characteristic function of $(0, a)$. Then

$$\varphi_a(x) \quad \text{and} \quad \frac{1}{\pi} \log \left| \frac{a+y}{a-y} \right|$$

are conjugate; so are the functions

$$\frac{1}{1+x^2} \quad \text{and} \quad \frac{y}{1+y^2},$$

$$\frac{\sin(ax)}{x} \quad \text{and} \quad \frac{\cos(ay)-1}{y}.$$

(iv) The Hilbert transform has the following properties:

    (a) It commutes with translations.

    (b) It commutes with dilations $x \to tx$, $t > 0$.

    (c) Anticommutes with $f(x) \to f(-x)$.

Conversely, a bounded operator on $\mathbf{L}^2(\mathbf{R})$ satisfying (a), (b) and (c) is a constant multiple of $\mathcal{H}$, (see e.g. Stein [1970], p. 55).

3. The Inversion Formula Interpreted in Terms of Distributions.

Given $\omega \in \mathbf{S}^{n-1}$, $p \in \mathbf{R}$ let $T_{p,\omega}$ denote the distribution on $\mathbf{R}^n$ given by

$$T_{p,\omega}(f) = \widehat{f}(\omega, p).$$

Then $T_{p,\omega}$ has support in the hyperplane $\langle \omega, x \rangle = p$ and $\frac{d}{dp}(T_{p,\omega})$ is the normal derivative of this distribution. For $p = 0$ we write this as $\frac{\partial}{\partial \nu} T_\omega$. For n odd Theorem 3.8 can be written

$$\delta = c \int_{\mathbf{S}^{n-1}} \left(\frac{\partial^{n-1}}{\partial \nu^{n-1}} T_\omega \right) d\omega$$

which is a decomposition of the delta function into plane supported distributions.

4. Convolutions.

With the method of proof of (5) show that with $\times$ denoting the convolution in (5),

$$\check{\varphi} * f = (\varphi \times \widehat{f})^\vee$$

for $\varphi \in \mathcal{D}(\mathbf{P}^n)$, $f \in \mathcal{D}(\mathbf{R}^n)$. (Natterer [1986].)

5. Exterior Problem.

The transform $f \to \widehat{f}$ is a one-to-one mapping of $L^2(\mathbf{R}^n, |x|^{n-1} dx)$ into $L^2(\mathbf{S}^{n-1} \times \mathbf{R})$. If B is the exterior of a ball with center 0 in $\mathbf{R}^n$ ($n \geq 3$), then the null space of $f \to \widehat{f}$ on $L^2(B, |x|^{n-1} dx)$ is the closure of the span of functions of the form $f(x) = |x|^{-n-k} Y_\ell(x/|x|)$ where $0 \leq k \leq \ell$, $\ell - k$ is even, and Y_ℓ is a homogeneous spherical harmonic of degree ℓ (Quinto [1982]).

6.   **Support Property.**

   This is a refinement of Lemma 2.7 (Quinto [2008]).

(i)   Let $f \in C^\infty(\mathbf{R}^n)$ be rapidly decreasing. Fix $x_0 \in \mathbf{R}^n$ and a neighborhood U of x_0. Let $r_0 > 0$ and assume

$$(M^r f)(x) = 0 \quad \text{for all} \quad x \in U, \quad \text{and all} \quad r > r_0.$$

Then $f \equiv 0$ outside $B_{r_0}(x_0)$.

(ii)   The statement holds for an analytic Riemannian manifold M with infinite injectivity radius, provided $f \in C_c^\infty(M)$.

7.   **Geometric Inversion.**

   For d even prove the inversion formula in Theorem 6.2 by the geometric method used for the hyperplane case (Theorem 3.1).

8.   **Density in $\mathbf{P}^2$.**

   When parametrizing the set of lines in $\mathbf{R}^2$ by using $ux + vy = 1$, the $\mathbf{M}(2)$ invariant measure is given by

$$\frac{du\, dv}{(u^2 + v^2)^{3/2}}.$$

(Cartan [1896].)

9.   **Generalized Radon Transform.**

   Let $\mu \in C^\infty(\mathbf{R}^n \times \mathbf{S}^{n-1} \times \mathbf{R})$ be a strictly positive function such that $\mu(z, \omega, p) = \mu(z, -\omega, -p)$, and R_μ the corresponding Radon transform

$$(R_\mu f)(\omega, p) = \int_{\langle \omega, x \rangle = p} f(x)\mu(x, \omega, p)\, dx.$$

Call R_μ *invariant* if for $f_a(x) = f(x + a)$

$$(R_\mu f_a)(\omega, p) = \nu(a, \omega, p)(R_\mu f)(\omega, p + \langle a, \omega \rangle) \qquad a \in \mathbf{R}^n,$$

where $\nu(a, \omega, p) = \mu(-a, \omega, p)/\mu(0, \omega, p + \langle a, \omega \rangle)$. Then R_μ is invariant if and only if

$$\mu(x, \omega, p) = \mu_1(\omega, p)e^{\langle \mu_2(\omega), x \rangle},$$

with $\mu_1 \in \mathcal{E}(\mathbf{S}^{n-1} \times \mathbf{R})$, $\mu_2 : \mathbf{S}^{n-1} \to \mathbf{R}$, C^∞ maps. In this case, R_μ is injective on $\mathcal{D}(\mathbf{R}^n)$ and if $(R_\mu f)(\omega, p) \equiv 0$ for $p > r$ then $\text{supp}(f) \subset \overline{B_r(0)}$ (Kurusa [1991c]).

10. Topological Properties of the Radon Transform.

(i)   The range $\mathcal{D}(\mathbf{R}^n)\widehat{}$ is closed in $\mathcal{D}_H(\mathbf{P}^n)$ and the range $\mathcal{E}'(\mathbf{R}^n)\widehat{}$ is closed in $\mathcal{E}'(\mathbf{P}^n)$ (Helgason [GGA], Chapter I, Ex. B4, B5).

(ii)   The transform $f \to \widehat{f}$ is a homeomorphism of $\mathcal{S}(\mathbf{R}^n)$ onto $\mathcal{S}_H(\mathbf{R}^n)$. (The closedness follows as in (i) and since both sides are Fréchet spaces, the result follows.)

(iii) $\mathcal{E}(\mathbf{P}^n)^{\vee} = \mathcal{E}(\mathbf{R}^n)$ (Hertle [1984a]). (This follows from (i) and the Fréchet space result that if $\alpha : E \to F$ is a continuous mapping and E and F Fréchet spaces then α is surjective iff $\,{}^t\alpha(F')$ is weak* closed in E' and $\alpha' : F' \to E'$ injective; take α as $\vee$ and ${}^t\alpha$ as $\widehat{}$.)

(iv)   The mapping $S \to \check{S}$ of $\mathcal{D}'(\mathbf{S}^{n-1} \times \mathbf{R})$ into $\mathcal{D}'(\mathbf{R}^n)$ is not surjective. In fact, Hertle shows [1984a] that the distribution

$$T(x) = \sum_{j=1}^{\infty} \Big(\frac{\partial}{\partial x_2}\Big)^{jn} \delta(x - x_j) \qquad (x_j = j, 0, \ldots 0)$$

is not in the range. Parts (iv) and (v) have not been verified by the author.

(v)   The mapping $f \to \widehat{f}$ is not a homeomorphism of $\mathcal{D}(\mathbf{R}^n)$ onto its image. (Hertle [1984].) (By the results under (iii) for the LF-space $\mathcal{D}$, the injectivity of $f \to \widehat{f}$ implies that the dual map $S \to \check{S}$ of $\mathcal{D}'(\mathbf{P}^n) \to \mathcal{D}'(\mathbf{R}^n)$ has a dense range. Thus by (iv) the range of $S \to \check{S}$ is not closed. Thus by Schaefer [1971, Ch. IV, 7.5], $f \to \widehat{f}$ is not a homeomorphism.)

Bibliographical Notes

§§*1-2.*   The inversion formulas

(i) $f(x) = \frac{1}{2}(2\pi i)^{1-n} L_x^{(n-1)/2} \int_{\mathbf{S}^{n-1}} J(\omega, \langle \omega, x \rangle)\, , d\omega$ (n odd)

(ii) $f(x) = \frac{1}{2}(2\pi i)^{-n} L_x^{(n-2)/2} \int_{\mathbf{S}^{n-1}} d\omega \int_{-\infty}^{\infty} \frac{dJ(\omega,p)}{p-\langle\omega,x\rangle}$ (n even)

for a function $f \in \mathcal{D}(X)$ in terms of its plane integrals $J(\omega, p)$ go back to Radon [1917] and John [1934], [1955]. According to Bockwinkel [1906] the case $n = 3$ had been proved before 1906 by H.A. Lorentz, but fortunately, both for Lorentz and Radon, the transformation $f(x) \to J(\omega, p)$ was not baptized "Lorentz transformation". In John [1955] the proofs are based on the Poisson equation $Lu = f$. Other proofs, using distributions, are given in Gelfand-Shilov [1960]. See also Nievergelt [1986]. The dual transforms, $f \to \widehat{f}, \varphi \to \check{\varphi}$, the unified inversion formula and its dual,

$$cf = L^{(n-1)/2}((\widehat{f})^{\vee}), \quad c\varphi = \square^{(n-1)/2}((\check{\varphi})\widehat{}),$$

were given by the author in [1964]. The first proof of Theorem 3.1 in the text is from the author's paper [1959]. It is valid for constant curvature spaces as well. The version in Theorem 3.8 is also proved in Ludwig [1966].

The support theorem, the Paley-Wiener theorem and the Schwartz theorem (Theorems 2.4,2.6, 2.10) are from Helgason [1964], [1965a]. The example in Remark 2.9 was also found by D.J. Newman, cf. Weiss' paper [1967], which gives another proof of the support theorem. See also Droste [1983]. The local result in Corollary 2.12 goes back to John [1935]; our derivation is suggested by the proof of a similar lemma in Flensted-Jensen [1977], p. 83. Another proof is in Ludwig [1966].See Palamodov and Denisjuk [1988] for a related inversion formula.

The simple geometric Lemma 2.7 is from the author's paper [1965a] and is extended to hyperbolic spaces in [1980b]. In the Proceedings containing [1966a] the author raised the problem (p. 174) to extend Lemma 2.7 to each complete simply connected Riemannian manifold M of negative curvature. If in addition M is analytic this was proved by Quinto [1993b] and Grinberg and Quinto [2000]. This is an example of injectivity and support results obtained by use of the techniques of microlocal analysis and wave front sets. As further samples involving very general Radon transforms we mention Boman [1990], [1992], [1993], Boman and Quinto [1987], [1993], Quinto [1983], [1992], [1993b], [1994a], [1994b], Agranovsky and Quinto [1996], Gelfand, Gindikin and Shapiro [1979].

Corollary 2.8 is derived by Ludwig [1966] in a different way. There he proposes alternative proofs of the Schwartz- and Paley-Wiener theorems by expanding $\widehat{f}(\omega, p)$ in spherical harmonics in ω. However, the proof fails because the principal point—the smoothness of the function F in the proof of Theorem 2.4 (the Schwartz theorem)—is overlooked. Theorem 2.4 is from the author's papers [1964] [1965a].

This proof of the Schwartz theorem in [1965a] is adopted in Palamodov [2004] and Gelfand–Gindikin–Graev [2003]. These versions do not seem to me to take sufficiently into account the needed relationship in (11) and (12), §2. See also Carton–Lebrun [1984] for a generalization.

Since the inversion formula is rather easy to prove for odd n it is natural to try to prove Theorem 2.4 for this case by showing directly that if $\varphi \in \mathcal{S}_H(\mathbf{P}^n)$ then the function $f = cL^{(n-1)/2}(\check{\varphi})$ for n odd belongs to $\mathcal{S}(\mathbf{R}^n)$ (in general $\check{\varphi} \notin \mathcal{S}(\mathbf{R}^n)$). This approach is taken in Gelfand-Graev-Vilenkin [1966], pp. 16-17. However, this method seems to offer some unresolved technical difficulties. For some generalizations see Kuchment and Lvin [1990], Aguilar, Ehrenpreis and Kuchment [1996] and Katsevich [1997]. Cor. 2.5 is stated in Semyanisty [1960].

§5.   The approach to Radon transforms of distributions adopted in the text is from the author's paper [1966a]. Other methods are proposed in Gelfand-Graev-Vilenkin [1966] and in Ludwig [1966]. See also Ramm [1995].

§6. Formula (60) for the d-plane transform was proved by Fuglede [1958]. The inversion in Theorem 6.2 is from Helgason [1959], p. 284. The range characterization for the d-plane transform in Theorem 6.3 is from our book [1980c] and was used skillfully by Kurusa [1991] to prove Theorem 6.5, which generalizes John's range theorem for the X-ray transform in $\mathbf{R}^3$ [1938]. The geometric range characterization (Corollary 6.12) is also due to John [1938]. Papers devoted to the d-plane range question for $\mathcal{S}(\mathbf{R}^n)$ are Gelfand-Gindikin and Graev [1982], Grinberg [1987], Richter [1986] and Gonzalez [1991]. This last paper gives the range as the kernel of a single 4th order differential operator on the space of d-planes. As shown by Gonzalez, the analog of Theorem 6.3 fails to hold for $\mathcal{S}(\mathbf{R}^n)$. An L^2-version of Theorem 6.3 was given by Solmon [1976], p. 77. Proposition 6.13 was communicated to me by Schlichtkrull.

Some difficulties with the d-plane transform on $L^2(\mathbf{R}^n)$ are pointed out by Smith and Solmon [1975] and Solmon [1976], p. 68. In fact, the function $f(x) = |x|^{-\frac{1}{2}n}(\log|x|)^{-1}$ $(|x| \geq 2)$, 0 otherwise, is square integrable on $\mathbf{R}^n$ but is not integrable over any plane of dimension $\geq \frac{n}{2}$. Nevertheless, see for example Rubin [1998a], Strichartz [1981], Markoe [2006] for L^p-extensions of the d-plane transform.

§7. The applications to partial differential equations go in part back to Herglotz [1931]; see John [1955]. Other applications of the Radon transform to partial differential equations with constant coefficients can be found in Courant-Lax [1955], Gelfand-Shapiro [1955], John [1955], Borovikov [1959], Gårding [1961] and Ludwig [1966]. Our discussion of the wave equation (Theorem 7.3 and Corollary 7.4) is closely related to the treatment in Lax-Phillips [1967], Ch. IV, where however, the dimension is assumed to be odd. Applications to general elliptic equations were given by John [1955].

While the Radon transform on $\mathbf{R}^n$ can be used to "reduce" partial differential equations to ordinary differential equations one can use a Radon type transform on a symmetric space X to "reduce" an invariant differential operator D on X to a partial differential operator with constant coefficients. For an account of these applications see the author's monograph [1994b], Chapter V.

While the applications to differential equations are perhaps the most interesting to mathematicians, the tomographic applications of the X-ray transform (see §7, B) have revolutionized medicine. These applications originated with Cormack [1963], [1964] and Hounsfield [1973]. See §7, B for the medical relevance of the support theorem. For the approximate reconstruction problem, including Propositions 7.8 and 7.9 and refinements of Theorems 7.10, 7.11 see Smith, Solmon and Wagner [1977], Solmon [1976] and Hamaker and Solmon [1978]. Theorem 7.11 is due to Halperin [1962], the proof in the text to Amemiya and Ando [1965]. For an account of some of those applications see e.g. Deans [1983], Natterer [1986], Markoe [2006] and Ramm and Katsevich [1996]. Applications in radio astronomy appear in Bracewell and Riddle [1967].

CHAPTER II

A DUALITY IN INTEGRAL GEOMETRY

§1   Homogeneous Spaces in Duality

The inversion formulas in Theorems 3.1, 3.7, 3.8 and 6.2, Ch. I suggest the general problem of determining a function on a manifold by means of its integrals over certain submanifolds. This is essentially the title of Radon's paper. In order to provide a natural framework for such problems we consider the Radon transform $f \to \widehat{f}$ on $\mathbf{R}^n$ and its dual $\varphi \to \check{\varphi}$ from a group-theoretic point of view, motivated by the fact that the isometry group $\mathbf{M}(n)$ acts transitively both on $\mathbf{R}^n$ and on the hyperplane space $\mathbf{P}^n$. Thus

(1) $$\mathbf{R}^n = \mathbf{M}(n)/\mathbf{O}(n), \quad \mathbf{P}^n = \mathbf{M}(n)/\mathbf{Z}_2 \times \mathbf{M}(n-1),$$

where $\mathbf{O}(n)$ is the orthogonal group fixing the origin $0 \in \mathbf{R}^n$ and $\mathbf{Z}_2 \times \mathbf{M}(n-1)$ is the subgroup of $\mathbf{M}(n)$ leaving a certain hyperplane ξ_0 through 0 stable. ($\mathbf{Z}_2$ consists of the identity and the reflection in this hyperplane.)

We observe now that a point $g_1 \mathbf{O}(n)$ in the first coset space above lies on a plane $g_2(\mathbf{Z}_2 \times \mathbf{M}(n-1))$ in the second if and only if these cosets, considered as subsets of $\mathbf{M}(n)$, have a point in common. In fact

$$g_1 \cdot 0 \subset g_2 \cdot \xi_0 \quad \Leftrightarrow \quad g_1 \cdot 0 = g_2 h \cdot 0 \text{ for some } h \in \mathbf{Z}_2 \times \mathbf{M}(n-1)$$
$$\Leftrightarrow \quad g_1 k = g_2 h \text{ for some } k \in \mathbf{O}(n).$$

This leads to the following general setup.

Let G be a locally compact group, X and Ξ two left coset spaces of G,

(2) $$X = G/K, \quad \Xi = G/H,$$

where K and H are closed subgroups of G. Let $L = K \cap H$. We assume that the subset $KH \subset G$ is **closed**. This is automatic if one of the groups K or H is compact.

Two elements $x \in X, \xi \in \Xi$ are said to be *incident* if as cosets in G they intersect. We put (see Fig. II.1)

$$\check{x} = \{\xi \in \Xi : x \text{ and } \xi \text{ incident}\}$$
$$\widehat{\xi} = \{x \in X : x \text{ and } \xi \text{ incident}\}.$$

Let $x_0 = \{K\}$ and $\xi_0 = \{H\}$ denote the origins in X and Ξ, respectively. If $\Pi : G \to G/H$ denotes the natural mapping then since $\check{x}_0 = K \cdot \xi_0$ we have

$$\Pi^{-1}(\Xi - \check{x}_0) = \{g \in G : gH \notin KH\} = G - KH.$$

S. Helgason, *Integral Geometry and Radon Transforms*,
DOI 10.1007/978-1-4419-6055-9_2, © Springer Science+Business Media, LLC 2010

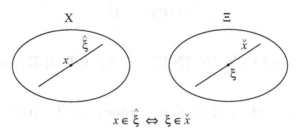

$$x \in \widehat{\xi} \iff \xi \in \check{x}$$

FIGURE II.1.

In particular $\Pi(G - KH) = \Xi - \check{x}_0$ so since Π is an open mapping, $\check{x}_0$ is a closed subset of Ξ. This proves the following:

Lemma 1.1. *Each $\check{x}$ and each $\widehat{\xi}$ is closed.*

Using the notation $A^g = gAg^{-1}$ ($g \in G, A \subset G$) we have the following lemma.

Lemma 1.2. *Let $g, \gamma \in G$, $x = gK$, $\xi = \gamma H$. Then*

$$\check{x} \text{ is an orbit of } K^g, \ \widehat{\xi} \text{ is an orbit of } H^\gamma,$$

and

$$\check{x} = K^g/L^g, \quad \widehat{\xi} = H^\gamma/L^\gamma.$$

Proof. By definition

(3) $$\check{x} = \{\delta H \ : \ \delta H \cap gK \neq \emptyset\} = \{gkH \ : \ k \in K\},$$

which is the orbit of the point gH under gKg^{-1}. The subgroup fixing gH is $gKg^{-1} \cap gHg^{-1} = L^g$. Also (3) implies

$$\check{x} = g \cdot \check{x}_0, \qquad \widehat{\xi} = \gamma \cdot \widehat{\xi}_0,$$

where the dot $\cdot$ denotes the action of G on X and Ξ.

We often write $\tau(g)$ for the maps $x \to g \cdot x, \ \xi \to g \cdot \xi$ and

$$f^{\tau(g)}(x) = f(g^{-1} \cdot x), \quad S^{\tau(g)}(f) = S(f^{\tau(g^{-1})})$$

for f a function, S a distribution.

Lemma 1.3. *Consider the subgroups*

$$\begin{aligned} K_H &= \{k \in K \ : \ kH \cup k^{-1}H \subset HK\}, \\ H_K &= \{h \in H \ : \ hK \cup h^{-1}K \subset KH\}. \end{aligned}$$

The following properties are equivalent:

(a) $K \cap H = K_H = H_K$.

(b)   *The maps* $x \to \check{x} \, (x \in X)$ *and* $\xi \to \hat{\xi} \, (\xi \in \Xi)$ *are injective.*

We think of property (a) as a kind of **transversality** of K and H.

Proof. Suppose $x_1 = g_1 K$, $x_2 = g_2 K$ and $\check{x}_1 = \check{x}_2$. Then by (3) $g_1 \cdot \check{x}_0 = g_1 \cdot \check{x}_0$ so $g \cdot \check{x}_0 = \check{x}_0$ if $g = g_1^{-1} g_2$. In particular $g \cdot \xi_0 \subset \check{x}_0$ so $g \cdot \xi_0 = k \cdot \xi_0$ for some $k \in K$. Hence $k^{-1} g = h \in H$ so $h \cdot \check{x}_0 = \check{x}_0$, that is $hK \cdot \xi_0 = K \cdot \xi_0$. As a relation in G, this means $hKH = KH$. In particular $hK \subset KH$. Since $h \cdot \check{x}_0 = \check{x}_0$ implies $h^{-1} \cdot \check{x}_0 = \check{x}_0$ we have also $h^{-1} K \subset KH$ so by (a) $h \in K$ which gives $x_1 = x_2$.

On the other hand, suppose the map $x \to \check{x}$ is injective and suppose $h \in H$ satisfies $h^{-1} K \cup hK \subset KH$. Then

$$hK \cdot \xi_0 \subset K \cdot \xi_0 \text{ and } h^{-1} K \cdot \xi_0 \subset K \cdot \xi_0.$$

By Lemma 1.2, $h \cdot \check{x}_0 \subset \check{x}_0$ and $h^{-1} \cdot \check{x}_0 \subset \check{x}_0$. Thus $h \cdot \check{x}_0 = \check{x}_0$ whence by the assumption, $h \cdot x_0 = x_0$ so $h \in K$.

Thus we see that under the transversality assumption a) the elements ξ can be viewed as the subsets $\hat{\xi}$ of X and the elements x as the subsets $\check{x}$ of Ξ. We say X and Ξ are **homogeneous spaces in duality**.

The maps are also conveniently described by means of the following **double fibration**

(4)

$$
\begin{array}{ccc}
 & G/L & \\
{}^{p}\swarrow & & \searrow^{\pi} \\
G/K & & G/H
\end{array}
$$

where $p(gL) = gK$, $\pi(\gamma L) = \gamma H$. In fact, by (3) we have

$$\check{x} = \pi(p^{-1}(x)) \qquad \hat{\xi} = p(\pi^{-1}(\xi)).$$

We now prove some group-theoretic properties of the incidence, supplementing Lemma 1.3.

Theorem 1.4. *(i)   We have the identification*

$$G/L = \{(x, \xi) \in X \times \Xi : x \text{ and } \xi \text{ incident}\}$$

via the bijection $\tau : gL \to (gK, gH)$.

(ii)   The property

$$KHK = G$$

is equivalent to the property:

Any two $x_1, x_2 \in X$ *are incident to some* $\xi \in \Xi$. *A similar statement holds for* $HKH = G$.

(iii) The property

$$HK \cap KH = K \cup H$$

is equivalent to the property:

For any two $x_1 \neq x_2$ in X there is at most one $\xi \in \Xi$ incident to both. By symmetry, this is equivalent to the property:

For any $\xi_1 \neq \xi_2$ in Ξ there is at most one $x \in X$ incident to both.

Proof. (i)   The map is well-defined and injective. The surjectivity is clear because if $gK \cap \gamma H \neq \emptyset$ then $gk = \gamma h$ and $\tau(gkL) = (gK, \gamma H)$.

(ii)   We can take $x_2 = x_0$. Writing $x_1 = gK$, $\xi = \gamma H$ we have

$$
\begin{aligned}
x_0, \xi \text{ incident} &\Leftrightarrow \gamma h &=& \quad k \quad (\text{some } h \in H, k \in K) \\
x_1, \xi \text{ incident} &\Leftrightarrow \gamma h_1 &=& \quad g_1 k_1 \quad (\text{some } h_1 \in H, k_1 \in K).
\end{aligned}
$$

Thus if x_0, x_1 are incident to ξ we have $g_1 = kh^{-1}h_1 k_1^{-1}$. Conversely if $g_1 = k'h'k''$ we put $\gamma = k'h'$ and then x_0, x_1 are incident to $\xi = \gamma H$.

(iii) Suppose first $KH \cap HK = K \cup H$. Let $x_1 \neq x_2$ in X. Suppose $\xi_1 \neq \xi_2$ in Ξ are both incident to x_1 and x_2. Let $x_i = g_i K$, $\xi_j = \gamma_j H$. Since x_i is incident to ξ_j there exist $k_{ij} \in K$, $h_{ij} \in H$ such that

$$(5) \qquad g_i k_{ij} = \gamma_j h_{ij} \qquad i = 1,2; \; j = 1,2.$$

Eliminating g_i and γ_j we obtain

$$(6) \qquad k_{22}^{-1} k_{21} h_{21}^{-1} h_{11} = h_{22}^{-1} h_{12} k_{12}^{-1} k_{11}.$$

This being in $KH \cap HK$ it lies in $K \cup H$. If the left hand side is in K then $h_{21}^{-1} h_{11} \in K$, so

$$g_2 K = \gamma_1 h_{21} K = \gamma_1 h_{11} K = g_1 K,$$

contradicting $x_2 \neq x_1$. Similarly if expression (6) is in H, then $k_{12}^{-1} k_{11} \in H$, so by (5) we get the contradiction

$$\gamma_2 H = g_1 k_{12} H = g_1 k_{11} H = \gamma_1 H.$$

Conversely, suppose $KH \cap HK \neq K \cup H$. Then there exist h_1, h_2, k_1, k_2 such that $h_1 k_1 = k_2 h_2$ and $h_1 k_1 \notin K \cup H$. Put $x_1 = h_1 K$, $\xi_2 = k_2 H$. Then $x_1 \neq x_0$, $\xi_0 \neq \xi_2$, yet both ξ_0 and ξ_2 are incident to both x_0 and x_1.

Examples

(i) **Points outside hyperplanes.** We saw before that if in the coset space representation (1) $\mathbf{O}(n)$ is viewed as the isotropy group of 0 and $\mathbf{Z}_2\mathbf{M}(n-1)$ is viewed as the isotropy group of a hyperplane *through* 0 then the abstract incidence notion is equivalent to the naive one: $x \in \mathbf{R}^n$ is incident to $\xi \in \mathbf{P}^n$ if and only if $x \in \xi$.

On the other hand we can also view $\mathbf{Z}_2\mathbf{M}(n-1)$ as the isotropy group of a hyperplane ξ_δ at a distance $\delta > 0$ from 0. (This amounts to a different embedding of the group $\mathbf{Z}_2\mathbf{M}(n-1)$ into $\mathbf{M}(n)$.) Then we have the following generalization.

Proposition 1.5. *The point* $x \in \mathbf{R}^n$ *and the hyperplane* $\xi \in \mathbf{P}^n$ *are incident if and only if distance* $(x, \xi) = \delta$.

Proof. Let $x = gK$, $\xi = \gamma H$ where $K = \mathbf{O}(n), H = \mathbf{Z}_2\mathbf{M}(n-1)$. Then if $gK \cap \gamma H \neq \emptyset$, we have $gk = \gamma h$ for some $k \in K, h \in H$. Now the orbit $H \cdot 0$ consists of the two planes ξ'_δ and ξ''_δ parallel to ξ_δ at a distance δ from ξ_δ. The relation

$$g \cdot 0 = \gamma h \cdot 0 \in \gamma \cdot (\xi'_\delta \cup \xi''_\delta)$$

together with the fact that g and γ are isometries shows that x has distance δ from $\gamma \cdot \xi_\delta = \xi$.

On the other hand if distance $(x, \xi) = \delta$, we have $g \cdot 0 \in \gamma \cdot (\xi'_\delta \cup \xi''_\delta) = \gamma H \cdot 0$, which means $gK \cap \gamma H \neq \emptyset$.

(ii) **Unit spheres.** Let σ_0 be a sphere in $\mathbf{R}^n$ of radius one passing through the origin. Denoting by Σ the set of all *unit* spheres in $\mathbf{R}^n$, we have the dual homogeneous spaces

(7) $$\mathbf{R}^n = \mathbf{M}(n)/\mathbf{O}(n); \quad \Sigma = \mathbf{M}(n)/\mathbf{O}^*(n)$$

where $\mathbf{O}^*(n)$ is the set of rotations around the center of σ_0. Here a point $x = g\mathbf{O}(n)$ is incident to $\sigma_0 = \gamma\mathbf{O}^*(n)$ if and only if $x \in \sigma$.

§2 The Radon Transform for the Double Fibration

With K, H and L as in §1 we assume now that K/L and H/L have positive measures $d\mu_0 = dk_L$ and $dm_0 = dh_L$ invariant under K and H, respectively. This is for example guaranteed if L is compact.

Lemma 2.1. *Assume the transversality condition* (a). *Then there exists a measure on each* $\check{x}$ *coinciding with* $d\mu_0$ *on* $K/L = \check{x}_0$ *such that whenever* $g \cdot \check{x}_1 = \check{x}_2$ *the measures on* $\check{x}_1$ *and* $\check{x}_2$ *correspond under* g. *A similar statement holds for* dm *on* $\hat{\xi}$.

Proof. If $\tilde{x} = g \cdot \tilde{x}_0$ we transfer the measure $d\mu_0 = dk_L$ over on $\tilde{x}$ by the map $\xi \to g \cdot \xi$. If $g \cdot \tilde{x}_0 = g_1 \cdot \tilde{x}_0$ then $(g \cdot x_0)^\vee = (g_1 \cdot x_0)^\vee$ so by Lemma 1.3, $g \cdot x_0 = g_1 \cdot x_0$ so $g = g_1 k$ with $k \in K$. Since $d\mu_0$ is K-invariant the lemma follows.

The measures defined on each $\tilde{x}$ and $\hat{\xi}$ under condition (a) are denoted by $d\mu$ and dm, respectively.

Definition. The Radon transform $f \to \hat{f}$ and its dual $\varphi \to \check{\varphi}$ are defined by

$$(8) \qquad \hat{f}(\xi) = \int_{\hat{\xi}} f(x)\, dm(x), \qquad \check{\varphi}(x) = \int_{\tilde{x}} \varphi(\xi)\, d\mu(\xi),$$

whenever the integrals converge. Because of Lemma 1.1, this will always happen for $f \in C_c(X)$, $\varphi \in C_c(\Xi)$.

In the setup of Proposition 1.5, $\hat{f}(\xi)$ is the integral of f over the two hyperplanes at distance δ from ξ and $\check{\varphi}(x)$ is the average of φ over the set of hyperplanes at distance δ from x. For $\delta = 0$ we recover the transforms of Ch. I, §1.

Formula (8) can also be written in the group-theoretic terms,

$$(9) \qquad \hat{f}(\gamma H) = \int_{H/L} f(\gamma h K)\, dh_L, \qquad \check{\varphi}(gK) = \int_{K/L} \varphi(gkH)\, dk_L.$$

Note that (9) serves as a definition even if condition (a) in Lemma 1.3 is not satisfied. In this abstract setup the spaces X and Ξ have equal status. The theory in Ch. I, in particular Lemma 2.1, Theorems 2.4, 2.10, 3.1 thus raises the following problems:

Principal Problems:

A.   Relate function spaces on X and on Ξ by means of the transforms $f \to \hat{f}$, $\varphi \to \check{\varphi}$. In particular, determine their ranges and kernels.

B.   Invert the transforms $f \to \hat{f}$, $\varphi \to \check{\varphi}$ on suitable function spaces.

C.   In the case when G is a Lie group so X and Ξ are manifolds let $\mathbf{D}(X)$ and $\mathbf{D}(\Xi)$ denote the algebras of G-invariant differential operators on X and Ξ, respectively. Is there a map $D \to \hat{D}$ of $\mathbf{D}(X)$ into $\mathbf{D}(\Xi)$ and a map $E \to \check{E}$ of $\mathbf{D}(\Xi)$ into $\mathbf{D}(X)$ such that

$$(Df)\hat{} = \hat{D}\hat{f}, \qquad (E\varphi)^\vee = \check{E}\check{\varphi}?$$

D.   Support Property: Does $\hat{f}$ of compact support imply that f has compact support?

Although weaker assumptions would be sufficient, we assume now that the groups G, K, H and L all have bi-invariant Haar measures dg, dk, dh and $d\ell$. These will then generate invariant measures dg_K, dg_H, dg_L, dk_L, dh_L on G/K, G/H, G/L, K/L, H/L, respectively. This means that

$$(10) \qquad \int\limits_G F(g)\, dg = \int\limits_{G/K} \left(\int\limits_K F(gk)\, dk \right) dg_K$$

and similarly dg and dh determine dg_H, etc. Then

$$(11) \qquad \int\limits_{G/L} Q(gL)\, dg_L = c \int\limits_{G/K} dg_K \int\limits_{K/L} Q(gkL)\, dk_L$$

for $Q \in C_c(G/L)$ where c is a constant. In fact, the integrals on both sides of (11) constitute invariant measures on G/L and thus must be proportional. However,

$$(12) \qquad \int\limits_G F(g)\, dg = \int\limits_{G/L} \left(\int\limits_L F(g\ell)\, d\ell \right) dg_L$$

and

$$(13) \qquad \int\limits_K F(k)\, dk = \int\limits_{K/L} \left(\int\limits_L F(k\ell)\, d\ell \right) dk_L \,.$$

We use (13) on (10) and combine with (11) taking $Q(gL) = \int F(g\ell)\, d\ell$. Then we see that from (12) the constant c equals 1.

We shall now prove that $f \to \widehat{f}$ and $\varphi \to \widecheck{\varphi}$ are adjoint operators. We write dx for dg_K and $d\xi$ for dg_H.

Proposition 2.2. Let $f \in C_c(X)$, $\varphi \in C_c(\Xi)$. Then $\widehat{f}$ and $\widecheck{\varphi}$ are continuous and

$$\int\limits_X f(x)\widecheck{\varphi}(x)\, dx = \int\limits_\Xi \widehat{f}(\xi)\varphi(\xi)\, d\xi \,.$$

Proof. The continuity statement is immediate from (9). We consider the function

$$P = (f \circ p)(\varphi \circ \pi)$$

on G/L. We integrate it over G/L in two ways using the double fibration (4). This amounts to using (11) and its analog with G/K replaced by G/H with $Q = P$. Since $P(gk\, L) = f(gK)\varphi(gkH)$, the right hand side of (11) becomes

$$\int\limits_{G/K} f(gK)\widecheck{\varphi}(gK)\, dg_K \,.$$

If we treat G/H similarly, the lemma follows.

The result shows how to define the Radon transform and its dual for measures and, in case G is a Lie group, for distributions.

Definition. Let s be a measure on X of compact support. Its Radon transform is the functional $\widehat{s}$ on $C_c(\Xi)$ defined by

$$(14) \qquad\qquad \widehat{s}(\varphi) = s(\widecheck{\varphi}).$$

Similarly $\widecheck{\sigma}$ is defined by

$$(15) \qquad\qquad \widecheck{\sigma}(f) = \sigma(\widehat{f}), \qquad f \in C_c(X),$$

if σ is a compactly supported measure on Ξ.

Lemma 2.3. *(i) If s is a compactly supported measure on X, $\widehat{s}$ is a measure on Ξ.*

(ii) If s is a bounded measure on X and if $\widecheck{x}_0$ has finite measure then $\widehat{s}$ as defined by (14) is a bounded measure.

Proof. (i)   The measure s can be written as a difference $s = s^+ - s^-$ of two positive measures, each of compact support. Then $\widehat{s} = \widehat{s}^+ - \widehat{s}^-$ is a difference of two positive **functionals** on $C_c(\Xi)$.

Since a positive functional is necessarily a measure, $\widehat{s}$ is a measure.

(ii)   We have

$$\sup_{x} |\widecheck{\varphi}(x)| \leq \sup_{\xi} |\varphi(\xi)| \, \mu_0(\widecheck{x}_0),$$

so for a constant K,

$$|\widehat{s}(\varphi)| = |s(\widecheck{\varphi})| \leq K \sup |\widecheck{\varphi}| \leq K\mu_0(\widecheck{x}_0) \sup |\varphi|,$$

and the boundedness of $\widehat{s}$ follows.

If G is a Lie group then (14), (15) with $f \in \mathcal{D}(X)$, $\quad \varphi \in \mathcal{D}(\Xi)$ serve to define the Radon transform $s \rightarrow \widehat{s}$ and the dual $\sigma \rightarrow \widecheck{\sigma}$ for distributions s and σ of compact support. We consider the spaces $\mathcal{D}(X)$ and $\mathcal{E}(X) \, (= \mathcal{C}^{\infty}(X))$ with their customary topologies (Chapter VII, §1). The duals $\mathcal{D}'(X)$ and $\mathcal{E}'(X)$ then consist of the distributions on X and the distributions on X of compact support, respectively.

Proposition 2.4. *The mappings*

$$f \in \mathcal{D}(X) \quad \rightarrow \quad \widehat{f} \in \mathcal{E}(\Xi)$$
$$\varphi \in \mathcal{D}(\Xi) \quad \rightarrow \quad \widecheck{\varphi} \in \mathcal{E}(X)$$

are continuous. In particular,

$$s \in \mathcal{E}'(X) \quad \Rightarrow \quad \widehat{s} \in \mathcal{D}'(\Xi)$$
$$\sigma \in \mathcal{E}'(\Xi) \quad \Rightarrow \quad \widecheck{\sigma} \in \mathcal{D}'(X).$$

Proof. We have

(16)
$$\widehat{f}(g \cdot \xi_0) = \int_{\widehat{\xi}_0} f(g \cdot x) \, dm_0(x).$$

Let g run through a local cross section through e in G over a neighborhood of ξ_0 in Ξ. If $(t_1, \ldots, t_n)$ are coordinates of g and $(x_1, \ldots, x_m)$ the coordinates of $x \in \widehat{\xi}_0$ then (16) can be written in the form

$$\widehat{F}(t_1, \ldots, t_n) = \int F(t_1, \ldots, t_n ; x_1, \ldots, x_m) \, dx_1 \ldots dx_m.$$

Now it is clear that $\widehat{f} \in \mathcal{E}(\Xi)$ and that $f \to \widehat{f}$ is continuous, proving the proposition.

The result has the following refinement.

Proposition 2.5. *Assume K compact. Then*

(i) $f \to \widehat{f}$ *is a continuous mapping of $\mathcal{D}(X)$ into $\mathcal{D}(\Xi)$.*

(ii) $\varphi \to \check{\varphi}$ *is a continuous mapping of $\mathcal{E}(\Xi)$ into $\mathcal{E}(X)$.*

A self-contained proof is given in the author's book [1994b], Ch. I, § 3. The result has the following consequence.

Corollary 2.6. *Assume K compact. Then $\mathcal{E}'(X)\widehat{}\subset \mathcal{E}'(\Xi)$, $\mathcal{D}'(\Xi)^\vee \subset \mathcal{D}'(X)$.*

Ranges and Kernels. General Features

It is clear from Proposition 2.2 that the range $\mathcal{R}$ of $f \to \widehat{f}$ is orthogonal to the kernel $\mathcal{N}$ of $\varphi \to \check{\varphi}$. When $\mathcal{R}$ is closed one can often conclude $\mathcal{R} = \mathcal{N}^\perp$, also when $\widehat{}$ is extended to distributions (Helgason [1994b], Chapter IV, §2, Chapter I, §2). Under fairly general conditions one can also deduce that the range of $\varphi \to \check{\varphi}$ equals the annihilator of the kernel of $T \to \widehat{T}$ for distributions (*loc. cit.*, Ch. I, §3).

In Chapter I we have given solutions to Problems A, B, C, D in some cases. Further examples will be given in § 4 of this chapter and Chapter III will include their solution for the antipodal manifolds for compact two-point homogeneous spaces.

The variety of the results for these examples make it doubtful that the individual results could be captured by a general theory. Our abstract setup in terms of homogeneous spaces in duality is therefore to be regarded as a framework for examples rather than as axioms for a general theory.

Nevertheless, certain general features emerge from the study of these examples. If $\dim X = \dim \Xi$ and $f \to \hat{f}$ is injective the range consists of functions which are either arbitrary or at least subjected to rather weak conditions. As the difference $\dim \Xi - \dim X$ increases more conditions are imposed on the functions in the range. (See the example of the d-plane transform in $\mathbf{R}^n$.)

In case G is a Lie group there is a group-theoretic explanation for this. Let X be a manifold and Ξ a manifold whose points ξ are submanifolds of X. We assume each $\xi \in \Xi$ to have a measure dm and that the set $\{\xi \in \Xi : \xi \ni x\}$ has a measure $d\mu$. We can then consider the transforms

$$(17) \qquad \hat{f}(\xi) = \int_{\xi} f(x)\, dm(x), \quad \check{\varphi}(x) = \int_{\xi \ni x} \varphi(\xi)\, d\mu(\xi).$$

If G is a Lie transformation group of X permuting the members of Ξ including the measures dm and $d\mu$, the transforms $f \to \hat{f}, \varphi \to \check{\varphi}$ commute with the G-actions on X and Ξ

$$(18) \qquad (\hat{f})^{\tau(g)} = (f^{\tau(g)})\hat{\ } \qquad (\varphi^{\tau(g)})^{\vee} = (\check{\varphi})^{\tau(g)}.$$

Let λ and Λ be the homomorphisms

$$\lambda : \mathbf{D}(G) \to \mathbf{E}(X)$$
$$\Lambda : \mathbf{D}(G) \to \mathbf{E}(\Xi)$$

in Ch. VIII, §2. Using (13) in Ch. VIII we derive

$$(19) \qquad (\lambda(D)f)\hat{\ } = \Lambda(D)\hat{f}, \ (\Lambda(D)\varphi)^{\vee} = \lambda(D)\check{\varphi}.$$

Therefore $\Lambda(D)$ annihilates the range of $f \to \hat{f}$ if $\lambda(D) = 0$. In some cases, including the case of the d-plane transform in $\mathbf{R}^n$, the range is characterized as the null space of these operators $\Lambda(D)$ (with $\lambda(D) = 0$). This is illustrated by Theorems 6.5 and 6.8 in Ch. I and even more by theorems of Richter, Gonzalez which characterized the range as the null space of certain explicit invariant operators ([GSS, I, §3]). Much further work in this direction has been done by Gonzalez and Kakehi (see Part I in Ch. II, §4). Examples of (17)–(18) would occur with G a group of isometries of a Riemannian manifold, Ξ a suitable family of geodesics. The framework (8) above fits here too but goes further in that Ξ does not have to consist of subsets of X. We shall see already in the next Theorem 4.1 that this feature is significant.

The Inversion Problem. General Remarks

In Theorem 3.1 and 6.2 in Chapter I as well as in several later results the Radon transform $f \to \hat{f}$ is inverted by a formula

$$(20) \qquad f = D((\hat{f})^{\vee}),$$

where D is a specific operator on X, often a differential operator. Rouvière has in [2001] outlined an effective strategy for producing such a D.

Consider the setup $X = G/K$, $\Xi = G/H$ from §1 and assume G, K and H are unimodular Lie groups and K compact. On G we have a convolution (in the style of Ch. VII),

$$(u * v)(h) = \int\limits_G u(hg^{-1})v(g)\,dg = \int\limits_G u(g)v(g^{-1}h)\,dg,$$

provided one of the functions u, v has compact support. Here dg is Haar measure. More generally, if s, t are two distributions on G at least one of compact support the tensor product $s \otimes t$ is a distribution on $G \times G$ given by

$$(s \otimes t)(u(x, y)) = \int\limits_{G \times G} u(x, y)\,ds(x)\,dt(y) \quad u \in \mathcal{D}(G \times G).$$

Note that $s \otimes t = t \otimes s$ because they agree on the space spanned by functions of the type $\varphi(x)\psi(y)$ which is dense in $\mathcal{D}(G \times G)$. The convolution $s * t$ is defined by

$$(s * t)(v) = \int\limits_G \int\limits_G v(xy)\,ds(x)\,dt(y).$$

Lifting a function f on X to G by $\tilde{f} = f \circ \pi$ where $\pi : G \to G/K$ is the natural map we lift a distribution S on X to a $\tilde{S} \in \mathcal{D}'(G)$ by $\tilde{S}(u) = S(\dot{u})$ where

$$\dot{u}(gK) = \int\limits_K u(gk)\,dk.$$

Thus $\tilde{S}(\tilde{f}) = S(f)$ for $f \in \mathcal{D}(X)$. If $S, T \in \mathcal{D}'(X)$, one of compact support the convolution $\times$ on X is defined by

(21) $$(S \times T)(f) = (\tilde{S} * \tilde{T})(\tilde{f}).$$

If one of these is a function f, we have

(22) $$(f \times S)(g \cdot x_0) = \int\limits_G f(gh^{-1} \cdot x_0)\,d\tilde{S}(h),$$

(23) $$(S \times f)(g \cdot x_0) = \int\limits_G f(h^{-1}g \cdot x_0)\,d\tilde{S}(h).$$

The first formula can also be written

(24) $$f \times S = \int\limits_G f(g \cdot x_0)S^{\tau(g)}\,dg$$

as distributions. In fact, let $\varphi \in \mathcal{D}(X)$. Then

$$(f \times S)(\varphi) = \int_G (f \times S)(g \cdot x_0)\varphi(g \cdot x_0)\, dg$$

$$= \int_G \left(\int_G f(gh^{-1} \cdot x_0)\, d\tilde{S}(h) \right)\varphi(g \cdot x_0)\, dg$$

$$= \int_G \left(\int_G f(g \cdot x_0)\tilde{\varphi}(gh)\, dg \right) d\tilde{S}(h)$$

$$= \int_G \int_G f(g \cdot x_0)(\varphi^{\tau(g^{-1})})^{\sim}(h)\, dg\, d\tilde{S}(h)$$

$$= \int_G f(g \cdot x_0)S(\varphi^{\tau(g^{-1})})\, dg = \int_G f(g \cdot x_0)S^{\tau(g)}(\varphi)\, dg\,.$$

Now let D be a G-invariant differential operator on X and D^* its adjoint. It is also G-invariant. If $\varphi = \mathcal{D}(X)$ then the invariance of D^* and (24) imply

$$(D(f \times S))(\varphi) = (f \times S)(D^*\varphi) = \int_G f(g \cdot x_0)S^{\tau(g)}(D^*\varphi)\, dg$$

$$= \int_G f(g \cdot x_0)S(D^*(\varphi \circ \tau(g)))\, dg = \int_G f(g \cdot x_0)(DS)^{\tau(g)}(\varphi)\, dg,$$

so

(25) $$\qquad\qquad\qquad D(f \times S) = f \times DS\,.$$

Let ϵ_D denote the distribution $f \to (D^*f)(x_0)$. Then

$$Df = f \times \epsilon_D\,,$$

because by (24)

$$(f \times \epsilon_D)(\varphi) = \int_G f(g \cdot x_0)\epsilon_D^{\tau(g)}(\varphi)$$

$$= \int_G f(g \cdot x_0)D^*(\varphi^{\tau(g^{-1})})(x_0)\, dg = \int_G f(g \cdot x_0)(D^*\varphi)(g \cdot x_0)\, dg$$

$$= \int_X f(x)(D^*\varphi)(x)\, dx = \int_X (Df)(x)\varphi(x)\, dx\,.$$

We consider now the situation where the elements ξ of Ξ are subsets of X (cf. Lemma 1.3).

Theorem 2.7 (Rouvière). *Under the assumptions above (K compact) there exists a distribution S on X such that*

(26) $$(\widehat{f})^{\vee} = f \times S, \quad f \in \mathcal{D}(X).$$

Proof. Define a functional S on $C_c(X)$ by

$$S(f) = (\widehat{f})^{\vee}(x_0) = \int_K \left(\int_H f(kh \cdot x_0) \, dh \right) dk.$$

Then S is a measure because if f has compact support C the set of $h \in H$ for which $kh \cdot x_0 \in C$ for some k is compact. The restriction of S to $\mathcal{D}(X)$ is a distribution which is clearly K-invariant. By (24) we have for $\varphi \in \mathcal{D}(X)$

$$(f \times S)(\varphi) = \int_G f(g \cdot x_0) S(\varphi^{\tau(g^{-1})}) \, dg,$$

which, since the operations $\widehat{}$ and $\vee$ commute with the G action, becomes

$$\int_G f(g \cdot x_0)(\widehat{\varphi})^{\vee}(g \cdot x_0) \, dg = \int_X (\widehat{f})^{\vee}(x)\varphi(x) \, dx,$$

because of Proposition 2.2. This proves the theorem.

Corollary 2.8. *If D is a G-invariant differential operator on X such that $DS = \delta$ (delta function at x_0) then we have the inversion formula*

(27) $$f = D((\widehat{f})^{\vee}), \quad f \in \mathcal{D}(X).$$

This follows from (26) and $f \times \delta = f$.

§3   Orbital Integrals

As before let $X = G/K$ be a homogeneous space with origin $o = (K)$. Given $x_0 \in X$ let G_{x_0} denote the subgroup of G leaving x_0 fixed, i.e., the isotropy subgroup of G at x_0.

Definition. A **generalized sphere** is an orbit $G_{x_0} \cdot x$ in X of some point $x \in X$ under the isotropy subgroup at some point $x_0 \in X$.

Examples. (i) If $X = \mathbf{R}^n$, $G = \mathbf{M}(n)$ then the generalized spheres are just the spheres.

(ii)   Let X be a locally compact subgroup L and G the product group $L \times L$ acting on L on the right and left, the element $(\ell_1, \ell_2) \in L \times L$ inducing action $\ell \to \ell_1 \ell \ell_2^{-1}$ on L. Let ΔL denote the diagonal in $L \times L$. If $\ell_0 \in L$ then the isotropy subgroup of ℓ_0 is given by

$$(28) \qquad\qquad (L \times L)_{\ell_0} = (\ell_0, e) \Delta L(\ell_0^{-1}, e)$$

and the orbit of ℓ under it by

$$(L \times L)_{\ell_0} \cdot \ell = \ell_0 (\ell_0^{-1} \ell)^L ,$$

which is the left translate by ℓ_0 of the conjugacy class of the element $\ell_0^{-1} \ell$. Thus the *generalized spheres in the group L are the left (or right) translates of its conjugacy classes*.

Coming back to the general case $X = G/K = G/G_0$ we assume that G_0, and therefore each G_{x_0}, is unimodular. But $G_{x_0} \cdot x = G_{x_0}/(G_{x_0})_x$ so $(G_{x_0})_x$ unimodular implies the orbit $G_{x_0} \cdot x$ has an invariant measure determined up to a constant factor. We can now consider the following general problem (following Problems A, B, C, D above).

E.   *Determine a function f on X in terms of its integrals over generalized spheres.*

Remark 3.1. In this problem it is of course significant how the invariant measures on the various orbits are normalized.

(a)   If G_0 is compact the problem above is rather trivial because each orbit $G_{x_0} \cdot x$ has finite invariant measure so $f(x_0)$ is given as the limit as $x \to x_0$ of the average of f over $G_{x_0} \cdot x$.

(b)   Suppose that for each $x_0 \in X$ there is a G_{x_0}-invariant open set $C_{x_0} \subset X$ containing x_0 in its closure such that for each $x \in C_{x_0}$ the isotropy group $(G_{x_0})_x$ is compact. The invariant measure on the orbit $G_{x_0} \cdot x$ $(x_0 \in X, x \in C_{x_0})$ can then be consistently normalized as follows: Fix a Haar measure dg_0 on G_0. If $x_0 = g \cdot o$ we have $G_{x_0} = gG_0 g^{-1}$ and can carry dg_0 over to a measure dg_{x_0} on G_{x_0} by means of the conjugation $z \to gzg^{-1}$ $(z \in G_0)$. Since dg_0 is bi-invariant, dg_{x_0} is independent of the choice of g satisfying $x_0 = g \cdot o$, and is bi-invariant. Since $(G_{x_0})_x$ is compact it has a unique Haar measure $dg_{x_0,x}$ with total measure 1 and now dg_{x_0} and $dg_{x_0,x}$ determine canonically an invariant measure μ on the orbit $G_{x_0} \cdot x = G_{x_0}/(G_{x_0})_x$. We can therefore state Problem E in a more specific form.

E'.   *Express $f(x_0)$ in terms of integrals*

$$(29) \qquad\qquad \int_{G_{x_0} \cdot x} f(p) \, d\mu(p), \qquad x \in C_{x_0} .$$

For the case when X is an **isotropic Lorentz manifold** the assumptions above are satisfied (with C_{x_0} consisting of the "timelike" rays from x_0) and we shall obtain in Ch. V an explicit solution to Problem E' (Theorem 4.1, Ch. V).

(c)   If in Example (ii) above L is a semisimple Lie group Problem E is a basic step (Gelfand–Graev [1955], Harish-Chandra [1954], [1957]) in proving the Plancherel formula for the Fourier transform on L.

§4   Examples of Radon Transforms for Homogeneous Spaces in Duality

In this section we discuss some examples of the abstract formalism and problems set forth in the preceding sections §1–§2.

A.   The Funk Transform

This case goes back to Funk [1913], [1916] (preceding Radon's paper [1917]) where he proved, inspired by Minkowski [1911], that a symmetric function on $\mathbf{S}^2$ is determined by its great circle integrals. This is carried out in more detail and in greater generality in Chapter III, §1. Here we state the solution of Problem B for $X = \mathbf{S}^2$, Ξ the set of all great circles, both as homogeneous spaces of $\mathbf{O}(3)$. Given $p \geq 0$ let $\xi_p \in \Xi$ have distance p from the North Pole o, $H_p \subset \mathbf{O}(3)$ the subgroup leaving ξ_p invariant and $K \subset \mathbf{O}(3)$ the subgroup fixing o. Then in the double fibration

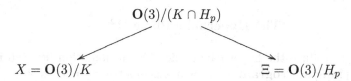

$$\mathbf{O}(3)/(K \cap H_p)$$

$$X = \mathbf{O}(3)/K \qquad\qquad \Xi = \mathbf{O}(3)/H_p$$

$x \in X$ and $\xi \in \Xi$ are incident if and only if $d(x, \xi) = p$. The proof is the same as that of Proposition 1.5. We denote by $\widehat{f}_p$ and $\check{\varphi}_p$ the Radon transforms (9) for the double fibration. Then $\widehat{f}_p(\xi)$ the integral of f over two circles at distance p from ξ and $\check{\varphi}_p$ is the average of $\check{\varphi}(x)$ over the great circles ξ that have distance p from x. (See Fig. II.2.) We need $\widehat{f}_p$ only for $p = 0$ and put $\widehat{f} = \widehat{f}_0$. Note that $(\widehat{f})_p^\vee(x)$ is the average of the integrals of f over the great circles ξ at distance p from x (see Figure II.2). As a special case of Theorem 1.22, Chapter III, we have the following inversion.

Theorem 4.1. *The Funk transform* $f \to \widehat{f}$ *is (for f even) inverted by*

$$(30) \qquad f(x) = \frac{1}{2\pi}\left\{ \frac{d}{du} \int_0^u (\widehat{f})_{\cos^{-1}(v)}^\vee(x)\, v(u^2 - v^2)^{-\frac{1}{2}}\, dv \right\}_{u=1}.$$

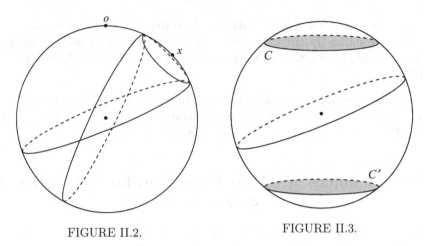

FIGURE II.2.   FIGURE II.3.

We shall see later that this formula can also be written

$$(31) \qquad f(x) = \int_{E_x} f(w)\, dw - \frac{1}{2\pi} \int_0^{\frac{\pi}{2}} \frac{d}{dp}\left((\widehat{f})_p^{\vee}(x) \right) \frac{dp}{\sin p},$$

where dw is the normalized measure on the equator E_x corresponding to x. In this form the formula holds in all dimensions.

Also Theorem 1.26, Ch. III shows that if f is even and if all its derivatives vanish on the equator then f vanishes outside the "arctic zones" C and C' if and only if $\widehat{f}(\xi) = 0$ for all great circles ξ disjoint from C and C' (Fig. II.3).

The Hyperbolic Plane $\mathbf{H^2}$

We now introduce the hyperbolic plane. This formulation fits well into **Klein's Erlanger Program** under which geometric properties of a space should be understood in terms of a suitable transformation group of the space.

Theorem 4.2. *On the unit disk $D : |z| < 1$ there exists a Riemannian metric g which is invariant under all conformal transformations of D. Also g is unique up to a constant factor.*

For this consider a point $a \in D$. The mapping $\varphi : z \to \frac{a-z}{1-\bar{a}z}$ is a conformal transformation of D and $\varphi(a) = 0$. The invariance of g requires

$$g_a(u, u) = g_0(d\varphi(u), d\varphi(u))$$

for each $u \in D_a$ (the tangent space to D at a) $d\varphi$ denoting the differential of φ. Since g_0 is invariant under rotations around 0, $g_0(z, z) = c|z|^2$, where c is a constant. Here $z \in D_0 \,(= \mathbf{C})$. Let $t \to z(t)$ be a curve with $z(0) = a$,

$z'(0) = u \in \mathbf{C}$. Then $d\varphi(u)$ is the tangent vector

$$\left\{ \frac{d}{dt}\varphi(z(t)) \right\}_{t=0} = \left(\frac{d\varphi}{dz}\right)_a \left(\frac{dz}{dt}\right)_0 = \left\{ \frac{|a|^2 - 1}{(1 - \bar{a}z)^2} \right\}_{z=a} u,$$

so

$$g_a(u, u) = c\frac{1}{(1 - |a|^2)^2}|u|^2,$$

and the proof shows that g is indeed invariant.

Thus we take the hyperbolic plane $\mathbf{H}^2$ as the disk D with the Riemannian structure

(32)
$$ds^2 = \frac{|dz|^2}{(1 - |z|^2)^2}.$$

This remarkable object enters into several fields in mathematics. In particular, it offers at least two interesting cases of Radon transforms. The Laplace-Beltrami operator for (32) is given by

$$L = (1 - x^2 - y^2)^2 \left(\frac{\partial^2}{\partial x^2} + \frac{\partial^2}{\partial y^2} \right).$$

The group $G = \mathbf{SU}(1, 1)$ of matrices

$$\left\{ \begin{pmatrix} a & b \\ b & a \end{pmatrix} : |a|^2 - |b|^2 = 1 \right\}$$

acts transitively on the unit disk by

(33)
$$\begin{pmatrix} a & b \\ b & a \end{pmatrix} \cdot z = \frac{az + b}{\bar{b}z + \bar{a}}$$

and leaves the metric (32) invariant. The length of a curve $\gamma(t)$ ($\alpha \leq t \leq \beta$) is defined by

(34)
$$L(\gamma) = \int_\alpha^\beta (\langle \gamma'(t), \gamma'(t) \rangle_{\gamma(t)})^{1/2} \, dt.$$

In particular take $\gamma(t) = (x(t), y(t))$ such that $\gamma(\alpha) = 0$, $\gamma(\beta) = x$ ($0 < x < 1$), and let $\gamma_0(\tau) = \tau x$, $0 \leq \tau \leq 1$, so γ and γ_0 have the same endpoints. Then

$$L(\gamma) \geq \int_\alpha^\beta \frac{|x'(t)|}{1 - x(t)^2} \, dt \geq \int_\alpha^\beta \frac{x'(t)}{1 - x(t)^2} \, dt,$$

which by $\tau = x(t)/x$, $d\tau/dt = x'(t)/x$ becomes

$$\int_0^1 \frac{x}{1 - \tau^2 x^2} \, d\tau = L(\gamma_0).$$

Thus $L(\gamma) \geq L(\gamma_0)$ so γ_0 is a geodesic and the distance d satisfies

(35)
$$d(o, x) = \int_0^1 \frac{|x|}{1 - t^2 x^2} \, dt = \tfrac{1}{2} \log \frac{1 + |x|}{1 - |x|}.$$

Since G acts conformally on D the *geodesics* in $\mathbf{H}^2$ are the circular arcs in $|z| < 1$ perpendicular to the boundary $|z| = 1$.

We consider now the following subgroups of G where $\operatorname{sh} t = \sinh t$ etc.:

$$K = \left\{ k_\theta = \begin{pmatrix} e^{i\theta} & 0 \\ 0 & e^{-i\theta} \end{pmatrix} : 0 \leq \theta < 2\pi \right\}$$

$$M = \{k_0, k_\pi\}, \qquad M' = \{k_0, k_\pi, k_{-\frac{\pi}{2}}, k_{\frac{\pi}{2}}\}$$

$$A = \left\{ a_t = \begin{pmatrix} \operatorname{ch} t & \operatorname{sh} t \\ \operatorname{sh} t & \operatorname{ch} t \end{pmatrix} : t \in \mathbf{R} \right\},$$

$$N = \left\{ n_x = \begin{pmatrix} 1 + ix, & -ix \\ ix, & 1 - ix \end{pmatrix} : x \in \mathbf{R} \right\}$$

$$\Gamma = C\mathbf{SL}(2, \mathbf{Z})C^{-1},$$

where C is the transformation $w \to (w - i)/(w + i)$ mapping the upper half-plane onto the unit disk.

The orbits of K are the circles around 0. To identify the orbit $A \cdot z$ we use this simple argument by Reid Barton:

$$a_t \cdot z = \frac{\operatorname{ch} t \, z + \operatorname{sh} t}{\operatorname{sh} t \, z + \operatorname{ch} t} = \frac{z + \operatorname{th} t}{\operatorname{th} t \, z + 1}.$$

Under the map $w \to \frac{z+w}{zw+1}$ ($w \in \mathbf{C}$) lines go into circles and lines. Taking $w = \operatorname{th} t$ we see that $A \cdot z$ is the circular arc through -1, z and 1. Barton's argument also gives the orbit $n_x \cdot t$ ($x \in \mathbf{R}$) as the image of $i\mathbf{R}$ under the map

$$w \to \frac{w(t - 1) + t}{w(t - 1) + 1}.$$

They are circles tangential to $|z| = 1$ at $z = 1$. Clearly $NA \cdot 0$ is the whole disk D so $G = NAK$ (and also $G = KAN$).

B. The X-ray Transform in $\mathbf{H}^2$

The (unoriented) geodesics for the metric (32) were mentioned above. Clearly the group G permutes these geodesics transitively (Fig. II.4). Let

Ξ be the set of all these geodesics. Let o denote the origin in $\mathbf{H}^2$ and ξ_o the horizontal geodesic through o. Then

(36)
$$X = G/K, \quad \Xi = G/M'A.$$

We can also fix a geodesic ξ_p at distance p from o and write

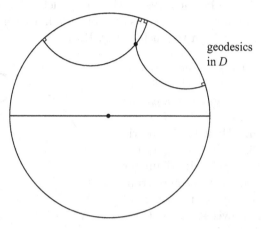

geodesics in D

(37)
$$X = G/K, \quad \Xi = G/H_p,$$

where H_p is the subgroup of G leaving ξ_p stable. Then for the homogeneous spaces (37), x and ξ are incident if and only if $d(x, \xi) = p$. The transform

FIGURE II.4.

$f \to \widehat{f}$ is inverted by means of the dual transform $\varphi \to \check{\varphi}_p$ for (37). The inversion below is a special case of Theorem 1.11, Chapter III, and is the analog of (30). Observe also that the metric ds is renormalized by the factor 2 (so curvature is -1).

Theorem 4.3. *The X-ray transform in $\mathbf{H}^2$ with the metric*

$$ds^2 = \frac{4|dz|^2}{(1 - |z|^2)^2}$$

is inverted by

(38)
$$f(z) = -\left\{ \frac{d}{dr} \int_r^\infty (t^2 - r^2)^{-\frac{1}{2}} t(\widehat{f})^\vee_{s(t)}(z)\, dt \right\}_{r=1},$$

where $s(t) = \cosh^{-1}(t)$.

Another version of this formula is

(39)
$$f(z) = -\frac{1}{\pi} \int_0^\infty \frac{d}{dp}\left((\widehat{f})^\vee_p(z) \right) \frac{dp}{\sinh p}$$

and in this form it is valid in all dimensions (Theorem 1.12, Ch. III).
    One more inversion formula is

(40)
$$f = -\frac{1}{4\pi} LS((\widehat{f})^\vee),$$

where S is the operator of convolution on $\mathbf{H}^2$ with the function $x \to \coth(d(x, o)) - 1$, (Theorem 1.16, Chapter III).

C.   The Horocycles in $\mathbf{H}^2$

Consider a family of geodesics with the same limit point on the boundary B. The **horocycles** in $\mathbf{H}^2$ are by definition the orthogonal trajectories of such families of geodesics. Thus the horocycles are the circles tangential to $|z| = 1$ from the inside (Fig. II.5).

One such horocycle is $\xi_0 = N \cdot o$, the orbit of the origin o under the action of N. Now we take $\mathbf{H}^2$ with the metric (32). Since $a_t \cdot \xi$ is the horocycle with diameter $(\tanh t, 1)$ G acts transitively on the set Ξ of horocycles. Since $G = KAN$ it is easy to see that MN is the subgroup leaving ξ_o invariant. Thus we have here (41)
$$X = G/K, \quad \Xi = G/MN.$$

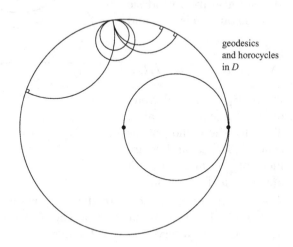

geodesics and horocycles in D

FIGURE II.5.

Furthermore each horocycle has the form $\xi = ka_t \cdot \xi_0$ where $kM \in K/M$ and $t \in \mathbf{R}$ are unique. Thus $\Xi \sim K/M \times A$, which is also evident from the figure.

We observe now that the maps
$$\psi : t \to a_t \cdot o, \quad \varphi : x \to n_x \cdot o$$

of $\mathbf{R}$ onto γ_0 and ξ_0, respectively, are isometries. The first statement follows from (35) because
$$d(o, a_t \cdot o) = d(o, \tanh t) = t.$$

For the second we note that
$$\varphi(x) = x(x + i)^{-1}, \quad \varphi'(x) = i(x + i)^{-2}$$

so
$$\langle \varphi'(x), \varphi'(x) \rangle_{\varphi(x)} = (x^2 + 1)^{-2}(1 - |x(x + i)^{-1}|^2)^{-2} = 1.$$

Thus we give A and N the Haar measures $d(a_t) = dt$ and $d(n_x) = dx$.

Geometrically, the Radon transform on X relative to the horocycles is defined by

(42)
$$\widehat{f}(\xi) = \int_\xi f(x) \, dm(x),$$

where dm is the measure on ξ induced by (32). Because of our remarks about φ, (42) becomes

$$(43) \qquad \widehat{f}(g \cdot \xi_0) = \int_N f(gn \cdot o) \, dn \, ,$$

so the geometric definition (42) coincides with the group-theoretic one in (9). The dual transform is given by

$$(44) \qquad \check{\varphi}(g \cdot o) = \int_K \varphi(gk \cdot \xi_o) \, dk \, , \qquad (dk = d\theta/2\pi) \, .$$

In order to invert the transform $f \to \widehat{f}$ we introduce the non-Euclidean analog of the operator Λ in Chapter I, §3. Let T be the distribution on $\mathbf{R}$ given by

$$(45) \qquad T\varphi = \tfrac{1}{2} \int_{\mathbf{R}} (\operatorname{sh} t)^{-1} \varphi(t) \, dt \, , \quad \varphi \in \mathcal{D}(\mathbf{R}) \, ,$$

considered as the Cauchy principal value, and put $T' = dT/dt$. Let Λ be the operator on $\mathcal{D}(\Xi)$ given by

$$(46) \qquad (\Lambda\varphi)(ka_t \cdot \xi_0) = \int_{\mathbf{R}} \varphi(ka_{t-s} \cdot \xi_0) e^{-s} \, dT'(s) \, .$$

Theorem 4.4. *The Radon transform $f \to \widehat{f}$ for horocycles in $\mathbf{H}^2$ is inverted by*

$$(47) \qquad f = \frac{1}{\pi}(\Lambda\widehat{f})^{\vee} \, , \quad f \in \mathcal{D}(\mathbf{H}^2) \, .$$

We begin with a simple lemma.

Lemma 4.5. *Let τ be a distribution on $\mathbf{R}$. Then the operator $\widetilde{\tau}$ on $\mathcal{D}(\Xi)$ given by the convolution*

$$(\widetilde{\tau}\varphi)(ka_t \cdot \xi_0) = \int_{\mathbf{R}} \varphi(ka_{t-s} \cdot \xi_0) \, d\tau(s)$$

is invariant under the action of G.

Proof. To understand the action of $g \in G$ on $\Xi \sim (K/M) \times A$ we write $gk = k' a_{t'} n'$. Since each $a \in A$ normalizes N we have

$$gka_t \cdot \xi_0 = gka_t N \cdot o = k' a_{t'} n' a_t N \cdot o = k' a_{t+t'} \cdot \xi_0 \, .$$

Thus the action of g on $\Xi \simeq (K/M) \times A$ induces this fixed translation $a_t \to a_{t+t'}$ on A. This translation commutes with the convolution by τ, so the lemma follows.

Since the operators $\Lambda, \hat{\ }, \vee$ in (47) are all G-invariant, it suffices to prove the formula at the origin o. We first consider the case when f is K-invariant, i.e., $f(k \cdot z) \equiv f(z)$. Then by (43),

$$(48) \qquad \hat{f}(a_t \cdot \xi_0) = \int_{\mathbf{R}} f(a_t n_x \cdot o) \, dx \,.$$

Because of (35) we have

$$(49) \qquad |z| = \tanh d(o, z), \quad \cosh^2 d(o, z) = (1 - |z|^2)^{-1} \,.$$

Since

$$a_t n_x \cdot o = (\operatorname{sh} t - ix \, e^t)/(\operatorname{ch} t - ix \, e^t)$$

(49) shows that the distance $s = d(o, a_t n_x \cdot o)$ satisfies

$$(50) \qquad \operatorname{ch}^2 s = \operatorname{ch}^2 t + x^2 e^{2t} \,.$$

Thus defining F on $[1, \infty)$ by

$$(51) \qquad F(\operatorname{ch}^2 s) = f(\tanh s) \,,$$

we have

$$F'(\operatorname{ch}^2 s) = f'(\tanh s)(2\operatorname{sh} s \operatorname{ch}^3 s)^{-1}$$

so, since $f'(0) = 0$, $\lim_{u \to 1} F'(u)$ exists. The transform (48) now becomes (with $xe^t = y$)

$$(52) \qquad e^t \hat{f}(a_t \cdot \xi_0) = \int_{\mathbf{R}} F(\operatorname{ch}^2 t + y^2) \, dy \,.$$

We put

$$\varphi(u) = \int_{\mathbf{R}} F(u + y^2) \, dy$$

and invert this as follows:

$$\int_{\mathbf{R}} \varphi'(u + z^2) \, dz \quad = \quad \int_{\mathbf{R}^2} F'(u + y^2 + z^2) \, dy \, dz$$

$$= \quad 2\pi \int_0^\infty F'(u + r^2) r \, dr = \pi \int_0^\infty F'(u + \rho) \, d\rho \,,$$

so

$$-\pi F(u) \quad = \quad \int_{\mathbf{R}} \varphi'(u + z^2) \, dz \,.$$

In particular,

$$
f(o) = -\frac{1}{\pi} \int_{\mathbf{R}} \varphi'(1 + z^2)\, dz = -\frac{1}{\pi} \int_{\mathbf{R}} \varphi'(\mathrm{ch}^2 \tau)\mathrm{ch}\,\tau\, d\tau ,
$$

$$
= -\frac{1}{\pi} \int_{\mathbf{R}} \int_{\mathbf{R}} F'(\mathrm{ch}^2 t + y^2)\, dy\, \mathrm{ch}\,t\, dt ,
$$

so

$$
f(o) = -\frac{1}{2\pi} \int_{\mathbf{R}} \frac{d}{dt}(e^t \widehat{f}(a_t \cdot \xi_0))\frac{dt}{\mathrm{sh}\,t} .
$$

Since $(e^t \widehat{f})(a_t \cdot \xi_0)$ is even (cf. (52)), its derivative vanishes at $t = 0$, so the integral is well defined. With T as in (45), the last formula can be written

(53)
$$
f(o) = \frac{1}{\pi} T'_t(e^t \widehat{f}(a_t \cdot \xi_0)) ,
$$

the prime indicating derivative. If f is not necessarily K-invariant we use (53) on the average

$$
f^\natural(z) = \int_K f(k \cdot z)\, dk = \frac{1}{2\pi} \int_0^{2\pi} f(k_\theta \cdot z)\, d\theta .
$$

Since $f^\natural(o) = f(o)$, (53) implies

(54)
$$
f(o) = \frac{1}{\pi} \int_{\mathbf{R}} [e^t (f^\natural)\widehat{\,}(a_t \cdot \xi_0)]\, dT'(t) .
$$

This can be written as the convolution at $t = 0$ of $(f^\natural)\widehat{\,}(a_t \cdot \xi_0)$ with the image of the distribution $e^t T'_t$ under $t \to -t$. Since T' is even the right hand side of (54) is the convolution at $t = 0$ of $\widehat{f^\natural}$ with $e^{-t}T'_t$. Thus by (46),

$$
f(o) = \frac{1}{\pi}(\Lambda \widehat{f^\natural})(\xi_0) .
$$

Since Λ and $\widehat{}$ commute with the K action this implies

$$
f(o) = \frac{1}{\pi} \int_K (\Lambda \widehat{f})(k \cdot \xi_0) = \frac{1}{\pi}(\Lambda \widehat{f})^\vee(o)
$$

and this proves the theorem.

Theorem 4.4 is of course the exact analog to Theorem 3.6 in Chapter I, although we have not specified the decay conditions for f needed in generalizing Theorem 4.4.

D.   The Poisson Integral as a Radon Transform

Here we preserve the notation introduced for the hyperbolic plane $\mathbf{H}^2$. Now we consider the homogeneous spaces

$$(55) \qquad\qquad X = G/MAN\,, \quad \Xi = G/K\,.$$

Then Ξ is the disk $D : |z| < 1$. On the other hand, X is identified with the boundary $B : |z| = 1$, because when G acts on B, MAN is the subgroup fixing the point $z = 1$. Since $G = KAN$, each coset $gMAN$ intersects eK. Thus each $x \in X$ is incident to each $\xi \in \Xi$. Our abstract Radon transform (9) now takes the form

$$(56) \qquad \widehat{f}(gK) \;=\; \int_{K/M} f(gkMAN)\, dk_M = \int_B f(g \cdot b)\, db\,,$$

$$\;=\; \int_B f(b)\,\frac{d(g^{-1} \cdot b)}{db}\, db\,.$$

Writing g^{-1} in the form

$$g^{-1} : \zeta \to \frac{\zeta - z}{-\overline{z}\zeta + 1}\,, \quad g^{-1} \cdot e^{i\theta} = e^{i\varphi}\,,$$

we have

$$e^{i\varphi} = \frac{e^{i\theta} - z}{-\overline{z}e^{i\theta} + 1}\,, \quad \frac{d\varphi}{d\theta} = \frac{1 - |z|^2}{|z - e^{i\theta}|}\,,$$

and this last expression is the classical Poisson kernel. Since $gK = z$, (56) becomes the classical Poisson integral

$$(57) \qquad\qquad \widehat{f}(z) = \int_B f(b)\,\frac{1 - |z|^2}{|z - b|^2}\, db\,.$$

Theorem 4.6. *The Radon transform $f \to \widehat{f}$ for the homogeneous spaces (55) is the classical Poisson integral (57). The inversion is given by the classical Schwarz theorem*

$$(58) \qquad\qquad f(b) = \lim_{z \to b} \widehat{f}(z)\,, \quad f \in C(B)\,,$$

solving the Dirichlet problem for the disk.

We repeat the geometric proof of (58) from our booklet [1981] since it seems little known and is considerably shorter than the customary solution in textbooks of the Dirichlet problem for the disk. In (58) it suffices to consider the case $b = 1$. Because of (56),

$$\widehat{f}(\tanh t) \quad = \quad \widehat{f}(a_t \cdot 0) = \frac{1}{2\pi} \int_0^{2\pi} f(a_t \cdot e^{i\theta}) \, d\theta$$

$$= \quad \frac{1}{2\pi} \int_0^{2\pi} f\left(\frac{e^{i\theta} + \tanh t}{\tanh t \, e^{i\theta} + 1} \right) \, d\theta \, .$$

Letting $t \to +\infty$, (58) follows by the dominated convergence theorem.

The range question A for $f \to \widehat{f}$ is also answered by classical results for the Poisson integral; for example, the classical characterization of the Poisson integrals of bounded functions now takes the form

(59) $$L^\infty(B)\widehat{} = \{\varphi \in L^\infty(\Xi) : L\varphi = 0\} \, .$$

The range characterization (59) is of course quite analogous to the range characterization for the X-ray transform described in Theorem 6.9, Chapter I. Both are realizations of the general expectations at the end of §2 that when $\dim X < \dim \Xi$ the range of the transform $f \to \widehat{f}$ should be given as the kernel of some differential operators. The analogy between (59) and Theorem 6.9 is even closer if we recall Gonzalez' theorem [1990b] that if we view the X-ray transform as a Radon transform between two homogeneous spaces of $\mathbf{M}(3)$ (see next example) then the range (91) in Theorem 6.9, Ch. I, can be described as the null space of a differential operator which is invariant under $\mathbf{M}(3)$. Furthermore, the dual transform $\varphi \to \check{\varphi}$ maps $\mathcal{E}(\Xi)$ onto $\mathcal{E}(X)$. (See Corollary 4.8 below.)

Furthermore, John's mean value theorem for the X-ray transform (Corollary 6.12, Chapter I) now becomes the exact analog of Gauss' mean-value theorem for harmonic functions.

From a non-Euclidean point of view, Godement's mean-value theorem (Ch. VI, §1) is even closer analog to John's theorem. Because of the special form of the Laplace–Beltrami operator in $\mathbf{H}^2$ non-Euclidean harmonic functions are the same as the usual ones (this fails for $\mathbf{H}^n$ $n > 2$). Also non-Euclidean circles are Euclidean circles (because the map (33) sends circles into circles). However, the mean-value theorem is different, namely,

$$u(z) = \int_S u(\zeta) \, d\mu(\zeta)$$

for a harmonic function u, z being the non-Euclidean center of the circle S and μ being the normalized non-Euclidean arc length measure on X, according to (32). However, this follows readily from the Gauss' mean-value theorem using a conformal map of D.

What is the dual transform $\varphi \rightarrow \check{\varphi}$ for the pair (55)? The invariant measure on $MAN/M = AN$ is the functional

$$(60) \qquad \varphi \rightarrow \int_{AN} \varphi(an \cdot o) \, da \, dn \, .$$

The right hand side is just $\check{\varphi}(b_0)$ where $b_0 = eMAN$. If $g = a'n'$ the measure (58) is seen to be invariant under g. Thus it is a constant multiple of the surface element $dz = (1 - x^2 - y^2)^{-2} \, dx \, dy$ defined by (32). Since the maps $t \rightarrow a_t \cdot o$ and $x \rightarrow n_x \cdot o$ were seen to be isometries, this constant factor is 1. Thus the measure (60) is invariant under each $g \in G$. Writing $\varphi_g(z) = \varphi(g \cdot z)$ we know $(\varphi_g)^\vee = \check{\varphi}_g$ so

$$\check{\varphi}(g \cdot b_0) = \int_{AN} \varphi_g(an) \, da \, dn = \check{\varphi}(b_0) \, .$$

Thus the dual transform $\varphi \rightarrow \check{\varphi}$ assigns to each $\varphi \in \mathcal{D}(\Xi)$ its integral over the disk.

Table II.1 summarizes the various results mentioned above about the Poisson integral and the X-ray transform. The inversion formulas and the ranges show subtle analogies as well as strong differences. The last item in the table comes from Corollary 4.8 below for the case $n = 3$, $d = 1$.

E.   The d-plane Transform

We now review briefly the d-plane transform from a group theoretic stand-point. As in (1) we write

$$(61) \ \ X = \mathbf{R}^n = \mathbf{M}(n)/\mathbf{O}(n), \quad \Xi = \mathbf{G}(d, n) = \mathbf{M}(n)/(\mathbf{M}(d) \times \mathbf{O}(n-d)),$$

where $\mathbf{M}(d) \times \mathbf{O}(n-d)$ is the subgroup of $\mathbf{M}(n)$ preserving a certain d-plane ξ_0 through the origin. Since the homogeneous spaces

$$\mathbf{O}(n)/\mathbf{O}(n) \cap (\mathbf{M}(d) \times \mathbf{O}(n - d)) = \mathbf{O}(n)/(\mathbf{O}(d) \times \mathbf{O}(n - d))$$

and

$$(\mathbf{M}(d) \times \mathbf{O}(n - d))/\mathbf{O}(n) \cap (\mathbf{M}(d) \times \mathbf{O}(n - d)) = \mathbf{M}(d)/\mathbf{O}(d)$$

have unique invariant measures the group-theoretic transforms (9) reduce to the transforms (57), (58) in Chapter I. The range of the d-plane transform is described by Theorem 6.3 and the equivalent Theorem 6.5 in Chapter I. It was shown by Richter [1986a] that the differential operators in Theorem 6.5 could be replaced by $\mathbf{M}(n)$-induced second order differential

	Poisson Integral	X-ray Transform
Coset spaces	$X = \mathbf{SU}(1,1)/MAN$ $\Xi = \mathbf{SU}(1,1)/K$	$X = \mathbf{M}(3)/\mathbf{O}(3)$ $\Xi = \mathbf{M}(3)/(\mathbf{M}(1) \times \mathbf{O}(2))$
$f \to \widehat{f}$	$\widehat{f}(z) = \int_B f(b) \frac{1-|z|^2}{|z-b|^2}\, db$	$\widehat{f}(\ell) = \int_\ell f(p)\, dm(p)$
$\varphi \to \check{\varphi}$	$\check{\varphi}(x) = \int_\Xi \varphi(\xi)\, d\xi$	$\check{\varphi}(x)$ = average of φ over set of ℓ through x
Inversion	$f(b) = \lim_{z \to b} \widehat{f}(z)$	$f = \frac{1}{\pi}(-L)^{1/2}((\widehat{f})^\vee)$
Range of $f \to \widehat{f}$	$L^\infty(X)\widehat{\;} =$ $\{\varphi \in L^\infty(\Xi) : L\varphi = 0\}$	$\mathcal{D}(X)\widehat{\;} =$ $\{\varphi \in \mathcal{D}(\Xi) : \Lambda(|\xi - \eta|^{-1}\varphi) = 0\}$
Range characterization	Gauss' mean value theorem	Mean value property for hyperboloids of revolution
Range of $\varphi \to \check{\varphi}$	$\mathcal{E}(\Xi)^\vee = \mathbf{C}$	$\mathcal{E}(\Xi)^\vee = \mathcal{E}(X)$

TABLE II.1. Analogies between the Poisson Integral and the X-ray Transform.

operators and then Gonzalez [1990b] showed that the whole system could be replaced by a single fourth order $\mathbf{M}(n)$-invariant differential operator on Ξ.

Writing (61) for simplicity in the form

(62) $X = G/K, \quad \Xi = G/H$

we shall now discuss the range question for the dual transform $\varphi \to \check{\varphi}$ by invoking the d-plane transform on $\mathcal{E}'(X)$.

Theorem 4.7. *Let $\mathcal{N}$ denote the kernel of the dual transform on $\mathcal{E}(\Xi)$. Then the range of $S \to \widehat{S}$ on $\mathcal{E}'(X)$ is given by*

$$\mathcal{E}'(X)\widehat{\;} = \{\Sigma \in \mathcal{E}'(\Xi) : \Sigma(\mathcal{N}) = 0\}.$$

The inclusion $\subset$ is clear from the definitions (14),(15) and Proposition 2.5. The converse is proved by the author in [1983a] and [1994b], Ch. I, §2 for $d = n - 1$; the proof is also valid for general d.

For Fréchet spaces E and F one has the following classical result. A continuous mapping $\alpha : E \to F$ is surjective if the transpose ${}^t\alpha : F' \to E'$ is injective and has a closed image. Taking $E = \mathcal{E}(\Xi)$, $F = \mathcal{E}(X)$, α as the dual transform $\varphi \to \check{\varphi}$, the transpose ${}^t\alpha$ is the Radon transform on

$\mathcal{E}'(X)$. By Theorem 4.7, $^t\alpha$ does have a closed image and by Theorem 5.5, Ch. I (extended to any d) $^t\alpha$ is injective. Thus we have the following result (Hertle [1984] for $d = n - 1$) expressing the surjectivity of α.

Corollary 4.8. *Every* $f \in \mathcal{E}(\mathbf{R}^n)$ *is the dual transform* $f = \check{\varphi}$ *of a smooth d-plane function* φ.

F.   Grassmann Manifolds

We consider now the (affine) Grassmann manifolds $\mathbf{G}(p, n)$ and $\mathbf{G}(q, n)$ where $p + q = n - 1$. If $p = 0$ we have the original case of points and hyperplanes. Both are homogeneous spaces of the group $\mathbf{M}(n)$ and we represent them accordingly as coset spaces

$$(63) \qquad\qquad X = \mathbf{M}(n)/H_p, \quad \Xi = \mathbf{M}(n)/H_q.$$

Here we take H_p as the isotropy group of a p-plane x_0 through the origin $0 \in \mathbf{R}^n$, H_q as the isotropy group of a q-plane ξ_0 through 0, *perpendicular to* x_0. Then

$$H_p = \mathbf{M}(p) \times \mathbf{O}(n - p), \quad H_q = \mathbf{M}(q) \times \mathbf{O}(n - q).$$

Also

$$H_q \cdot x_0 = \{x \in X : x \perp \xi_0, x \cap \xi_0 \neq \emptyset\},$$

the set of p-planes intersecting ξ_0 orthogonally. It is then easy to see that

$$x \text{ is incident to } \xi \Leftrightarrow x \perp \xi, \quad x \cap \xi \neq \emptyset.$$

Consider as in Chapter I, §6 the mapping

$$\pi : \mathbf{G}(p, n) \to \mathbf{G}_{p,n}$$

given by parallel translating a p-plane to one such through the origin. If $\sigma \in \mathbf{G}_{p,n}$, the fiber $F = \pi^{-1}(\sigma)$ is naturally identified with the Euclidean space $\sigma^\perp$. Consider the linear operator $\square_p$ on $\mathcal{E}(\mathbf{G}(p, n))$ given by

$$(64) \qquad\qquad (\square_p f)|F = L_F(f|F).$$

Here L_F is the Laplacian on F and bar denotes restriction. Then one can prove that $\square_p$ is a differential operator on $\mathbf{G}(p, n)$ which is invariant under the action of $\mathbf{M}(n)$. Let $f \to \hat{f}$, $\varphi \to \check{\varphi}$ be the Radon transform and its dual corresponding to the pair (61). Then $\hat{f}(\xi)$ represents the integral of f over all p-planes x intersecting ξ under a right angle. For n odd this is inverted as follows (Gonzalez [1984, 1987]).

Theorem 4.9. *Let* $p, q \in \mathbf{Z}^+$ *such that* $p + q + 1 = n$ *is odd. Then the transform* $f \to \hat{f}$ *from* $\mathbf{G}(p, n)$ *to* $\mathbf{G}(q, n)$ *is inverted by the formula*

$$C_{p,q} f = ((\square_q)^{(n-1)/2} \hat{f})^\vee, \quad f \in \mathcal{D}(\mathbf{G}(p, n))$$

where $C_{p,q}$ *is a constant.*

If $p = 0$ this reduces to Theorem 3.6, Ch. I.

G.   Half-lines in a Half-plane

In this example X denotes the half-plane $\{(a,b) \in \mathbf{R}^2 : a > 0\}$ viewed as a subset of the plane $\{(a,b,1) \in \mathbf{R}^3\}$. The group G of matrices

$$(\alpha, \beta, \gamma) = \begin{pmatrix} \alpha & 0 & 0 \\ \beta & 1 & \gamma \\ 0 & 0 & 1 \end{pmatrix} \in \mathbf{GL}(3, \mathbf{R}), \quad \alpha > 0$$

acts transitively on X with the action

$$(\alpha, \beta, \gamma) \odot (a,b) = (\alpha a, \beta a + b + \gamma).$$

This is the restriction of the action of $\mathbf{GL}(3, \mathbf{R})$ on $\mathbf{R}^3$. The isotropy group of the point $x_0 = (1,0)$ is the group

$$K = \{(1, \beta, -\beta) : \beta \in \mathbf{R}\}.$$

Let Ξ denote the set of half-lines in X which end on the boundary $\partial X = 0 \times \mathbf{R}$. These lines are given by

$$\xi_{v,w} = \{(t, v + tw) : t > 0\}$$

for arbitrary $v, w \in \mathbf{R}$. Thus Ξ can be identified with $\mathbf{R} \times \mathbf{R}$. The action of G on X induces a transitive action of G on Ξ which is given by

$$(\alpha, \beta, \gamma) \Diamond (v, w) = (v + \gamma, \frac{w + \beta}{\alpha}).$$

(Here we have for simplicity written (v, w) instead of $\xi_{v,w}$.) The isotropy group of the point $\xi_{(0,0)}$ (the x-axis) is

$$H = \{(\alpha, 0, 0) : \alpha > 0\} = \mathbf{R}_+^\times,$$

the multiplicative group of the positive real numbers. Thus we have the identifications

(65) $$X = G/K, \quad \Xi = G/H.$$

The group $K \cap H$ is now trivial so the Radon transform and its dual for the double fibration in (63) are defined by

(66) $$\widehat{f}(gH) = \int_H f(ghK)\, dh,$$

(67) $$\check{\varphi}(gK) = \chi(g) \int_K \varphi(gkH)\, dk,$$

where χ is the homomorphism $(\alpha, \beta, \gamma) \to \alpha^{-1}$ of G onto $\mathbf{R}_+^\times$. The reason for the presence of χ is that we wish Proposition 2.2 to remain valid even if G is not unimodular. In (66) and (67) we have the Haar measures

(68) $$dk_{(1,\beta-\beta)} = d\beta\,, \quad dh_{(\alpha,0,0)} = d\alpha/\alpha\,.$$

Also, if $g = (\alpha, \beta, \gamma)$, $h = (a, 0, 0)$, $k = (1, b, -b)$ then

$$gH = (\gamma, \beta/\alpha)\,, \qquad ghK = (\alpha a, \beta a + \gamma)$$
$$gK = (\alpha, \beta + \gamma)\,, \qquad gkH = (-b + \gamma, \tfrac{b+\beta}{\alpha})$$

so (66)–(67) become

$$\widehat{f}(\gamma, \beta/\alpha) = \int_{\mathbf{R}+} f(\alpha a, \beta a + \gamma) \frac{da}{a}$$

$$\check{\varphi}(\alpha, \beta + \gamma) = \alpha^{-1} \int_{\mathbf{R}} \varphi(-b + \gamma, \tfrac{b+\beta}{\alpha})\, db\,.$$

Changing variables these can be written

(69) $$\widehat{f}(v, w) = \int_{\mathbf{R}+} f(a, v + aw) \frac{da}{a}\,,$$

(70) $$\check{\varphi}(a, b) = \int_{\mathbf{R}} \varphi(b - as, s)\, ds \qquad a > 0\,.$$

Note that in (69) the integration takes place over all points on the line $\xi_{v,w}$ and in (70) the integration takes place over the set of lines $\xi_{b-as,s}$ all of which pass through the point (a, b). This is an *a posteriori* verification of the fact that our incidence for the pair (65) amounts to $x \in \xi$.

From (69)–(70) we see that $f \to \widehat{f}, \varphi \to \check{\varphi}$ are adjoint relative to the measures $\frac{da}{a}\, db$ and $dv\, dw$:

(71) $$\int_{\mathbf{R}} \int_{\mathbf{R}_+^\times} f(a, b)\check{\varphi}(a, b) \frac{da}{a}\, db = \int_{\mathbf{R}} \int_{\mathbf{R}} \widehat{f}(v, w)\varphi(v, w)\, dv\, dw\,.$$

The proof is a routine computation.

We recall (Chapter VII) that $(-L)^{1/2}$ is defined on the space of rapidly decreasing functions on $\mathbf{R}$ by

(72) $$((-L)^{1/2}\psi)^\sim (\tau) = |\tau|\widetilde{\psi}(\tau)$$

and we define Λ on $\mathcal{S}(\Xi)(= \mathcal{S}(\mathbf{R}^2))$ by having $(-L)^{1/2}$ only act on the second variable:

(73) $$(\Lambda\varphi)(v, w) = ((-L)^{1/2}\varphi(v, \cdot))(w)\,.$$

Viewing $(-L)^{1/2}$ as the Riesz potential I^{-1} on $\mathbf{R}$ (Chapter VII, §6) it is easy to see that if $\varphi_c(v, w) = \varphi(v, \frac{w}{c})$ then

(74) $$\Lambda\varphi_c = |c|^{-1}(\Lambda\varphi)_c .$$

The Radon transform (66) is now inverted by the following theorem.

Theorem 4.10. *Let* $f \in \mathcal{D}(X)$. *Then*

$$f = \frac{1}{2\pi}(\Lambda\widehat{f})^{\vee} .$$

Proof. In order to use the Fourier transform $F \to \widetilde{F}$ on $\mathbf{R}^2$ and on $\mathbf{R}$ we need functions defined on all of $\mathbf{R}^2$. Thus we define

$$f^*(a, b) = \begin{cases} \frac{1}{a}f(\frac{1}{a}, \frac{-b}{a}) & a > 0, \\ 0 & a \leq 0. \end{cases}$$

Then

$$
\begin{aligned}
f(a, b) &= \frac{1}{a}f^*\left(\frac{1}{a}, -\frac{b}{a}\right) \\
&= a^{-1}(2\pi)^{-2} \iint \widetilde{f^*}(\xi, \eta)e^{i(\frac{\xi}{a} - \frac{b\eta}{a})} \, d\xi \, d\eta \\
&= (2\pi)^{-2} \iint \widetilde{f^*}(a\xi + b\eta, \eta)e^{i\xi} \, d\xi \, d\eta \\
&= a(2\pi)^{-2} \iint |\xi|\widetilde{f^*}((a + ab\eta)\xi, a\eta\xi)e^{i\xi} \, d\xi \, d\eta .
\end{aligned}
$$

Next we express the Fourier transform in terms of the Radon transform. We have

$$
\begin{aligned}
\widetilde{f^*}((a + ab\eta)\xi, a\eta\xi) &= \iint f^*(x, y)e^{-ix(a+ab\eta)\xi}e^{-iya\eta\xi} \, dx \, dy \\
&= \int_{\mathbf{R}}\int_{x \geq 0} \frac{1}{x}f\left(\frac{1}{x}, -\frac{y}{x}\right) e^{-ix(a+ab\eta)\xi}e^{-iya\eta\xi} \, dx \, dy \\
&= \int_{\mathbf{R}}\int_{x \geq 0} f\left(\frac{1}{x}, b + \frac{1}{\eta} + \frac{z}{x}\right) e^{iza\eta\xi} \frac{dx}{x} \, dz .
\end{aligned}
$$

This last expression is

$$\int_{\mathbf{R}} \widehat{f}(b + \eta^{-1}, z)e^{iza\eta\xi} \, dz = (\widehat{f})^{\sim}(b + \eta^{-1}, -a\eta\xi) ,$$

where $\sim$ denotes the 1-dimensional Fourier transform (in the second variable). Thus

$$f(a, b) = a(2\pi)^{-2} \iint |\xi|(\widehat{f})^{\sim}(b + \eta^{-1}, -a\eta\xi)e^{i\xi} \, d\xi \, d\eta .$$

However $\widetilde{F}(c\xi) = |c|^{-1}(F_c)^\sim(\xi)$, so by (74)

$$
\begin{aligned}
f(a,b) &= a(2\pi)^{-2} \iint |\xi|((\widehat{f})_{a\eta})^\sim(b + \eta^{-1}, -\xi)e^{i\xi}\, d\xi|a\eta|^{-1}\, d\eta \\
&= (2\pi)^{-1} \int \Lambda((\widehat{f})_{a\eta})(b + \eta^{-1}, -1)|\eta|^{-1}\, d\eta \\
&= (2\pi)^{-1} \int |a\eta|^{-1}(\Lambda\widehat{f})_{a\eta}(b + \eta^{-1}, -1)|\eta|^{-1}\, d\eta \\
&= a^{-1}(2\pi)^{-1} \int (\Lambda\widehat{f})(b + \eta^{-1}, -(a\eta)^{-1})\eta^{-2}\, d\eta,
\end{aligned}
$$

so

$$
\begin{aligned}
f(a,b) &= (2\pi)^{-1} \int_{\mathbf{R}} (\Lambda\widehat{f})(b - av, v)\, dv \\
&= (2\pi)^{-1}(\Lambda\widehat{f})^\vee(a,b).
\end{aligned}
$$

proving the theorem.

Remark 4.11. It is of interest to compare this theorem with Theorem 3.8, Ch. I. If $f \in \mathcal{D}(X)$ is extended to all of $\mathbf{R}^2$ by defining it 0 in the left half plane then Theorem 3.8 does give a formula expressing f in terms of its integrals over half-lines in a strikingly similar fashion. Note however that while the operators $f \to \widehat{f}, \varphi \to \check{\varphi}$ are in the two cases defined by integration over the same sets (points on a half-line, half-lines through a point) the measures in the two cases are different. Thus it is remarkable that the inversion formulas look exactly the same.

H.   Theta Series and Cusp Forms

Let G denote the group $\mathbf{SL}(2, \mathbf{R})$ of 2×2 matrices of determinant one and Γ the *modular group* $\mathbf{SL}(2, \mathbf{Z})$. Let N denote the unipotent group $\left(\begin{smallmatrix} 1 & n \\ 0 & 1 \end{smallmatrix}\right)$ where $n \in \mathbf{R}$ and consider the homogeneous spaces

$$
(75) \qquad\qquad X = G/N, \qquad \Xi = G/\Gamma.
$$

Under the usual action of G on $\mathbf{R}^2$, N is the isotropy subgroup of $(1, 0)$ so X can be identified with $\mathbf{R}^2 - (0)$, whereas Ξ is of course 3-dimensional.

In number theory one is interested in decomposing the space $L^2(G/\Gamma)$ into G-invariant irreducible subspaces. We now give a rough description of this by means of the transforms $f \to \widehat{f}$ and $\varphi \to \check{\varphi}$.

As customary we put $\Gamma_\infty = \Gamma \cap N$; our transforms (9) then take the form

$$
\widehat{f}(g\Gamma) = \sum_{\Gamma/\Gamma_\infty} f(g\gamma N), \quad \check{\varphi}(gN) = \int_{N/\Gamma_\infty} \varphi(gn\Gamma)\, dn_{\Gamma_\infty}.
$$

Since N/Γ_∞ is the circle group, $\check{\varphi}(gN)$ is just the constant term in the Fourier expansion of the function $n\Gamma_\infty \to \varphi(gn\Gamma)$. The null space $L_d^2(G/\Gamma)$ in $L^2(G/\Gamma)$ of the operator $\varphi \to \check{\varphi}$ is called the space of **cusp forms** and the series for $\hat{f}$ is called **theta series**. According to Prop. 2.2 they constitute the orthogonal complement of the image $C_c(X)\hat{\;}$.

We have now the G-invariant decomposition

(76) $$L^2(G/\Gamma) = L_c^2(G/\Gamma) \oplus L_d^2(G/\Gamma),$$

where ($-$ denoting closure)

(77) $$L_c^2(G/\Gamma) = (C_c(X)\hat{\;})^-$$

and as mentioned above,

(78) $$L_d^2(G/\Gamma) = (C_c(X)\hat{\;})^\perp.$$

It is known (cf. Selberg [1962], Godement [1966]) that the representation of G on $L_c^2(G/\Gamma)$ is the *continuous* direct sum of the irreducible representations of G from the principal series whereas the representation of G on $L_d^2(G/\Gamma)$ is the *discrete* direct sum of irreducible representations each occurring with finite multiplicity.

I.   The Plane-to-Line Transform in $\mathbf{R}^3$. The Range

Now we consider the set $\mathbf{G}(2,3)$ of planes in $\mathbf{R}^3$ and the set $\mathbf{G}(1,3)$ of lines. The group $G = \mathbf{M}^+(3)$ of orientation preserving isometries of $\mathbf{R}^3$ acts transitively on both $\mathbf{G}(2,3)$ and $\mathbf{G}(1,3)$. The group $\mathbf{M}^+(3)$ can be viewed as the group of 4×4 matrices

$$\left(\begin{array}{c|c} \mathbf{SO}(3) & \begin{array}{c} x_1 \\ x_2 \\ x_3 \end{array} \\ \hline & 1 \end{array} \right),$$

whose Lie algebra $\mathfrak{g}$ has basis

$$E_i = E_{i4} \; (1 \le i \le 3), \quad X_{ij} = E_{ij} - E_{ji}, \quad 1 \le i \le j \le 3.$$

We have bracket relations

(79) $[E_i, X_{jk}] = 0$ if $i \ne j, k$, $[E_i, X_{ij}] = E_j - E_i$,

(80) $[X_{ij}, X_{k\ell}] = -\delta_{ik}X_{j\ell} + \delta_{jk}X_{i\ell} + \delta_{i\ell}X_{jk} - \delta_{j\ell}X_{ik}.$

We represent $\mathbf{G}(2,3)$ and $\mathbf{G}(1,3)$ as coset spaces

(81) $$\mathbf{G}(2,3) = G/H, \quad \mathbf{G}(1,3) = G/K,$$

where

$$H = \text{ stability of group of } \tau_0 \ (x_1, x_2\text{-plane}),$$
$$K = \text{ stability group of } \sigma_0 \ (x_1\text{-axis}).$$

We have $\mathbf{G} = \mathbf{SO}(3)\,\mathbf{R}^3$, $H = \mathbf{SO}_3(2)\,\mathbf{R}^2$, $K = \mathbf{SO}_1(2) \times \mathbf{R}$ the first two being semi-direct products. The subscripts indicate fixing of the x_3-axis and x_1-axis, respectively. The intersection $L = H \cap K = \mathbf{R}$ (the translations along the x_1-axis).

The elements $\tau_0 = eH$ and $\sigma_0 = eK$ are incident for the pair G/H, G/K and $\sigma_0 \subset \tau_0$. Since the inclusion notion is preserved by G we see that

$$\tau = \gamma H \text{ and } \sigma = gK \text{ are incident } \Leftrightarrow \sigma \subset \tau.$$

In the double fibration

(82)
$$G/L = \{(\sigma, \tau) | \sigma \subset \tau\}$$

$$\mathbf{G}(2,3) = G/H \qquad\qquad G/K = \mathbf{G}(1,3)$$

we see that the transform $\varphi \to \check{\varphi}$ in (9) (Chapter II,§2) is the plane-to-line transform which sends a function on $\mathbf{G}(2,3)$ into a function on lines:

(83)
$$\check{\varphi}(\sigma) = \int_{\tau \ni \sigma} \varphi(\tau)\, d\mu(\tau),$$

the measure $d\mu$ being the normalized measure on the circle.

For the study of the range of (83) it turns out to be simpler to replace G/L by another homogeneous space of G, namely the space of unit vectors $\omega \in \mathbf{S}^2$ with an initial point $x \in \mathbf{R}^3$. We denote this pair by ω_x. The action of G on this space $\mathbf{S}^2 \times \mathbf{R}^3$ is the obvious geometric action of $(u, y) \in \mathbf{SO}(3)\mathbf{R}^3$ on ω_x:

(84)
$$(u, y) \cdot \omega_x = (u \cdot \omega)_{(u \cdot x + y)} .$$

The subgroup fixing the North Pole ω_0 on $\mathbf{S}^2$ equals $\mathbf{SO}_3(2)$ so $\mathbf{S}^2 \times \mathbf{R}^3 = G/\mathbf{SO}(2)$. Instead of (82) we consider

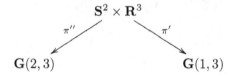

$$\mathbf{S}^2 \times \mathbf{R}^3$$

$$\pi'' \qquad\qquad \pi'$$

$$\mathbf{G}(2,3) \qquad\qquad \mathbf{G}(1,3)$$

the maps π' and π'' being given by

$$\pi'(\omega_x) = \mathbf{R}\omega + x \quad \text{(line through x in direction ω)},$$
$$\pi''(\omega_x) = \omega^\perp + x \quad \text{(plane through $x \perp \omega$)}.$$

The geometric nature of the action (84) shows that π' and π'' commute with the action of G.

For analysis on $\mathbf{S}^2 \times \mathbf{R}^3$ it will be convenient to write ω_x as the pair (ω, x). Note that

$$(85) \qquad (\pi')^{-1}(\mathbf{R}\omega + x) = \{(\omega, y) : y - x \in \mathbf{R}\omega\}$$

or equivalently, the set of translates of ω along the line $x + \mathbf{R}\omega$. Also

$$(86) \qquad (\pi'')^{-1}(\omega^\perp + x) = \{(\omega, z) : x - z \in \omega^\perp\},$$

the set of translates of ω with initial point on the plane through x perpendicular to ω.

Let ∇_x denote the gradient $(\partial/\partial x_1, \partial/\partial x_2, \partial/\partial x_3)$. Let $F \in \mathcal{E}(\mathbf{S}^2 \times \mathbf{R}^3)$. Then if θ is a unit vector and $\langle\,,\,\rangle$ the standard inner product on $\mathbf{R}^3$,

$$(87) \qquad \frac{d}{dt} F(\omega, x + t\theta) = \langle (\nabla_x F(\omega, x + t\theta)), \theta \rangle.$$

Thus for $\Psi \in \mathcal{E}(\mathbf{S}^2 \times \mathbf{R}^3)$,

$$(88) \qquad \Psi(\omega, x + t\omega) = \Psi(\omega, x) \quad (t \in \mathbf{R}) \Leftrightarrow (\nabla_x \Psi)(\omega, x) \perp \omega.$$

Lemma 4.12. *A function* $\Psi \in \mathcal{E}(\mathbf{S}^2 \times \mathbf{R}^3)$ *has the form* $\Psi = \psi \circ \pi'$ *with* $\psi \in \mathcal{E}(\mathbf{G}(1,3))$ *if and only if*

$$(89) \qquad \Psi(\omega, x) = \Psi(-\omega, x), \quad \nabla_x \Psi(\omega, x) \perp \omega.$$

Proof. Clearly, if $\psi \in \mathcal{E}(\mathbf{G}(1,3))$ then Ψ has the property stated. Conversely, if Ψ satisfies the conditions (89) it is constant on each set (85).

Lemma 4.13. *A function* $\Phi \in \mathcal{E}(\mathbf{S}^2 \times \mathbf{R}^3)$ *has the form* $\Phi = \varphi \circ \pi''$ *with* $\varphi \in \mathcal{E}(\mathbf{G}(2,3))$ *if and only if*

$$(90) \qquad \Phi(\omega, x) = \Phi(-\omega, x), \quad \nabla_x \Phi(\omega, x) \in \mathbf{R}\omega.$$

Proof. If $\varphi \in \mathcal{E}(\mathbf{G}(2,3))$ then (87) for $F = \Phi$ implies

$$\frac{d}{dt} \Phi(\omega, x + t\theta) = 0 \text{ for each } \theta \in \omega^\perp$$

so (90) holds. Conversely, if Φ satisfies (90) then by (87) for $F = \Phi$, Φ is constant on each set (86).

We consider now the action of G on $\mathbf{S}^2 \times \mathbf{R}^3$. The Lie algebra $\mathfrak{g}$ is $\mathfrak{so}(3) + \mathbf{R}^3$, where $\mathfrak{so}(3)$ consists of the 3×3 real skew-symmetric matrices.

For $X \in \mathfrak{so}(3)$ and $\Psi \in \mathcal{E}(\mathbf{S}^2 \times \mathbf{R}^3)$ we have by Ch. VIII, (12),

$$(\lambda(X)\Psi)(\omega, x) = \left\{ \frac{d}{dt} \Psi(\exp(-tX) \cdot \omega, \exp(-tX) \cdot \omega) \right\}_{t=0}$$

$$= \left\{ \frac{d}{dt} \Psi(\exp(-tX) \cdot \omega, x) \right\}_{t=0}$$

$$+ \left\{ \frac{d}{dt} \Psi(\omega, \exp(-tX) \cdot x) \right\}_{t=0}$$

so

(91) $$\lambda(X)\Psi(\omega, x) = X_\omega \Psi(\omega, x) + X_x \Psi(\omega, x),$$

where X_ω and X_x are tangent vectors to the circles $\exp(-tX) \cdot \omega$ and $\exp(-tX) \cdot x$ in $\mathbf{S}^2$ and $\mathbf{R}^3$, respectively.

For $v \in \mathbf{R}^3$ acting on $\mathbf{S}^2 \times \mathbf{R}^3$ we have

(92) $$(\lambda(v)\Psi)(\omega, x) = \left\{ \frac{d}{dt} \Psi(\omega, x - tv) \right\}_{t=0} = -\langle \nabla_x \Psi(\omega, x), v \rangle.$$

For $X_{12} = E_{12} - E_{21}$ we have

$$\exp t X_{12} = \begin{pmatrix} \cos t & \sin t & 0 \\ -\sin t & \cos t & 0 \\ 0 & 0 & 1 \end{pmatrix} \quad \text{etc.}$$

so if $f \in \mathcal{E}(\mathbf{R}^3)$

(93) $$(\lambda(X_{ij})f)(x) = \left\{ \frac{d}{dt} f(\exp(-tX_{ij}) \cdot x) \right\} = x_i \frac{\partial f}{\partial x_j} - x_j \frac{\partial f}{\partial x_i}.$$

Given $\ell \in \mathbf{G}(1,3)$ let $\check{\ell}$ denote the set of 2-planes in $\mathbf{R}^3$ containing it. If $\ell = \pi'(\sigma, x)$ then $\check{\ell} = \{\pi''(\omega, x) : \omega \in \mathbf{S}^2, \omega \perp \sigma\}$, which is identified with the great circle $A(\sigma) = \sigma^\perp \cap \mathbf{S}^2$. We give $\check{\ell}$ the measure μ_ℓ corresponding to the arc-length measure on $A(\sigma)$. In this framework, the plane-to-line transform (83) becomes

(94) $$(R\varphi)(\ell) = \int_{\xi \in \check{\ell}} \varphi(\xi) \, d\mu_\ell(\xi)$$

for $\varphi \in \mathcal{E}(\mathbf{G}(2,3))$, $\ell \in \mathbf{G}(1,3)$. Expressing this on $\mathbf{S}^2 \times \mathbf{R}^3$ we have with $\Phi = \varphi \circ \pi''$

(95) $$(R\varphi \circ \pi')(\sigma, x) = \frac{1}{2\pi} \int_{A(\sigma)} \Phi(\omega, x) \, d\sigma(\omega),$$

where d_σ represents the arc-length measure on $A(\sigma)$.

We consider now the basis E_i, X_{jk} of the Lie algebra $\mathfrak{g}$. For simplicity we drop the tilde in $\widetilde{E}_i$ and $\widetilde{X}_{jk}$.

Lemma 4.14. (Richter.) *The operator*

$$D = E_1 X_{23} - E_2 X_{13} + E_3 X_{12}$$

belongs to the center $\mathbf{Z}(G)$ *of* $\mathbf{D}(G)$.

Proof. First note by the above commutation relation that the factors in each summand commute. Thus D commutes with E_i. The commutation with each X_{ij} follows from the above commutation relations (79)–(80).

Because of Propositions 1.7 and 2.3 in Ch. VIII, D induces G-invariant operators $\lambda(D)$, $\lambda'(D)$ and $\lambda''(D)$ on $\mathbf{S}^2 \times \mathbf{R}^3$, $\mathbf{G}(1,3)$ and $\mathbf{G}(2,3)$, respectively.

Lemma 4.15. *(i)* $\lambda(D) = 0$ *on* $\mathcal{E}(\mathbf{R}^3)$.

(ii) $\lambda''(D) = 0$ *on* $\mathcal{E}(\mathbf{G}(2,3))$.

Proof. Part (i) follows from $(\lambda(E_i)f)(x) = -\partial f/\partial x_i$ and the formula (93). For (ii) we take $\varphi \in \mathcal{E}(\mathbf{G}(2,3))$ and put $\Phi = \varphi \circ \pi''$. Since π'' commutes with the G-action, we have

$$\Phi(g \cdot (\omega, x)) = \varphi(g \cdot \pi''(\omega, x))$$

so by (13) in Ch.VIII,

$$(96) \qquad \lambda(D)\Phi = \lambda''(D)\varphi \circ \pi''.$$

By (91)–(92) we have

$$(97) \qquad \lambda(D)\Phi(\omega, x) = (\lambda(E_1 X_{23} - E_2 X_{13} + E_3 X_{12}))_x \Phi(\omega, x)$$
$$+ \left[\lambda(E_1)_x \lambda(X_{23})_\omega - \lambda(E_2)_x \lambda(X_{13})_\omega + \lambda(E_3)_x \lambda(X_{12})_\omega\right]\Phi(\omega, x).$$

By Part (i) the first of the two terms vanishes. In the second term we exchange E_i and X_{jk}. Recalling that $\nabla_x \Phi(\omega, x)$ equals $h(\omega, x)\,\omega$ (h a scalar) we have

$$\lambda(E_i)_x \Phi(\omega, x) = h(\omega, x)\,\omega_i, \quad 1 \le i \le 3.$$

Since $\exp t X_{23}$ fixes ω_1 we have $\lambda(X_{23})\omega_1 = 0$ etc. Putting this together we deduce

$$(98) \qquad \lambda(D)\Phi(\omega, x) = -\omega_1 \lambda(X_{23})_\omega\, h(\omega, x) + \omega_2 \lambda(X_{13})_\omega\, h(\omega, x)$$
$$- \omega_3 \lambda(X_{12})_\omega\, h(\omega, x).$$

Part (ii) will now follow from the following.

Lemma 4.16. *Let $u \in \mathcal{E}(\mathbf{S}^2)$. Let $\mu(X_{ij})$ denote the restriction of the vector field $\lambda(X_{ij})$ to the sphere. Then*

$$(\omega_1\,\mu(X_{23}) - \omega_2\,\mu(X_{13}) + \omega_3\,\mu(X_{12}))u = 0\,.$$

Proof. For a fixed $\epsilon > 0$ extend u to a smooth function $\tilde{u}$ on the shell $S_\epsilon^2 : 1 - \epsilon < \|x\| < 1 + \epsilon$ in $\mathbf{R}^3$. The group $\mathbf{SO}(3)$ acts on S_ϵ^2 by rotation so by (12), Ch. VIII, the vector fields $\mu(X_{ij})$ extend to vector fields $\tilde{\mu}(X_{ij})$ on S_ϵ^2. But these are just the restrictions of the vector fields $x_i \frac{\partial}{\partial x_j} - x_j \frac{\partial}{\partial x_i}$ to S_ϵ^2. These vector fields satisfy

$$x_1\left(x_2\frac{\partial}{\partial x_3} - x_3\frac{\partial}{\partial x_2}\right) - x_2\left(x_1\frac{\partial}{\partial x_3} - x_3\frac{\partial}{\partial x_1}\right) + x_3\left(x_1\frac{\partial}{\partial x_2} - x_2\frac{\partial}{\partial x_1}\right) = 0,$$

so the lemma holds.

We can now state Gonzalez's main theorem describing the range of R.

Theorem 4.17. *The plane-to-line transform R maps $\mathcal{E}(\mathbf{G}(2,3))$ onto the kernel of D:*

$$R\big(\mathcal{E}(\mathbf{G}(2,3))\big) = \{\psi \in \mathcal{E}(\mathbf{G}(1,3)) : \lambda'(D)\psi = 0\}\,.$$

Proof. The operator R obviously commutes with the action of G. Thus by (13) in Ch. VIII, we have for each $E \subset \mathbf{D}(G)$,

(99) $$R(\lambda''(E)\varphi) = 0 = \lambda'(E)R\varphi \qquad \varphi \in \mathcal{E}(\mathbf{G}(2,3))\,.$$

In particular, Lemma 4.15 implies

$$\lambda'(D)(R\varphi) = 0 \qquad \text{for } \varphi \in \mathcal{E}(\mathbf{G}(2,3))\,.$$

For the converse assume $\psi \in \mathcal{E}(\mathbf{G}(1,3))$ satisfies

$$\lambda'(D)\psi = 0\,.$$

Put $\Psi = \psi \circ \pi'$. Then by the analog of (96) $\lambda(D)\Psi = 0$. In analogy with the formula (97) for $\lambda(D)\Phi$ (where the first term vanished) we get for each $(\sigma, x) \in \mathbf{S}^2 \times \mathbf{R}^3$,
(100)
$$0 = \lambda(D)\Psi = \big[\lambda(E_1)_x\lambda(X_{23})_\sigma - \lambda(E_2)_x\lambda(X_{13})_\sigma + \lambda(E_3)_x\lambda(X_{12})_\sigma\big]\Psi(\sigma, x)\,.$$

Now $\Psi(\sigma, x) = \Psi(-\sigma, x)$ so by the surjectivity of the great circle transform (which is contained in Theorem 2.2 in Ch. III) there exists a unique even smooth function $\omega \to \Phi_x(\omega)$ on $\mathbf{S}^2$ such that

(101) $$\Psi(\sigma, x) = \frac{1}{2\pi} \int\limits_{A(\sigma)} \Phi_x(\omega)\,d\sigma(\omega)\,.$$

We put $\Phi(\omega, x) = \Phi_x(\omega)$. The task is now to prove that $\nabla_x \Phi(\omega, x)$ is a multiple of ω, because then by Lemma 4.13, $\Phi = \varphi \circ \pi''$ for some $\varphi \in \mathcal{E}(\mathbf{G}(1, 3))$. Then we would in fact have by (95), $R\varphi \circ \pi' = \Psi = \psi \circ \pi'$ so $R\varphi = \psi$.

Applying the formula (100) above and differentiating (101) under the integral sign, we deduce

(102)

$$0 = \int_{A(\sigma)} \left[\lambda(X_{23})_\omega \lambda(E_1)_x - \lambda(X_{13})_\omega \lambda(E_2)_x + \lambda(X_{12})_\omega(E_3)_x \right] \Phi(\omega, x) \, d_\sigma(\omega).$$

For x fixed the integrand is even in ω, so by injectivity of the great circle transform, the integrand vanishes. Consider the $\mathbf{R}^3$-valued vector field on $\mathbf{S}^2$ given by

$$\vec{G}(\omega) = -\nabla_x \Phi(\omega, x) = (\lambda(E_1)_x \Phi(\omega, x), \lambda(E_2)_x \Phi(\omega, x), \lambda(E_3)_x \Phi(\omega, x))$$
$$= (G_1(\omega), G_2(\omega), G_3(\omega)),$$

where each $G_i(\omega)$ is even. By the vanishing of the integrand in (102) we have

(103)
$$\lambda(X_{23})G_1 - \lambda(X_{13})G_2 + \lambda(X_{12})G_3 = 0.$$

We decompose $\vec{G}(\omega)$ into tangential and normal components, respectively, $\vec{G}(\omega) = \vec{T}(\omega) + \vec{N}(\omega)$, with components $T_i(\omega)$, $N_i(\omega)$, $1 \leq i \leq 3$. We wish to show that $\vec{G}(\omega)$ proportional to ω, or equivalently, $\vec{T}(\omega) = 0$. We substitute $G_i = T_i + N_i$ into (103) and observe that

(104)
$$\lambda(X_{23})(N_1) - \lambda(X_{13})(N_2) + \lambda(X_{12})(N_3) = 0,$$

because writing $\vec{N}(\omega) = n(\omega)\omega$, n is an odd function on $\mathbf{S}^2$ and (104) equals

$$\omega_1 \lambda(X_{23})n(\omega) - \omega_2 \lambda(X_{13})n(\omega) + \omega_3 \lambda(X_{12})n(\omega)$$
$$+ n(\omega)(\lambda(X_{23})(\omega_1) - \lambda(X_{13})\omega_2 + \lambda(X_{12})(\omega_3)) = 0$$

by Lemma 4.16 and $\lambda(X_{jk})\omega_i = 0$, $(i \neq j, k)$. Thus we have the equation

(105)
$$\lambda(X_{23})T_1 - \lambda(X_{13})T_2 + \lambda(X_{12})T_3 = 0.$$

From Lemma 4.12 $\langle \sigma, \nabla_x \Psi(\sigma, x) \rangle = 0$ and by (101) we get

$$0 = \int_{A(\sigma)} \langle \sigma, \nabla_x \Phi(\omega, x) \, d_\sigma(\omega) \rangle = - \int_{A(\sigma)} \langle \sigma, \vec{G}(\omega) \rangle \, d_\sigma \omega$$

$$= - \int_{A(\sigma)} \langle \sigma, \vec{T}(\omega) \rangle \, d_\sigma(\omega) - \int_{A(\sigma)} \langle \sigma, \vec{N}(\omega) \rangle \, d_\sigma \omega$$

$$= - \int_{A(\sigma)} \langle \sigma, \vec{T}(\omega) \rangle \, d_\sigma \omega,$$

since $\sigma \perp \vec{N}(\omega)$ on $A(\sigma)$. Thus $\vec{T}(\omega)$ is an even vector field on $\mathbf{S}^2$ satisfying (105) and

$$(106) \qquad \int_{A(\sigma)} \langle \sigma, \vec{T}(\omega) \rangle \, d_\sigma \omega = 0 \, .$$

We claim that $\vec{T}(\omega) = \operatorname{grad}_{\mathbf{S}^2} t(\omega)$, where $\operatorname{grad}_{\mathbf{S}^2}$ denotes the gradient on $\mathbf{S}^2$ and t is an odd function on $\mathbf{S}^2$. To see this, we extend $T(\omega)$ to a smooth vector field $\widetilde{T}$ on a shell $S_\epsilon^2 : 1 - \epsilon < \|x\| < 1 + \epsilon$ in $\mathbf{R}^3$ by $\vec{T}(r\omega) = \vec{T}(\omega)$ for $r \in (1 - \epsilon, 1 + \epsilon)$. Again the $\mathbf{SO}(3)$ action on S_ϵ^2 induces vector fields $\widetilde{\mu}(X_{ij})$ on S_ϵ^2, which are just $x_i \frac{\partial}{\partial x_j} - x_j \frac{\partial}{\partial x_i}$. Thus (105) becomes

$$\langle \operatorname{curl} \widetilde{T}(x), x \rangle = 0 \text{ on } S_\epsilon^2 \, .$$

By the classical Stokes' theorem for $\mathbf{S}^2$ this implies that the line integral

$$\int_\gamma T_1 \, dx_1 + T_2 \, dx_2 + T_3 \, dx_3 = 0$$

for each simple closed curve γ on $\mathbf{S}^2$. Let τ be the pull back of the form $\sum_i T_i \, dx_i$ to $\mathbf{S}^2$. By the Stokes' theorem for τ on $\mathbf{S}^2$ we deduce $d\tau = 0$ on $\mathbf{S}^2$, i.e., τ is closed. Since $\mathbf{S}^2$ is simply connected, τ is exact, i.e., $\tau = dt$, $t \in \mathcal{E}(\mathbf{S}^2)$. (This is an elementary case of deRham's theorem; t can be constructed as in complex variable theory.) For any vector field Z on $\mathbf{S}^2$ $dt(Z) = \langle \operatorname{grad}_{\mathbf{S}^2} t, Z \rangle$ so $T(\omega) = \operatorname{grad}_{\mathbf{S}^2} t(\omega)$. Decomposing $t(\omega)$ into odd and even components we see that the even component is constant so we can take $t(\omega)$ odd.

Let $H(\sigma)$ denote the hemisphere on the side of $A(\sigma)$ away from σ. Note that σ located at points of $A(\sigma)$ form the outward pointing normals of the boundary $A(\sigma)$ of $H(\sigma)$. With $\vec{T}(\omega) = \operatorname{grad}_{\mathbf{S}^2} t(\omega)$ the integral (106) equals

$$\int_{H(\sigma)} (L_{\mathbf{S}^2} t)(\omega) \, d\omega \, , \quad \sigma \in \mathbf{S}^2 \, ,$$

by the divergence theorem on $\mathbf{S}^2$. Since $L_{\mathbf{S}^2} t$ is odd the next lemma implies that $L_{\mathbf{S}^2} t = 0$ so t is a constant, hence $t \equiv 0$ (because t is odd).

Lemma 4.18. *Let τ denote the hemisphere transform on $\mathbf{S}^2$, $\tau(h) = \int_{H(\sigma)} h(\omega) \, d\omega$ for $h \in \mathcal{E}(\mathbf{S}^2)$. If $\tau(h) = 0$ then h is an even function.*

Proof. Let H_m denote the space of degree m spherical harmonics on $\mathbf{S}^2$ ($m = 0, 1, 2, \ldots$). Then $\mathbf{SO}(3)$ acts irreducibly on H_m. Since τ commutes with the action of $\mathbf{SO}(3)$ it must (by Schur's lemma) be a scalar operator

c_m on H_m. The value can be obtained by integrating a zonal harmonic P_m over the hemisphere

$$c_m = 2\pi \int_0^{\frac{\pi}{2}} P_m(\cos\theta) \sin\theta \, d\theta = 2\pi \int_0^1 P_m(x) \, dx.$$

According to Erdelyi et al. [1953], I p. 312, this equals

$$(107) \qquad c_m = 4\pi^{\frac{1}{2}} \frac{\Gamma(1 + \frac{m}{2})}{\Gamma\left(\frac{m+1}{2}\right)} \frac{1}{m(m+1)} \sin\frac{m\pi}{2},$$

which equals 2π for $m = 0$, is 0 for m even, and is $\neq 0$ for m odd. Since each $h \in \mathcal{E}(\mathbf{S}^2)$ has an expansion $h = \sum_0^\infty h_m$ with $h_m \in H_m$, $\tau(h) = 0$ implies $c_m = 0$ (so m is even) if $h_m \neq 0$. Thus h is even as claimed.

Remark. The value of c_m in (107) appears in an exercise in Whittaker–Watson [1927], p. 306, attributed to Clare, 1902.

J.   Noncompact Symmetric Space and Its Family of Horocycles

This example belongs to the realm of the theory of semisimple Lie groups G. See Chapter IX, §2 for orientation. To such a group with finite center is associated a coset space $X = G/K$ (a Riemannian symmetric space) where K is a maximal compact subgroup (unique up to conjugacy). The group G has an Iwasawa decomposition $G = NAK$ (generalizing the one in Example C for $\mathbf{H}^2$.) Here N is nilpotent and A abelian. The orbits in X of the conjugates gNg^{-1} to N are called **horocycles**. These are closed submanifolds of X and are permuted transitively by G. The set Ξ of those horocycles ξ is thus a coset space of G, in fact $\Xi = G/MN$, where M is the centralizer of A in K. To this pair

$$X = G/K, \qquad \Xi = G/MN$$

are associated a Radon transform $f \to \widehat{f}$ and its dual $\varphi \to \check{\varphi}$ as in formula (9). More explicitly,

$$(108) \qquad \widehat{f}(\xi) = \int_\xi f(x) \, dm(x), \qquad \check{\varphi}(x) = \int_{\xi \ni x} \varphi(\xi) \, d\mu(\xi),$$

where dm is the Riemannian measure on the submanifold ξ and $d\mu$ is the average over the (compact) set of horocycles passing through x.

   Problems A, B, C, D all have solutions here (with some open questions); there is injectivity of $f \to \widehat{f}$ (with inversion formulas), surjectivity of $\varphi \to \check{\varphi}$, determinations of ranges and kernels of these maps, support theorems and applications to differential equations and group representations.

The transform $f \rightarrow \widehat{f}$ has the following inversion

$$f = (\Lambda \widehat{f})^{\vee},$$

where Λ is a G-invariant pseudo-differential operator on Ξ. In the case when G has all Cartan subgroups conjugate one has a better formula

$$f = \square((\widehat{f})^{\vee}),$$

where $\square$ is an explicit differential operator on X.

The support theorem for $f \rightarrow \widehat{f}$ states informally, $B \subset X$ being any ball:

$$\widehat{f}(\xi) = 0 \quad \text{for} \quad \xi \cap B = \emptyset \Rightarrow f(x) = 0 \quad \text{for} \quad x \notin B.$$

Here f is assumed "rapidly decreasing" in a certain technical sense.

Thus the conjugacy of the Cartan subgroups corresponds to the case of odd dimension for the Radon transform on $\mathbf{R}^n$. For complete proofs of these results, with documentation, see my book [1994b] or [2008].

Exercises and Further Results

1.   The Discrete Case.

For a discrete group G, Proposition 2.2 (via diagram (4)) takes the form (# denoting incidence):

$$\sum_{x \in X} f(x) \check{\varphi}(x) = \sum_{(x,\xi) \in X \times \Xi, x \# \xi} f(x)\varphi(\xi) = \sum_{\xi \in X} \widehat{f}(\xi)\varphi(\xi).$$

2.   Linear Codes. (Boguslavsky [2001])

Let $\mathbf{F}_q$ be a finite field and $\mathbf{F}_q^n$ the n-dimensional vector space with its natural basis. The *Hamming metric* is the distance d given by $d(x,y) =$ number of distinct coordinate positions in x and y.

A linear $[n, k, d]$-code C is a k-dimensional subspace of $\mathbf{F}_k^n$ such that $d(x,y) \geq d$ for all $x, y \in C$. Let $\mathbf{P}C$ be the projectivization of C on which the projective group $G = \mathbf{PGL}(k-1, \mathbf{F}_q)$ acts transitively. Let $\ell \in \mathbf{P}C$ be fixed and π a hyperplane containing ℓ. Let K and H be the corresponding isotropy groups. Then $X = G/K$, and $\Xi = G/H$ satisfy Lemma 1.3 and the transforms

$$\widehat{f}(\xi) = \sum_{x \in \xi} f(x), \quad \check{\varphi}(x) = \sum_{\xi \ni x} \varphi(\xi)$$

are well defined. They are inverted as follows. Put

$$s(\varphi) = \sum_{\xi \in \Xi} \varphi(\xi), \quad \sigma(f) = \sum_{x \in X} f(x).$$

The projective space $\mathbf{P}^m$ over $\mathbf{F}_q$ has a number of points equal to $p_m = \frac{q^{m+1}-1}{q-1}$. Here $m = k - 1$ and we consider the operators D and Δ given by

$$(D\varphi)(\xi) = \varphi(\xi) - \frac{q^{k-2}-1}{(q^{k-1}-1)^2}s(\varphi), \quad (\Delta f)(x) = f(x) - \frac{q^{k-2}-1}{(q^{k-1}-1)^2}\sigma(f).$$

Then

$$f(x) = \frac{1}{q^{k-2}}(D\widehat{f})^\vee(x), \quad \varphi(\xi) = \frac{1}{q^{k-2}}(\Delta\check{\varphi})\widehat{\ }(\xi).$$

3.   Radon Transform on Loops. (Brylinski [1996])

Let M be a manifold and LM the free loop space in M. Fix a 1-form α on M. Consider the functional I_α on LM given by $I_\alpha(\gamma) = \int_\gamma \alpha$.

With the standard C^∞ structure on LM

$$(dI_\alpha, v)_\gamma = \int_0^1 d\alpha(v(t), \overset{\circ}{\gamma}(t))\, dt$$

for $v \in (LM)_\gamma$. Clearly $I_\alpha = 0$ if and only if α is exact.

Inversion, support theorem and range description of this transform are established in the cited reference. Actually I_α satisfies differential equations reminiscent of John's equations in Theorem 6.5, Ch. I.

4.   Theta Series and Cusp Forms.

This concerns Ch. II, §4, Example **H**. For the following results see Godement [1966].

(i)   In the identification $G/N \approx \mathbf{R}^2 - (0)$ (via $gN \to g\binom{1}{0}$), let $f \in \mathcal{D}(G/N)$ satisfy $f(x) = f(-x)$. Then, in the notation of Example **H**,

$$\tfrac{1}{2}(\widehat{f})^\vee(gN) = f(gN) + \sum_{(\gamma)} \int_N f(gn\gamma N)\, dn,$$

where $\sum_{(\gamma)}$ denotes summation over the nontrivial double cosets $\pm\Gamma_\infty\gamma\Gamma_\infty$ (γ and $-\gamma$ in Γ identified).

(ii)   Let $A = \left\{\left(\begin{smallmatrix} t & 0 \\ 0 & t^{-1} \end{smallmatrix}\right) : t > 0\right\}$ be the diagonal subgroup of G and $\beta(h) = t^2$ if $h = \left(\begin{smallmatrix} t & 0 \\ 0 & t^{-1} \end{smallmatrix}\right)$. Consider the Mellin transform

$$\widetilde{f}(gN, 2s) = \int_A f(ghN)\beta(h)^s\, dh$$

and (viewing G/N as $\mathbf{R}^2 - 0$) the twisted Fourier transform

$$f^*(x) = \int_{\mathbf{R}^2} f(y)e^{-2\pi i B(x,y)}\, dy,$$

where $B(x, y) = x_2 y_1 - x_1 y_2$ for $x = (x_1, x_2)$, $y = (y_1, y_2)$. The *Eisenstein series* is defined by

$$E_f(g, s) = \sum_{\gamma \in \Gamma/\Gamma_\infty} \tilde{f}(g\gamma N, 2s), \quad (\text{convergent for } \operatorname{Re} s > 1).$$

Theorem. *Assuming $f^*(0) = 0$, the function $s \to \zeta(2s)E_f(g, s)$ extends to an entire function on $\mathbf{C}$ and does not change under $s \to (1-s)$, $f \to f^*$.*

5.   Radon Transform on Minkowski Space. (Kumahara and Wakayama [1993], see Figure II,6)

Let X be an $(n+1)$-dimensional real vector space with inner product $\langle\ ,\ \rangle$ of signature $(1, n)$. Let $e_0, e_1, \ldots, e_n$ be a basis such that $\langle e_i, e_j \rangle = -1$ for $i = j = 0$ and 1 if $i = j > 0$ and 0 if $i \neq j$. Then if $x = \sum_0^n x_i e_i$, a hyperplane in X is given by

$$\sum_0^n a_i x_i = c, \quad a \in \mathbf{R}^{n+1}, a \neq 0$$

and $c \in \mathbf{R}$. We put

$$\omega_0 = -a_0/|\langle a, a \rangle|^{\frac{1}{2}}, \quad \omega_j = a_j/|\langle a, a \rangle|^{\frac{1}{2}}, j > 0, \quad p = c/|\langle a, a \rangle|^{\frac{1}{2}}$$

if $\langle a, a \rangle \neq 0$ and if $\langle a, a \rangle = 0$

$$\omega_0 = -a_0/|a_0|, \quad \omega_j = a_j/|a_0| j > 0, \quad p = c/|a_0|.$$

The hyperplane above is thus

$$\langle x, \omega \rangle = -x_0 \omega_0 + x_1 \omega_1 + \cdots + x_n \omega_n = p,$$

written $\xi(\omega, p)$. The semidirect product $\mathbf{M}(1, n)$ of the translations of X with the connected Lorentz group $G = \mathbf{SO}_0(1, n)$ acts transitively on X and $\mathbf{M}(1, n)/\mathbf{SO}(1, n) \cong X$.

To indicate how the light cone $\langle \omega, \omega \rangle = 0$ splits X we make the following definitions (see Figure II,6).

$$\begin{aligned}
X_-^+ &= \{\omega \in X : \langle \omega, \omega \rangle = -1, \omega_0 > 0\} \\
X_-^- &= \{\omega \in X : \langle \omega, \omega \rangle = -1, \omega_0 < 0\} \\
X_+ &= \{\omega \in X : \langle \omega, \omega \rangle = +1\} \\
X_0^+ &= \{\omega \in X : \langle \omega, \omega \rangle = 0, \omega_0 > 0\} \\
X_0^- &= \{\omega \in X : \langle \omega, \omega \rangle = 0, \omega_0 < 0\} \\
S_\pm &= \{\omega \in X : \langle \omega, \omega \rangle = 0, \omega_0 = \pm 1\}.
\end{aligned}$$

The scalar multiples of the X_i fill up X. The group $\mathbf{M}(1, n)$ acts on X as follows: If $(g, z) \in \mathbf{M}(1, n)$, $g \in G$, $z \in X$ then

$$(g, z) \cdot x = z + g \cdot x.$$

The action on the space Ξ of hyperplanes $\xi(\omega, p)$ is

$$(g, z) \cdot \xi(\omega, p) = \xi(g \cdot \omega, p + \langle z, g \cdot \omega \rangle).$$

Then Ξ has the $\mathbf{M}(1, n)$ orbit decomposition

$$\Xi = \mathbf{M}(1, n)\xi(e_0, 0) \cup \mathbf{M}(1, n)\xi(e_1, 0) \cup M(1, n)\xi(e_0 + e_1, 0)$$

into three homogeneous spaces of $\mathbf{M}(1, n)$.

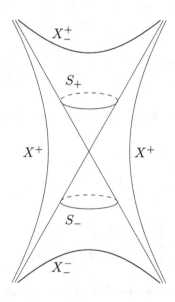

FIGURE II.6.

Minkowski space for dimension 3

Note that $\xi(-\omega, -p) = \xi(\omega, p)$ so in the definition of the Radon transform we assume $\omega_0 > 0$. The sets X_-^+, X_+ and X_0^+ have natural G-invariant measures $d\mu_-(\omega)$ and $d\mu_+(\omega)$ on $X_-^+ \cup X_-^-$ and X_+, respectively; in fact

$$d\mu_{\pm} = \frac{1}{|\omega_i|} \prod_{j \neq i} d\omega_j \qquad \text{where } \omega_i \neq 0.$$

Viewing $X_-^+ \cup X_-^- \cup X_+ \cup S_+ \cup S_-$ as a substitute for a "boundary" ∂X of X we define

$$\int_{\partial X} \psi(\omega) \, d\mu(\omega) = \int_{X_-^+ \cup X_-^-} \psi(\omega) \, d\mu_-(\omega) + \int_{X_+} \psi(\omega) \, d\mu_+(\omega)$$

for $\psi \in C_c(X)$. (S_+ and S_- have lower dimension.) The Radon transform $f \to \widehat{f}$ and its dual $\varphi \to \check{\varphi}$ are now defined by

$$\widehat{f}(\xi) = \int_\xi f(x)\, dm(x) = \int_{\langle x,\omega \rangle = p} f(x)\, dm(x) = \widehat{f}(\omega, p),$$

$$\check{\varphi}(x) = \int_{\xi \ni x} \varphi(\xi)\, d\sigma_x(\xi) = \int_{\partial X} \varphi(\omega, \langle x, \omega \rangle)\, d\mu(\xi).$$

Here dm is the Euclidean measure on the hyperplane ξ and a function φ on Ξ is identified with an even function $\varphi(\omega, p)$ on $\partial X \times \mathbf{R}$. The measure $d\sigma_x$ is defined by the last relation.

There are natural analogs $\mathcal{S}(X)$, $\mathcal{S}(\Xi)$ and $\mathcal{S}_H(\Xi)$ of the spaces $\mathcal{S}(\mathbf{R}^n)$, $\mathcal{S}(\mathbf{P}^n)$ and $\mathcal{S}_H(\mathbf{P}^n)$ defined in Ch. I, §2. The following analogs of the $\mathbf{R}^n$ theorems hold.

Theorem. $f \to \widehat{f}$ is a bijection of $\mathcal{S}(X)$ onto $\mathcal{S}_H(\Xi)$.

Theorem. For $f \in \mathcal{S}(X)$,

$$f = (\Lambda \widehat{f})^\vee,$$

where

$$(\Lambda \varphi)(\omega, p) = \begin{cases} \frac{1}{2(2\pi)^2} \left(\frac{\partial}{i\partial p} \right)^n \varphi(\omega, p) & n \text{ even} \\ \frac{1}{2(2\pi)^n} \mathcal{H}_p \left(\frac{\partial}{i\partial p} \right)^n \varphi(\omega, p) & n \text{ odd} \end{cases}.$$

6. John's Equation for the X-ray transform on $\mathbf{R}^3$.

According to Richter [1986b] the equation $\lambda'(D)\psi = 0$ in Gonzalez' Theorem 4.17 characterizes the range of the X-ray transform on $\mathbf{R}^3$. Relate this to John's equation $\Lambda \psi = 0$ in Theorem 6.9, Ch. I.

Bibliographical Notes

The Radon transform and its dual for a double fibration

(1)
$$Z = G/(K \cap H)$$

$$X = G/K \qquad\qquad \Xi = G/H$$

was introduced in the author's paper [1966a]. The results of §1–§2 are from there and from [1994b]. The definition uses the concept of *incidence* for

$X = G/K$ and $\Xi = G/H$ which goes back to Chern [1942]. Even when the elements of Ξ can be viewed as subsets of X and vice versa (Lemma 1.3) it can be essential for the inversion of $f \to \hat{f}$ not to restrict the incidence to the naive one $x \in \xi$. (See for example the classical case $X = \mathbf{S}^2, \Xi = $ set of great circles where in Theorem 4.1 a more general incidence is essential.) The double fibration in (1) was generalized in Gelfand, Graev and Shapiro [1969], by relaxing the homogeneity assumption.

For the case of geodesics in constant curvature spaces (Examples A, B in §4) see notes to Ch. III.

The proof of Theorem 4.4 (a special case of the author's inversion formula in [1964], [1965b]) makes use of a method by Godement [1957] in another context. Another version of the inversion (47) for $\mathbf{H}^2$ (and $\mathbf{H}^n$) is given in Gelfand-Graev-Vilenkin [1966]. A further inversion of the horocycle transform in $\mathbf{H}^2$ (and $\mathbf{H}^n$), somewhat analogous to (38) for the X-ray transform, is given by Berenstein and Tarabusi [1994].

The analogy suggested above between the X-ray transform and the horocycle transform in $\mathbf{H}^2$ goes even further in $\mathbf{H}^3$. There the 2-dimensional transform for totally geodesic submanifolds has *the same* inversion formula as the horocycle transform (Helgason [1994b], p. 209).

For a treatment of the horocycle transform on a Riemannian symmetric space see the author's paper [1963] and monograph [1994b], Chapter II, where Problems A, B, C, D in §2 are discussed in detail along with some applications to differential equations and group representations. See also Gelfand–Graev [1964] for a discussion and inversion for the case of complex G. See also Quinto [1993a] and Gonzalez and Quinto [1994] for new proofs of the support theorem.

Example G is from Hilgert's paper [1994], where a related Fourier transform theory is also established. It has a formal analogy to the Fourier analysis on $\mathbf{H}^2$ developed by the author in [1965b] and [1972].

Example I is from Gonzalez's beautiful paper [2001]. Higher dimensional versions have been proved by Gonzalez and Kakehi [2004]. The relationship between the operator D and John's operator Λ in Ch. I, §6 was established by Richter [1986b].

In conclusion we note that the determination of a function in $\mathbf{R}^n$ in terms of its integrals over unit spheres (John [1955]) can be regarded as a solution to the first half of Problem B in §2 for the double fibration (4) and (7). See Exercise 5 in Ch. VI.

CHAPTER III

THE RADON TRANSFORM ON TWO-POINT HOMOGENEOUS SPACES

Let X be a complete Riemannian manifold, x a point in X and X_x the tangent space to X at x. Let Exp_x denote the mapping of X_x into X given by $\mathrm{Exp}_x(u) = \gamma_u(1)$, where $t \to \gamma_u(t)$ is the geodesic in X through x with tangent vector u at $x = \gamma_u(0)$.

A connected submanifold S of a Riemannian manifold X is said to be **totally geodesic** if each geodesic in X which is tangential to S at a point lies entirely in S.

The totally geodesic submanifolds of $\mathbf{R}^n$ are the planes in $\mathbf{R}^n$. Therefore, in generalizing the Radon transform to Riemannian manifolds, it is natural to consider integration over totally geodesic submanifolds. In order to have enough totally geodesic submanifolds at our disposal we consider in this section Riemannian manifolds X which are **two-point homogeneous** in the sense that for any two-point pairs $p, q \in X$ $p', q' \in X$, satisfying $d(p,q) = d(p',q')$, (where $d = $ distance), there exists an isometry g of X such that $g \cdot p = p'$, $g \cdot q = q'$. We start with the subclass of Riemannian manifolds with the richest supply of totally geodesic submanifolds, namely the spaces of constant curvature.

While §1, which constitutes most of this chapter, is elementary, §2–§4 will involve a bit of Lie group theory.

§1 Spaces of Constant Curvature. Inversion and Support Theorems

Let X be a simply connected complete Riemannian manifold of dimension $n \geq 2$ and constant sectional curvature.

Lemma 1.1. *Let $x \in X$, V a subspace of the tangent space X_x. Then $\mathrm{Exp}_x(V)$ is a totally geodesic submanifold of X.*

Proof. For this we choose a specific embedding of X into $\mathbf{R}^{n+1}$, and assume for simplicity the curvature is $\epsilon(= \pm 1)$. Consider the quadratic form

$$B_\epsilon(x) = x_1^2 + \cdots + x_n^2 + \epsilon x_{n+1}^2$$

and the quadric Q_ϵ given by $B_\epsilon(x) = \epsilon$. The orthogonal group $\mathbf{O}(B_\epsilon)$ acts transitively on Q_ϵ. The form B_ϵ is positive definite on the tangent space $\mathbf{R}^n \times (0)$ to Q_ϵ at $x^0 = (0, \ldots, 0, 1)$; by the transitivity B_ϵ induces a positive definite quadratic form at each point of Q_ϵ, turning Q_ϵ into a

S. Helgason, *Integral Geometry and Radon Transforms*,
DOI 10.1007/978-1-4419-6055-9_3, © Springer Science+Business Media, LLC 2010

Riemannian manifold, on which $\mathbf{O}(B_\epsilon)$ acts as a transitive group of isometries. The isotropy subgroup at the point x^0 is isomorphic to $\mathbf{O}(n)$ and its acts transitively on the set of 2-dimensional subspaces of the tangent space $(Q_\epsilon)_{x^0}$. It follows that all sectional curvatures at x^0 are the same, namely ϵ, so by homogeneity, Q_ϵ has constant curvature ϵ. In order to work with connected manifolds, we replace Q_{-1} by its intersection Q_{-1}^+ with the half-space $x_{n+1} > 0$. Then Q_{+1} and Q_{-1}^+ are simply connected complete Riemannian manifolds of constant curvature. Since such manifolds are uniquely determined by the dimension and the curvature it follows that we can identify X with Q_{+1} or Q_{-1}^+.

The geodesic in X through x^0 with tangent vector $(1, 0, \ldots, 0)$ will be left point-wise fixed by the isometry

$$(x_1, x_2, \ldots, x_n, x_{n+1}) \to (x_1, -x_2, \ldots, -x_n, x_{n+1}).$$

This geodesic is therefore the intersection of X with the two-plane $x_2 = \cdots = x_n = 0$ in $\mathbf{R}^{n+1}$. By the transitivity of $\mathbf{O}(n)$ all geodesics in X through x^0 are intersections of X with two-planes through 0. By the transitivity of $\mathbf{O}(Q_\epsilon)$ it then follows that the geodesics in X are precisely the nonempty intersections of X with two-planes through the origin.

Now if $V \subset X_{x^0}$ is a subspace, $\mathrm{Exp}_{x^0}(V)$ is by the above the intersection of X with the subspace of $\mathbf{R}^{n+1}$ spanned by V and x^0. Thus $\mathrm{Exp}_{x^0}(V)$ is a quadric in $V + \mathbf{R}x^0$ and its Riemannian structure induced by X is the same as induced by the restriction $B_\epsilon|(V + \mathbf{R}x^0)$. Thus, by the above, the geodesics in $\mathrm{Exp}_{x^0}(V)$ are obtained by intersecting it with two-planes in $V + \mathbf{R}x^0$ through 0. Consequently, the geodesics in $\mathrm{Exp}_{x^0}(V)$ are geodesics in X so $\mathrm{Exp}_{x^0}(V)$ is a totally geodesic submanifold of X. By the homogeneity of X this holds with x^0 replaced by an arbitrary point $x \in X$. The lemma is proved.

It will be convenient in the following to use another model of the space $X = Q_-^+$, namely the generalization of the space $\mathbf{H}^2$ in Ch. II, Theorem 4.2. It is the hyperbolic space $\mathbf{H}^n$ which is the unit ball $|y| < 1$ in $\mathbf{R}^n$ with the Riemannian structure ds^2 related to the flat ds_0^2 by

$$(1) \qquad ds^2 = \frac{4(dy_1^2 + \cdots + dy_n^2)}{(1 - y_1^2 - \cdots - y_n^2)^2} = \rho^2 \, ds_0^2.$$

Consider the mapping $y = \Phi(x)$, $x \in Q_-^+$, given by

$$(2) \qquad (y_1, \ldots, y_n) = \frac{1}{x_{n+1} + 1}(x_1, \ldots, x_n),$$

with inverse

$$(3) \qquad (x_1, \ldots, x_n) = \frac{2}{1 - |y|^2}(y_1, \ldots, y_n), \qquad x_{n+1} = \frac{1 + |y|^2}{1 - |y|^2}.$$

Then Φ is a bijection of Q_-^+ onto $\mathbf{H}^n$ and we shall see that it is an isometry. For this we compute the pull-back $\Phi^*(ds^2)$. We have

$$\Phi^*(dy_i) = (x_{n+1}+1)^{-1}\, dx_i - x_i(x_{n+1}+1)^{-2}\, dx_{n+1},$$

so

$$\Phi^*(ds^2) = \sum_{i=1}^n dx_i^2 + (x_{n+1}+1)^{-2}\sum_1^n x_i^2\, dx_{n+1}^2$$

$$- 2(x_{n+1}+1)^{-1}\left(\sum_1^n x_i\, dx_i\right) dx_{n+1}.$$

On Q_-^+ we have

$$\sum_1^n x_i^2 = x_{n+1}^2 - 1, \quad \sum_1^n x_i\, dx_i = x_{n+1}\, dx_{n+1},$$

so if $I : Q_-^+ \to \mathbf{R}^{n+1}$ is the identity map

$$I^*(\Phi^*(ds^2)) = \sum_1^n dx_i^2 + \frac{x_{n+1}-1}{x_{n+1}+1}\, dx_{n+1}^2 - \frac{2x_{n+1}}{x_{n+1}+1}\, dx_{n+1}^2$$

$$= \sum_1^n dx_i^2 - dx_{n+1}^2,$$

which shows that Φ is an isometry.

Since the totally geodesic submanifolds of Q_-^+ are the intersections of Q_-^+ with plane sections through 0 the mapping Φ (which maps $(0,\ldots,0,1)$ into $(0,\ldots,0)$) shows that the totally geodesic submanifolds in $\mathbf{H}^n$ ($|y| < 1$) through $0 = (0,\ldots,0)$ are the plane sections in the ball $|y| < 1$ through 0. By the description of the geodesics in $\mathbf{H}^2$ (Ch. II, §4) we conclude that the totally geodesic submanifolds in $\mathbf{H}^n$ are the spherical caps of various dimensions perpendicular to the boundary $|y| = 1$.

Now let X stand for the spaces $\mathbf{R}^n$, $\mathbf{H}^n$ and $\mathbf{S}^n$ and $I(X)$ the group of isometries of X. For $1 \le k \le n-1$ let Ξ denote the space of k-dimensional totally geodesic submanifolds of X. Then G, the identity component of $I(X)$, acts transitively both on X and on Ξ. For $\mathbf{H}^n$ we use the model (1) above. We fix an origin o in X ($o = 0$ for $\mathbf{R}^n$ and $\mathbf{H}^n$, $o =$ the North Pole for $\mathbf{S}^n$) and let K denote the isotropy subgroup of G at o. Fix $p = \mathbf{R}^+$, fix $\xi_p \in \Xi$ at distance $d(o,\xi_p) = p$ from o and H_p the subgroup of G mapping ξ_p into itself. Then we have

(4) $$X = G/K, \quad \Xi = G/H_p.$$

Proposition 1.2. *The point $x \in X$ and the manifold $\xi \in \Xi$ are incident for (4) if and only if $d(x,\xi) = p$.*

For this we will first prove a simple lemma.

Lemma 1.3. *If ξ_1 and ξ_2 in Ξ have the same distance from o then $\xi_1 = k\,\xi_2$ for some $k \in K$.*

Proof. If $X = \mathbf{S}^n \subset \mathbf{R}^{n+1}$, ξ_1 and ξ_2 are the intersection of $\mathbf{S}^n$ with $(k+1)$-planes π_1 and π_2 through the center 0 of $\mathbf{S}^n$. Consider the normal planes of dimension $n + 1 - (k+1) = n - k$ to π_1 and π_2 passing through the North Pole o. These normal planes are conjugate under K and thus so are ξ_1 and ξ_2. For the case $\mathbf{H}^n$ let ξ_1 and ξ_2 be two spherical caps of dimension k perpendicular to the unit sphere $\mathbf{S}_1(0)$ Let π_1 and π_2 be the tangent spaces to ξ_1 and ξ_2 at the points closest to o. The corresponding normal planes through o are conjugate under K and so are π_1 and π_2 and ξ_1 and ξ_2. The same proof works for $X = \mathbf{R}^n$.

To prove Proposition 1.2 let $x = gK$, $\xi = \gamma H_p$. Then $gK \cap \gamma H_p \neq \emptyset$ implies $\gamma^{-1}g \cdot o = h_p \cdot o$ for some h_p so $d(x,\xi) = d(g \cdot o, \gamma \cdot \xi_p) = d(h_p \cdot o, \xi_p) = d(o, \xi_p) = p$. Conversely, if $d(x,\xi) = p$ so $d(o, g^{-1}\gamma \cdot \xi_p) = p$ then by Lemma 1.3

$$g^{-1}\gamma \cdot \xi_p = k \cdot \xi_p \text{ for some } k$$

so $g^{-1}\gamma = kh$ for some $h \in H_p$ so x and ξ are incident.

A.   The Euclidean Case $\mathbf{R}^n$

Let $f \to \widehat{f}$ where

$$(5) \qquad \widehat{f}(\xi) = \int_\xi f(x)\,dm(x)\,, \quad \xi \in \Xi,$$

denote the k-plane transform. In Ch. I, §6 we proved the support theorem and an inversion formula for this transform. We shall now prove another inversion formula, more geometric, in the spirit of the double fibration of Ch. II, §2 which here is

$$
\begin{array}{ccc}
 & G/L & \\
 \swarrow & & \searrow \\
G/K & & G/H_p,
\end{array}
\qquad L = K \cap H_p.
$$

We denote the corresponding Radon transforms (9), Ch. II, §2 by $\widehat{f}_p$ and $\check{\varphi}_p$. Then by Proposition 1.2,

$$(6) \qquad \widehat{f}_p(\xi) = \int_{d(x,\xi)=p} f(x)\,dm(x)\,, \qquad \check{\varphi}_p(x) = \int_{d(x,\xi)=p} \varphi(\xi)\,d\mu(\xi)\,,$$

where $d\mu$ is the invariant average over $\breve{x}$. Here $\widehat{f}_0 = \widehat{f}$ in (5). If $g \in G$ is such that $g \cdot o = x$ then $\xi = g \cdot \xi_p$ has $d(x, \xi) = p$ and

$$\breve{\varphi}_p(g \cdot o) = \int_K \varphi(gkg^{-1} \cdot \xi) \, dk \,,$$

where dk is the normalized Haar measure on K. If $z \in \mathbf{R}^n$ has distance r from 0 then as in (15), Ch.I,

$$(M^r f)(g \cdot o) = \int_K f(gk \cdot z) \, dk \,.$$

Thus since $d(o, g^{-1} \cdot y) = d(x, y)$,

$$(\widehat{f})_p^\vee(g \cdot o) = \int_K \widehat{f}(gkg^{-1} \cdot \xi) \, dk = \int_K dk \int_\xi f(gkg^{-1} \cdot y) \, dm(y)$$

$$= \int_\xi dm(y) \int_K f(gkg^{-1} \cdot y) \, dk = \int_\xi (M^{d(x,y)} f)(x) \, dm(y) \,.$$

Let x_o be the point in ξ at minimum distance from x. The integrand $(M^{d(x,y)} f)(x)$ is constant in y on each sphere in ξ with center x_o. Hence we have

(7)
$$(\widehat{f})_p^\vee(x) = \Omega_k \int_0^\infty (M^q f)(x) r^{k-1} \, dr \,,$$

where $r = d(x_0, y)$, $q = d(x, y)$. We have $q^2 = p^2 + r^2$ so putting

$$F(q) \quad = \quad (M^q f)(x), \quad \widehat{F}(p) = (\widehat{f})_p^\vee(x) \,,$$

we have

(8)
$$\widehat{F}(p) \quad = \quad \Omega_k \int_p^\infty F(q)(q^2 - p^2)^{(k-2)/2} q \, dq \,.$$

We multiply with $p(p^2 - r^2)^{(k-2)/2}$ and compute

$$\int_r^\infty \widehat{F}(p) p (p^2 - r^2)^{(k-2)/2} \, dp$$

$$= \Omega_k \int_r^\infty \left(\int_p^\infty F(q) \left[(q^2 - p^2)(p^2 - r^2) \right]^{(k-2)/2} q \, dq \right) p \, dp$$

$$= \Omega_k \int_{q=r}^\infty F(q) q \left(\int_{p=r}^q \left[(q^2 - p^2)(p^2 - r^2) \right]^{(k-2)/2} p \, dp \right) dq \,,$$

which by the proof of Theorem 2.6 in Chapter I and $\Omega_k = 2\pi^{k/2}/\Gamma(k/2)$ becomes

$$C_k \int_r^\infty F(q)q(q^2 - r^2)^{k-1}\,dq\,,$$

where

$$C_k = 2^{1-k}\pi^{(k+1)/2}/\Gamma((k+1)/2)\,.$$

Applying $\left(\frac{d}{d(r^2)}\right)^k$ this reduces to $\frac{1}{2}C_k(k-1)!(-F(r))$. Using the duplication formula $\Gamma(k) = 2^{k-1}\pi^{-\frac{1}{2}}\Gamma\left(\frac{k}{2}\right)\Gamma((k+1)/2)$ we get the inversion

$$(9) \qquad F(r) = -c(k)\left(\frac{d}{d(r^2)}\right)^k \int_r^\infty \widehat{F}(p)(p^2 - r^2)^{(k-2)/2}p\,dp\,,$$

where $c(k) = 2/(\pi^{k/2}\Gamma(k/2))$. Putting $r = 0$ we get the formula

$$(10) \qquad f(x) = -c(k)\left[\left(\frac{d}{d(r^2)}\right)^k \int_r^\infty p(p^2 - r^2)^{(k-2)/2}(\widehat{f})^{\vee}_p(x)\,dp\right]_{r=0}.$$

Since

$$p(p^2 - r^2)^{(k-2)/2} = \frac{d}{dp}(p^2 - r^2)^{k/2}\frac{1}{k}$$

we can in (9) integrate by parts and the integral becomes

$$-k^{-1}\int_r^\infty (p^2 - r^2)^{k/2}\widehat{F}'(p)\,dp\,.$$

Applying $d/d(r^2) = (2r)^{-1}d/dr$ to this integral reduces the exponent by 1. For k even we continue until the exponent is 0 and then replace $\int_r^\infty \widehat{F}'(p)\,dp$ by $-\widehat{F}(r)$. This $\widehat{F}(r)$ is an even function so taking $(d/d(r^2))^{k/2}$ at $r = 0$ amounts to taking a constant multiple of $(d/dr)^k$ at $r = 0$. For k odd we differentiate by $d/d(r^2)$ $\frac{k+1}{2}$ times and then the exponent in the integral becomes $-\frac{1}{2}$. Thus we obtain the following refinement of (10). We recall that $(\widehat{f})^{\vee}_p(x)$ is the average of the integrals of f over the k-planes tangent to $S_p(x)$.

Theorem 1.4 (Inversion formulas.). *The k-plane transform (5) on $\mathbf{R}^n$ is inverted as follows:*

(i) If k is even then

$$f(x) = C_1\left[\left(\frac{d}{dr}\right)^k \left((\widehat{f})^{\vee}_r(x)\right)\right]_{r=0}.$$

(ii) If k is odd then

$$f(x) = C_2 \left[\left(\frac{d}{d(r^2)} \right)^{(k-1)/2} \int\limits_{r}^{\infty} (p^2 - r^2)^{-1/2} \frac{d}{dp} (\widehat{f})_p^{\vee}(x) \, dp \right]_{r=0}.$$

(iii) If k = 1 then

$$f(x) = -\frac{1}{\pi} \int\limits_{0}^{\infty} \frac{1}{p} \frac{d}{dp} ((\widehat{f})_p^{\vee}(x)) \, dp.$$

Here C_1 and C_2 are constants depending only on k. In (iii) the constant comes from $c(1) = 2/\pi$.

Remark. Note that the integral in (ii) is convergent since $(p^2 - r^2)^{-\frac{1}{2}}$ behaves like $(p - r)^{-\frac{1}{2}}$ near $p = r$.

For k even there is an alternative inversion formula. For $p = 0$ formula (7) becomes

$$(\widehat{f})^{\vee}(x) = \Omega_k \int\limits_{0}^{\infty} (M^r f)(x) r^{k-1} \, dr$$

and the proof of Theorem 3.1, Ch. I, for n odd can be imitated. This gives another proof of Theorem 6.2, Ch. I for k even.

Remark. For $k = n - 1$ even the formula in (i) is equivalent to (31) in Ch. I, §3. For $\varphi \in \mathcal{E}(\Xi)$ we have

$$\check{\varphi}_r(x) = \int\limits_{\mathbf{S}^{n-1}} \varphi(w, r + \langle w, x \rangle) \, dw,$$

so

$$\left\{ \frac{d^2}{dr^2} \check{\varphi}_r(x) \right\}_{r=0} = (\Box \varphi)^{\vee}(x) = (L\check{\varphi})(x).$$

By iteration and putting $\varphi = \widehat{f}$ we recover (31) for $n - 1$ even.

During the proof of Theorem 1.4 we proved the following general relationship (8) and (9) between the spherical average $(M^r f)(x)$ and the average $(\widehat{f})_p^{\vee}(x)$ of the integrals of f over k planes at distance p from x.

Theorem 1.5. *The spherical averages and the k-plane integral averages are related by*

$$(11) \quad (\widehat{f})_p^{\vee}(x) = \Omega_k \int_p^{\infty} (M^q f)(x)(q^2 - p^2)^{(k-2)/2} q \, dq \,,$$

$$(12) \quad (M^r f)(x) = -c(k) \left(\frac{d}{d(r^2)} \right)^k \int_r^{\infty} (\widehat{f})_p^{\vee}(x)(p^2 - r^2)^{(k-2)/2} p \, dp \,,$$

where

$$c(k) = \frac{2}{\pi^{k/2} \Gamma \left(\frac{k}{2} \right)} \,.$$

Examples.

For $k = 2$ this reduces to

$$(\widehat{f})_p^{\vee}(x) = 2\pi \int_p^{\infty} (M^q f)(x) q \, dq \,, \quad (M^r f)(x) = \frac{1}{2\pi} \frac{1}{r} \frac{d}{dr} (\widehat{f})_r^{\vee}(x) \,.$$

For $k = 1$ we have instead

$$(\widehat{f})_p^{\vee}(x) = 2 \int_p^{\infty} (M^q f)(x)(q^2 - p^2)^{-\frac{1}{2}} q \, dq \,,$$

$$(M^r f)(x) = -\frac{1}{\pi} \int_r^{\infty} \frac{d}{dp} ((\widehat{f})_p^{\vee}(x)) \frac{dp}{(p^2 - r^2)^{1/2}} \,.$$

It is remarkable that the dimension n does not enter into the formulas (11) and (12) nor in Theorem 1.4(iii).

B.   The Hyperbolic Space

We take first the case of negative curvature, that is $\epsilon = -1$. The transform $f \to \widehat{f}$ is now given by

$$(13) \qquad\qquad \widehat{f}(\xi) = \int_{\xi} f(x) \, dm(x),$$

ξ being any k-dimensional totally geodesic submanifold of X ($1 \leq k \leq n-1$) with the induced Riemannian structure and dm the corresponding measure. From our description of the geodesics in X it is clear that any two points in

X can be joined by a unique geodesic. Let d be a distance function on X, and for simplicity we write o for the origin x^o in X. Consider now geodesic polar-coordinates for X at o; this is a mapping

$$\mathrm{Exp}_o Y \rightarrow (r, \theta_1, \ldots, \theta_{n-1}),$$

where Y runs through the tangent space X_o, $r = |Y|$ (the norm given by the Riemannian structure) and $(\theta_1, \ldots, \theta_{n-1})$ are coordinates of the unit vector $Y/|Y|$. Then the Riemannian structure of X is given by

$$(14) \qquad ds^2 = dr^2 + (\sinh r)^2 \, d\sigma^2 \,,$$

where $d\sigma^2$ is the Riemannian structure

$$\sum_{i,j=1}^{n-1} g_{ij}(\theta_1, \cdots, \theta_{n-1}) \, d\theta_i \, d\theta_j$$

on the unit sphere in X_o. The surface area $A(r)$ and volume $V(r) = \int_o^r A(t) \, dt$ of a sphere in X of radius r are thus given by

$$(15) \qquad A(r) = \Omega_n (\sinh r)^{n-1}, \quad V(r) = \Omega_n \int_o^r \sinh^{n-1} t \, dt,$$

so $V(r)$ increases like $e^{(n-1)r}$. This explains the growth condition in the next result where $d(o, \xi)$ denotes the distance of o to the manifold ξ.

Theorem 1.6. (The support theorem.) *Suppose $f \in C(X)$ satisfies*

(i)   *For each integer $m > 0$, $f(x) e^{md(o,x)}$ is bounded.*

(ii) *There exists a number $R > 0$ such that*

$$\widehat{f}(\xi) = 0 \quad \text{for } d(o, \xi) > R.$$

Then

$$f(x) = 0 \quad \text{for } d(o, x) > R.$$

Taking $R \rightarrow 0$ we obtain the following consequence.

Corollary 1.7. *The Radon transform $f \rightarrow \widehat{f}$ is one-to-one on the space of continuous functions on X satisfying condition (i) of "exponential decrease".*

Proof of Theorem 1.6. Using smoothing of the form

$$\int_G \varphi(g) f(g^{-1} \cdot x) \, dg$$

($\varphi \in \mathcal{D}(G)$, dg Haar measure on G) we can (as in Theorem 2.6, Ch. I) assume that $f \in \mathcal{E}(X)$.

We first consider the case when f in (2) is a radial function. Let P denote the point in ξ at the minimum distance $p = d(o, \xi)$ from o, let $Q \in \xi$ be arbitrary and let

$$q = d(o, Q), \quad r = d(P, Q).$$

Since ξ is totally geodesic $d(P, Q)$ is also the distance between P and Q in ξ. Consider now the totally geodesic plane π through the geodesics oP and oQ as given by Lemma 1.1 (Fig. III.1). Since a totally geodesic submanifold contains the geodesic joining any two of its points, π contains the geodesic PQ. The angle oPQ being 90° (see e.g. [DS], p. 77) we conclude by hyperbolic trigonometry, (see e.g. Coxeter [1957])

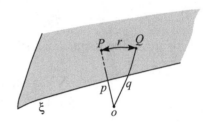

FIGURE III.1.

(16) $$\cosh q = \cosh p \, \cosh r.$$

Since f is radial it follows from (16) that the restriction $f|\xi$ is constant on spheres in ξ with center P. Since these have area $\Omega_k(\sinh r)^{k-1}$ formula (13) takes the form

(17) $$\widehat{f}(\xi) = \Omega_k \int_0^\infty f(Q)(\sinh r)^{k-1} \, dr.$$

Since f is a radial function it is invariant under the subgroup $K \subset G$ which fixes o. But K is not only transitive on each sphere $S_r(o)$ with center o, it is for each fixed k transitive on the set of k-dimensional totally geodesic submanifolds which are tangent to $S_r(o)$. Consequently, $\widehat{f}(\xi)$ depends only on the distance $d(o, \xi)$. Thus we can write

$$f(Q) = F(\cosh q), \quad \widehat{f}(\xi) = \widehat{F}(\cosh p)$$

for certain 1-variable functions F and $\widehat{F}$, so by (16) we obtain

(18) $$\widehat{F}(\cosh p) = \Omega_k \int_0^\infty F(\cosh p \, \cosh r)(\sinh r)^{k-1} \, dr.$$

Writing here $t = \cosh p$, $s = \cosh r$ this reduces to

(19)
$$\widehat{F}(t) = \Omega_k \int_1^\infty F(ts)(s^2 - 1)^{(k-2)/2}\, ds\,.$$

Here we substitute $u = (ts)^{-1}$ and then put $v = t^{-1}$. Then (19) becomes

$$v^{-1}\widehat{F}(v^{-1}) = \Omega_k \int_0^v \{F(u^{-1})u^{-k}\}(v^2 - u^2)^{(k-2)/2}\, du\,.$$

This integral equation is of the form (19), Ch. I so we get the following analog of (20), Ch. I:

(20)
$$F(u^{-1})u^{-k} = c\, u \left(\frac{d}{d(u^2)}\right)^k \int_o^u (u^2 - v^2)^{(k-2)/2}\widehat{F}(v^{-1})\, dv\,,$$

where c is a constant. Now by assumption (ii) $\widehat{F}(\cosh p) = 0$ if $p > R$. Thus

$$\widehat{F}(v^{-1}) = 0 \quad \text{if } 0 < v < (\cosh R)^{-1}\,.$$

From (20) we can then conclude

$$F(u^{-1}) = 0 \quad \text{if } u < (\cosh R)^{-1}$$

which means $f(x) = 0$ for $d(o,x) > R$. This proves the theorem for f radial.

Next we consider an arbitrary $f \in \mathcal{E}(X)$ satisfying (i), (ii) . Fix $x \in X$ and if dk is the normalized Haar measure on K consider the integral

$$F_x(y) = \int_K f(gk \cdot y)\, dk\,, \quad y \in X\,,$$

where $g \in G$ is an element such that $g \cdot o = x$. Clearly, $F_x(y)$ is the average of f on the sphere with center x, passing through $g \cdot y$. The function F_x satisfies the decay condition (i) and it is radial. Moreover,

(21)
$$\widehat{F}_x(\xi) = \int_K \widehat{f}(gk \cdot \xi)\, dk\,.$$

We now need the following estimate

(22)
$$d(o, gk \cdot \xi) \geq d(o, \xi) - d(o, g \cdot o)\,.$$

For this let x_o be a point on ξ closest to $k^{-1}g^{-1} \cdot o$. Then by the triangle inequality

$$d(o, gk \cdot \xi) = d(k^{-1}g^{-1} \cdot o, \xi) \;\geq\; d(o, x_o) - d(o, k^{-1}g^{-1} \cdot o)$$
$$\geq\; d(o, \xi) - d(o, g \cdot o)\,.$$

Thus it follows by (ii) that

$$\widehat{F}_x(\xi) = 0 \quad \text{if} \quad d(o, \xi) > d(o, x) + R.$$

Since F_x is radial this implies by the first part of the proof that

$$(23) \qquad\qquad \int_K f(gk \cdot y)\, dk = 0$$

if

$$(24) \qquad\qquad d(o, y) > d(o, g \cdot o) + R.$$

But the set $\{gk \cdot y : k \in K\}$ is the sphere $S_{d(o,y)}(g \cdot o)$ with center $g \cdot o$ and radius $d(o, y)$; furthermore, the inequality in (24) implies the inclusion relation

$$(25) \qquad\qquad B_R(o) \subset B_{d(o,y)}(g \cdot o)$$

for the balls. But considering the part in $B_R(o)$ of the geodesic through o and $g \cdot o$ we see that conversely relation (25) implies (24). Theorem 1.6 will therefore be proved if we establish the following lemma.

Lemma 1.8. *Let $f \in C(X)$ satisfy the conditions:*

(i)   For each integer $m > 0$, $f(x)e^{m\, d(o,x)}$ is bounded.

(ii) There exists a number $R > 0$ such that the surface integral

$$\int_S f(s)\, d\omega(s) = 0,$$

whenever the spheres S encloses the ball $B_R(o)$.

   Then

$$f(x) = 0 \quad \text{for} \quad d(o, x) > R.$$

Proof. This lemma is the exact analog of Lemma 2.7, Ch. I, whose proof, however, used the vector space structure of $\mathbf{R}^n$. By using the model (1) of the hyperbolic space we shall nevertheless adapt the proof to the present situation. As before we may assume that f is smooth, i.e., $f \in \mathcal{E}(X)$.

   This model is useful here because the spheres in X are the ordinary Euclidean spheres inside the ball. This fact is obvious for the spheres Σ with center 0. For the general statement it suffices to prove that if T is the geodesic symmetry with respect to a point (which we can take on the x_1-axis) then $T(\Sigma)$ is a Euclidean sphere. The unit disk D in the x_1x_2-plane is totally geodesic in X, hence invariant under T. Now the isometries of the non-Euclidean disk D are generated by the complex conjugation

$x_1 + ix_2 \rightarrow x_1 - ix_2$ and fractional linear transformations so they map Euclidean circles into Euclidean circles. In particular $T(\Sigma \cap D) = T(\Sigma) \cap D$ is a Euclidean circle. But T commutes with the rotations around the x_1-axis. Thus $T(\Sigma)$ is invariant under such rotations and intersects D in a circle; hence it is a Euclidean sphere.

After these preliminaries we pass to the proof of Lemma 1.8. Let $S = S_r(y)$ be a sphere in X enclosing $B_r(o)$ and let $B_r(y)$ denote the corresponding ball. Expressing the exterior $X - B_r(y)$ as a union of spheres in X with center y we deduce from assumption (ii)

$$(26) \qquad \int\limits_{B_r(y)} f(x)\,dx = \int\limits_{X} f(x)\,dx\,,$$

which is a constant for small variations in r and y. The Riemannian measure dx is given by

$$(27) \qquad dx = \rho^n\,dx_o\,,$$

where $dx_o = dx_1 \ldots dx_n$ is the Euclidean volume element. Let r_o and y_o, respectively, denote the Euclidean radius and Euclidean center of $S_r(y)$. Then $S_{r_o}(y_o) = S_r(y), B_{r_o}(y_o) = B_r(y)$ set-theoretically and by (26) and (27)

$$(28) \qquad \int\limits_{B_{r_0}(y_0)} f(x_0)\rho(x_0)^n\,dx_o = \text{const.}\,,$$

for small variations in r_o and y_o; thus by differentiation with respect to r_o,

$$(29) \qquad \int\limits_{S_{r_0}(y_0)} f(s_0)\rho(s_0)^n\,d\omega_o(s_0) = 0\,,$$

where $d\omega_o$ is the Euclidean surface element. Putting $f^*(x) = f(x)\rho(x)^n$ we have by (28)

$$\int\limits_{B_{r_0}(y_0)} f^*(x_0)\,dx_o = \text{const.}\,,$$

so by differentiating with respect to y_o, we get

$$\int\limits_{B_{r_0}(o)} (\partial_i f^*)(y_o + x_o)\,dx_o = 0\,.$$

Using the divergence theorem (26), Chapter I, §2, on the vector field $F(x_o) = f^*(y_o + x_o)\partial_i$ defined in a neighborhood of $B_{r_0}(0)$ the last equation implies

$$\int\limits_{S_{r_0}(0)} f^*(y_o + s)s_i\,d\omega_o(s) = 0\,,$$

which in combination with (29) gives

$$(30) \qquad \int\limits_{S_{r_o}(y_o)} f^*(s) s_i \, d\omega_o(s) = 0 \,.$$

The Euclidean and the non-Euclidean Riemannian structures on $S_{r_o}(y_o)$ differ by the factor ρ^2. It follows that $d\omega = \rho(s)^{n-1} \, d\omega_o$ so (30) takes the form

$$(31) \qquad \int\limits_{S_r(y)} f(s)\rho(s) s_i \, d\omega(s) = 0 \,.$$

We have thus proved that the function $x \to f(x)\rho(x)x_i$ satisfies the assumptions of the theorem. By iteration we obtain

$$(32) \qquad \int\limits_{S_r(y)} f(s)\rho(s)^k s_{i_1} \ldots s_{i_k} \, d\omega(s) = 0 \,.$$

In particular, this holds with $y = 0$ and $r > R$. Then $\rho(s) =$ constant and (32) gives $f \equiv 0$ outside $B_R(o)$ by the Weierstrass approximation theorem. Now Theorem 1.6 is proved.

Now let L denote the Laplace-Beltrami operator on X. (See Ch. V, §1 for the definition.) Because of formula (14) for the Riemannian structure of X, L is given by

$$(33) \qquad L = \frac{\partial^2}{\partial r^2} + (n-1) \coth r \frac{\partial}{\partial r} + (\sinh r)^{-2} L_S$$

where L_S is the Laplace-Beltrami operator on the unit sphere in X_0. We consider also for each $r \geq 0$ the mean-value operator M^r defined by

$$(M^r f)(x) = \frac{1}{A(r)} \int\limits_{S_r(x)} f(s) \, d\omega(s) \,.$$

As we saw before this can also be written

$$(34) \qquad (M^r f)(g \cdot o) = \int\limits_K f(gk \cdot y) \, dk$$

if $g \in G$ is arbitrary and $y \in X$ is such that $r = d(o, y)$. If f is an analytic function one can, by expanding it in a Taylor series, prove from (34) that M^r is a certain power series in L (e.g. Helgason [1959], pp. 270-272); see also remark below. In particular we have the commutativity

$$(35) \qquad M^r L = L M^r \,.$$

This in turn implies the "Darboux equation"

$$(36) \qquad L_x(F(x,y)) = L_y(F(x,y))$$

for the function $F(x,y) = (M^{d(o,y)}f)(x)$. In fact, using (34) and (35) we have if $g \cdot o = x$, $r = d(o,y)$

$$
\begin{aligned}
L_x(F(x,y)) &= (LM^r f)(x) = (M^r L f)(x) \\
&= \int_K (Lf)(gk \cdot y)\, dk = \int_K (L_y(f(gk \cdot y)))\, dk
\end{aligned}
$$

the last equation following from the invariance of the Laplacian under the isometry gk. But this last expression is $L_y(F(x,y))$.

We remark that the analog of Lemma 2.13 in Ch. VI which also holds here would give another proof of (35) and (36).

For a fixed integer $k(1 \le k \le n-1)$ let Ξ denote the manifold of all k-dimensional totally geodesic submanifolds of X. If φ is a continuous function on Ξ we denote by $\check{\varphi}$ the point function

$$\check{\varphi}(x) = \int_{x \in \xi} \varphi(\xi)\, d\mu(\xi),$$

where μ is the unique measure on the (compact) space of ξ passing through x, invariant under all rotations around x and having total measure one.

Theorem 1.9. (The inversion formula.) *For k even let Q_k denote the polynomial*

$$Q_k(z) = [z + (k-1)(n-k)][z + (k-3)(n-k+2)] \ldots [z+1 \cdot (n-2)]$$

of degree $k/2$. The k-dimensional Radon transform on X is then inverted by the formula

$$cf = Q_k(L)((\hat{f})^{\vee}), \quad f \in \mathcal{D}(X).$$

Here c is the constant

$$(37) \qquad c = (-4\pi)^{k/2} \Gamma(n/2)/\Gamma((n-k)/2).$$

The formula holds also if f satisfies the decay condition (i) in Corollary 2.1, Ch. IV.

Proof. Fix $\xi \in \Xi$ passing through the origin $o \in X$. If $x \in X$ fix $g \in G$ such that $g \cdot o = x$. As k runs through K, $gk \cdot \xi$ runs through the set of totally geodesic submanifolds of X passing through x and

$$\check{\varphi}(g \cdot o) = \int_K \varphi(gk \cdot \xi)\, dk.$$

Hence

$$(\widehat{f})^{\vee}(g \cdot o) = \int_K \left(\int_{\xi} f(gk \cdot y) \, dm(y) \right) dk = \int_{\xi} (M^r f)(g \cdot o) \, dm(y),$$

where $r = d(o, y)$. But since ξ is totally geodesic in X, it has also constant curvature -1 and two points in ξ have the same distance in ξ as in X. Thus we have

$$(38) \qquad (\widehat{f})^{\vee}(x) = \Omega_k \int_0^{\infty} (M^r f)(x)(\sinh r)^{k-1} \, dr .$$

We apply L to both sides and use (36). Then

$$(39) \qquad (L(\widehat{f})^{\vee})(x) = \Omega_k \int_0^{\infty} (\sinh r)^{k-1} L_r(M^r f)(x) \, dr ,$$

where L_r is the "radial part" $\frac{\partial^2}{\partial r^2} + (n-1) \coth r \frac{\partial}{\partial r}$ of L. Putting now $F(r) = (M^r f)(x)$ we have the following result.

Lemma 1.10. *Let m be an integer $0 < m < n = \dim X$. Then*

$$\int_0^{\infty} \sinh^m r L_r F \, dr$$

$$= (m+1-n) \left[m \int_0^{\infty} \sinh^m r F(r) \, dr + (m-1) \int_0^{\infty} \sinh^{m-2} r F(r) \, dr \right].$$

If $m = 1$ the term $(m-1) \int_0^{\infty} \sinh^{m-2} r F(r) \, dr$ should be replaced by $F(0)$.

This follows by repeated integration by parts.
From this lemma combined with the Darboux equation (36) in the form

$$(40) \qquad L_x(M^r f(x)) = L_r(M^r f(x))$$

we deduce

$$[L_x + m(n-m-1)] \int_0^{\infty} \sinh^m r(M^r f)(x) \, dr$$

$$= -(n-m-1)(m-1) \int_0^{\infty} \sinh^{m-2} r(M^r f)(p) \, dr .$$

Applying this repeatedly to (39) we obtain Theorem 1.9.

Unfortunately, this proof does not work for k odd, in particular not for the X-ray transform. Because of Proposition 1.2, the formula (6) remains valid for the hyperbolic space with the same proof.

We shall now invert the transform $f \to \widehat{f}$ by invoking the more general transform $\varphi \to \check{\varphi}_p$. Consider $x \in X, \xi \in \Xi$ with $d(x, \xi) = p$. Select $g \in G$ such that $g \cdot o = x$. Then $d(o, g^{-1}\xi) = p$ so $\{kg^{-1} \cdot \xi : k \in K\}$ is the set of $\eta \in \Xi$ at distance p from o and $\{gkg^{-1} \cdot \xi : k \in K\}$ is the set of $\eta \in \Xi$ at distance p from x. Hence

$$(\widehat{f})_p^{\vee}(g \cdot o) = \int_K \widehat{f}(gkg^{-1} \cdot \xi)\, dk = \int_K dk \int_\xi f(gkg^{-1} \cdot y)\, dm(y)$$

$$= \int_\xi \left(\int_K f(gkg^{-1} \cdot y)\, dk \right) dm(y)$$

so

(41)
$$(\widehat{f})_p^{\vee}(x) = \int_\xi (M^{d(x,y)} f)(x)\, dm(y).$$

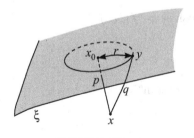

FIGURE III.2.

Let $x_0 \in \xi$ be a point at minimum distance (i.e., p) from x and let (Fig. III.2)

$$r = d(x_0, y), \quad q = d(x, y), \quad y \in \xi.$$

Since $\xi \subset X$ is totally geodesic, $d(x_0, y)$ is also the distance between x_o and y in ξ. In (41) the integrand $(M^{d(x,y)} f)(x)$ is constant in y on each sphere in ξ with center x_o.

Theorem 1.11 (General inversion). *The transform $f \to \widehat{f}$ is inverted by*

$$f(x) = -c_k \left[\left(\frac{d}{d(r^2)} \right)^k \int_r^\infty (t^2 - r^2)^{(k-2)/2} t^k (\widehat{f})_{s(t)}^{\vee}(x)\, dt \right]_{r=1}$$

where $s(p) = \cosh^{-1} p$ and

$$c_k = 2/(\pi^{k/2} \Gamma(k/2)).$$

As for the flat case the following improvements hold for k even and the case $k = 1$.

Theorem 1.12. *(i) If k is even the inversion can be written*

$$f(x) = C \left[\left(\frac{d}{d(r^2)} \right)^{k/2} (r^{k-1} (\widehat{f})^{\vee}_{s(r)}(x)) \right]_{r=1} \quad , C = \text{const.} .$$

(ii) If $k = 1$ then

$$f(x) = -\frac{1}{\pi} \int\limits_0^\infty \frac{1}{\sinh p} \frac{d}{dp} ((\widehat{f})^{\vee}_p(x)) \, dp \, .$$

Proof. Applying geodesic polar coordinates in ξ with center x_0 we obtain from (37),

$$(42) \qquad (\widehat{f})^{\vee}_p(x) = \Omega_k \int\limits_0^\infty (M^q f)(x) \sinh^{k-1} r \, dr \, .$$

Using the cosine relation on the right-angled triangle $(x x_0 y)$ we have by $r = d(x_0, y), q = d(x, y)$ and $d(x_0, x) = p$,

$$(43) \qquad\qquad\qquad \cosh q = \cosh p \cosh r \, .$$

With x fixed we define F and $\widehat{F}$ by

$$(44) \qquad F(\cosh q) = (M^q f)(x), \quad \widehat{F}(\cosh p) = (\widehat{f})^{\vee}_p(x) \, .$$

Then by (42),

$$(45) \qquad \widehat{F}(\cosh p) = \Omega_k \int\limits_0^\infty F(\cosh p \cosh r) \sinh^{k-1} r \, dr \, .$$

Putting here $t = \cosh p$, $s = \cosh r$ this becomes

$$\widehat{F}(t) = \Omega_k \int\limits_1^\infty F(ts)(s^2 - 1)^{\frac{k}{2} - 1} \, ds \, ,$$

Putting here $u = ts$, $ds = t^{-1} \, du$ we get the Abel type integral equation

$$(46) \qquad t^{k-1} \widehat{F}(t) = \Omega_k \int\limits_t^\infty u^{-1} F(u)(u^2 - t^2)^{(k-2)/2} u \, du \, .$$

This is inverted just as in the flat case giving

$$(47) \qquad r^{-1} F(r) = -c(k) \left(\frac{d}{d(r^2)} \right)^k \int\limits_r^\infty t^{k-1} \widehat{F}(t)(t^2 - r^2)^{(k-2)/2} t \, dt \, .$$

Here we put $r = 1$ and derive Theorem 1.11.

If k is even, each time we apply $d/d(r^2)$ the exponent $(k-2)/2$ is lowered by 1. After $\frac{k}{2} - 1$ steps the factor $(t^2 - r^2)^{(k-2)/2}$ has disappeared. Then $\frac{1}{2r}d/dr$ produces $r^{k-1}(\widehat{f})^{\vee}_{s(r)}$ and part (i) of Theorem 1.12 follows.

If $k = 1$ we note that in (47)

$$(t^2 - r^2)^{-\frac{1}{2}}t = \frac{d}{dt}(t^2 - r^2)^{\frac{1}{2}},$$

so

$$F(1) = +\frac{c(1)}{2}\left[\frac{d}{dr}\int_r^{\infty}(t^2 - r^2)^{\frac{1}{2}}\frac{d}{dt}\widehat{F}(t)\,dt\right]_{r=1}$$

$$= -\frac{c(1)}{2}\int_1^{\infty}(t^2 - 1)^{-\frac{1}{2}}\frac{d}{dt}\widehat{F}(t)\,dt.$$

Now part (ii) follows by putting $t = \cosh p$, $dt = \sinh p\,dp$.

In the proof we used (46)–(47) only for $r = 1$. More generally these formulas give an explicit relationship between spherical averages of f and averages of integrals of f over k-dimensional totally geodesic submanifolds.

Theorem 1.13. *With* $s(t) = \cosh^{-1}t$ *we have*

$$(\widehat{f})^{\vee}_{s(t)}(x) = \Omega_k t^{1-k}\int_t^{\infty}(M^{s(u)}f)(x)(u^2 - t^2)^{(k-2)/2}\,du,\quad t \geq 1,$$

$$(M^{s(r)}f)(x) = -c(k)r\left(\frac{d}{d(r^2)}\right)^k\int_r^{\infty}t^k(\widehat{f})^{\vee}_{s(t)}(t^2 - r^2)^{(k-2)/2}\,dt,$$

where

$$c(k) = \frac{2}{\pi^{1/2}\Gamma(k/2)}.$$

Example.

For $k = 2$ this reduces to

$$(\widehat{f})^{\vee}_{s(t)}(x) = 2\pi\frac{1}{t}\int_t^{\infty}(M^{s(u)}f)(x)\,du\qquad t \geq 1,$$

$$(M^{s(r)}f)(x) = \frac{1}{2\pi}\frac{d}{dr}\left(r(\widehat{f})^{\vee}_{s(r)}(x)\right),\qquad r \geq 1.$$

For $k = 1$ we have even more symmetric relations between spherical averages and averages of X-ray transforms:

$$(\widehat{f})^{\vee}_{s(t)}(x) = 2 \int_t^{\infty} (M^{s(u)} f)(x) \frac{du}{(u^2 - t^2)^{1/2}}, \qquad t \geq 1,$$

$$(M^{s(r)} f)(x) = -\frac{2}{\pi} \int_r^{\infty} \frac{d}{dt} \left((\widehat{f})^{\vee}_{s(t)}(x) \right) \frac{dt}{(t^2 - r^2)^{1/2}}, \qquad r \geq 1.$$

Again the dimension n does not enter explicitly in the formulas.

A special case. The case of the hyperbolic plane $\mathbf{H}^2$ being particularly interesting we shall now prove one more inversion formula ((31), Ch. II) which has more resemblance to Theorem 3.1, Ch. I, than the formula above does. For this we use a few facts about spherical functions on $\mathbf{H}^n$. A **spherical function** φ on $G/K = \mathbf{H}^n$ is by definition a K-invariant function which is an eigenfunction of the Laplacian L on X satisfying $\varphi(0) = 1$. Then by Helgason [1962] or [1984] the eigenspace of L containing φ consists of the functions f on X satisfying the functional equation (cf. [GGA], p. 64)

$$(48) \qquad \int_K f(gk \cdot x) \, dk = f(g \cdot o) \, \varphi(x).$$

Consider now the case $\mathbf{H}^2$. Then the spherical functions are the solutions $\varphi_\lambda(r)$ of the differential equation

$$(49) \qquad \frac{d^2 \varphi_\lambda}{dr^2} + \coth r \frac{d\varphi_\lambda}{dr} = -(\lambda^2 + \tfrac{1}{4})\varphi_\lambda, \quad \varphi_\lambda(o) = 1.$$

Here $\lambda \in \mathbf{C}$ and $\varphi_{-\lambda} = \varphi_\lambda$. The function φ_λ has the integral representation

$$(50) \qquad \varphi_\lambda(r) = \frac{1}{\pi} \int_0^{\pi} (\operatorname{ch} r - \operatorname{sh} r \cos \theta)^{-i\lambda + \frac{1}{2}} \, d\theta.$$

In fact, already the integrand is easily seen to be an eigenfunction of the operator L in (33) (for $n = 2$) with eigenvalue $-(\lambda^2 + 1/4)$. It is known that $|\varphi_\lambda(r)| \leq \text{const.} r e^{-\frac{r}{2}}$ for λ real and large r.

If $f \in C_c(X)$ and $\psi(x) = \Psi(d(o, x))$ the convolution $\times$ in Ch. II, §2 satisfies

$$(51) \qquad (f \times \psi)(x) = \int_X f(y) \Psi(d(x, y)) \, dy.$$

If f is a radial function on X its **spherical transform** $\widetilde{f}$ is defined by

$$(52) \qquad \widetilde{f}(\lambda) = \int_X f(x) \varphi_{-\lambda}(x) \, dx$$

for all $\lambda \in \mathbf{C}$ for which this integral exists. The continuous radial functions on X form a commutative algebra $C_c^\natural(X)$ under convolution and

$$(53) \qquad (f_1 \times f_2)^\sim(\lambda) = \tilde{f}_1(\lambda)\tilde{f}_2(\lambda).$$

In fact, using (48) for $f = \varphi = \varphi_{-\lambda}$,

$$
\begin{aligned}
(f_1 \times f_2)^\sim(\lambda) &= \int_G f_1(h \cdot o)\left(\int_G f_2(g \cdot o)\varphi_{-\lambda}(hg \cdot o)\,dg\right) dh \\
&= \int_G f_1(h \cdot o)\left(\int_G f_2(g \cdot o)\right)\left(\int_K \varphi_{-\lambda}(hkg \cdot o)\,dk\,dg\right) dh \\
&= \tilde{f}_1(\lambda)\tilde{f}_2(\lambda).
\end{aligned}
$$

We know already from Corollary 1.7 that the Radon transform on $\mathbf{H}^n$ is injective and is inverted in Theorems 1.10 and 1.11. For the case $n = 2, k = 1$ we shall now obtain another inversion formula based on (53).

The spherical function $\varphi_\lambda(r)$ in (50) is the classical Legendre function $P_v(\cosh r)$ with $v = i\lambda - \frac{1}{2}$ for which we shall need the following result ([Prudnikov, Brychkov and Marichev], Vol. III, 2.17.8(2)).

Lemma 1.14.

$$(54) \qquad 2\pi \int_0^\infty e^{-pr} P_v(\cosh r)\,dr = \pi\frac{\Gamma(\frac{p-v}{2})\Gamma(\frac{p+v+1}{2})}{\Gamma(1 + \frac{p+v}{2})\Gamma(\frac{1+p-v}{2})},$$

for

$$(55) \qquad \operatorname{Re}(p - v) > 0, \quad \operatorname{Re}(p + v) > -1.$$

We shall require this result for $p = 0, 1$ and λ real. In both cases, conditions (55) are satisfied.

Let τ and σ denote the functions

$$(56) \qquad \tau(x) = \sinh d(o, x)^{-1}, \quad \sigma(x) = \coth(d(o, x)) - 1, \quad x \in X.$$

Lemma 1.15. *For $f \in \mathcal{D}(X)$ we have*

$$(57) \qquad (\hat{f})^\vee(x) = \pi^{-1}(f \times \tau)(x).$$

Proof. In fact, the right hand side is, by (51),

$$\int_X \sinh d(x, y)^{-1} f(y)\,dy = \int_0^\infty dr(\sinh r)^{-1} \int_{S_r(x)} f(y)\,d\omega(y),$$

so the lemma follows from (38).

Similarly we have

(58) $$Sf = f \times \sigma,$$

where S is the operator

(59) $$(Sf)(x) = \int_X (\coth(d(x,y)) - 1) f(y) \, dy.$$

Theorem 1.16 (Another inversion of the X-ray transform on $\mathbf{H}^2$).
The operator $f \to \widehat{f}$ is inverted by

(60) $$LS((\widehat{f})^\vee) = -4\pi f, \quad f \in \mathcal{D}(X).$$

Proof. The operators $\widehat{}$, $^\vee$, S and L are all G-invariant so it suffices to
verify (60) at o. Let $f^\natural(x) = \int_K f(k \cdot x) \, dk$. Then

$$(f \times \tau)^\natural = f^\natural \times \tau, \quad (f \times \sigma)^\natural = f^\natural \times \sigma, \quad (Lf)(o) = (Lf^\natural)(o).$$

Thus by (57)–(58)

$$\begin{aligned} LS((\widehat{f})^\vee)(o) &= L(S((\widehat{f})^\vee))^\natural(o) = \pi^{-1} L(f \times \tau \times \sigma)^\natural(o) \\ &= LS(((f^\natural)^{\widehat{}})^\vee))(o). \end{aligned}$$

Now, if (60) is proved for a radial function this equals $c f^\natural(o) = c f(o)$. Thus
(60) would hold in general. Consequently, it suffices to prove

(61) $$L(f \times \tau \times \sigma) = -4\pi f, \quad f \text{ radial in } \mathcal{D}(X).$$

Now f, $\tau \varphi_\lambda$ (λ real) and σ are all integrable on X. By the proof of (53) we
have for $f_1 = \tau$, $f_2 = \sigma$, and λ real, the formula $(\tau \times \sigma)^\sim(\lambda) = \widetilde{\tau}(\lambda) \widetilde{\sigma}(\lambda)$;
repeating the argument with $f_1 = f$ and $f_2 = \tau \times \sigma$ we deduce that

(62) $$(f \times \tau \times \sigma)^\sim(\lambda) = \widetilde{f}(\lambda) \widetilde{\tau}(\lambda) \widetilde{\sigma}(\lambda).$$

Since $\coth r - 1 = e^{-r}/\sinh r$, and since $dx = \sinh r \, dr \, d\theta$, $\widetilde{\tau}(\lambda)$ and $\widetilde{\sigma}(\lambda)$
are given by the left hand side of (54) for $p = 0$ and $p = 1$, respectively.
Thus

$$\widetilde{\tau}(\lambda) = \pi \frac{\Gamma(\frac{1}{4} - \frac{i\lambda}{2})\Gamma(\frac{i\lambda}{2} + \frac{1}{4})}{\Gamma(\frac{i\lambda}{2} + \frac{3}{4})\Gamma(\frac{3}{4} - \frac{i\lambda}{2})},$$

$$\widetilde{\sigma}(\lambda) = \pi \frac{\Gamma(\frac{3}{4} - \frac{i\lambda}{2})\Gamma(\frac{i\lambda}{2} + \frac{3}{4})}{\Gamma(\frac{i\lambda}{2} + \frac{5}{4})\Gamma(\frac{5}{4} - \frac{i\lambda}{2})}.$$

Using the identity $\Gamma(x+1) = x\Gamma(x)$ on the denominator of $\widetilde{\sigma}(\lambda)$ we see that

(63) $$\widetilde{\tau}(\lambda)\widetilde{\sigma}(\lambda) = 4\pi^2(\lambda^2 + \tfrac{1}{4})^{-1}.$$

Now

$$L(f \times \tau \times \sigma) = (Lf \times \tau \times \sigma), \quad f \in \mathcal{D}^{\natural}(X),$$

and by (49), $(Lf)^{\sim}(\lambda) = -(\lambda^2 + \tfrac{1}{4})\widetilde{f}(\lambda)$. Using the decomposition $\tau = \varphi\tau + (1-\varphi)\tau$ where φ is the characteristic function of a ball $B(0)$ we see that $f \times \tau \in L^2(X)$ for $f \in \mathcal{D}^{\natural}(X)$. Since $\sigma \in L^1(X)$ we have $f \times \tau \times \sigma \in L^2(X)$. By the Plancherel theorem, the spherical transform $f(r) \to \widetilde{f}(\lambda)$ (λ real) is injective on $(L^2)^{\natural}(X)$ so we deduce from (62)–(63) that (60) holds with the constant -4π.

C.   The Spheres and the Elliptic Spaces

Now let X be the unit sphere $\mathbf{S}^n(0) \subset \mathbf{R}^{n+1}$ and Ξ the set of k-dimensional totally geodesic submanifolds of X. Each $\xi \in \Xi$ is a k-sphere. We shall now invert the Radon transform

$$\widehat{f}(\xi) = \int\limits_{\xi} f(x)\,dm(x), \quad f \in \mathcal{E}(X)$$

where dm is the measure on ξ given by the Riemannian structure induced by that of X. In contrast to the hyperbolic space, each geodesic X through a point x also passes through the antipodal point A_x. As a result, $\widehat{f} = (f \circ A)\widehat{}$ and our inversion formula will reflect this fact. Although we state our result for the sphere, it is really a result for the **elliptic space**, that is the sphere with antipodal points identified. The functions on this space are naturally identified with symmetric functions on the sphere.

Again let

$$\check{\varphi}(x) = \int\limits_{x \in \xi} \varphi(\xi)\,d\mu(\xi)$$

denote the average of a continuous function on Ξ over the set of ξ passing through x.

Theorem 1.17 (Inversion on $\mathbf{S}^n$). *Let k be an integer, $1 \leq k < n = \dim X$.*

(i)   *The mapping $f \to \widehat{f}$ ($f \in \mathcal{E}(X)$) has kernel consisting of the skew functions (the functions f satisfying $f + f \circ A = 0$).*

(ii)   *Assume k even and let P_k denote the polynomial*

$$P_k(z) = [z - (k-1)(n-k)][z - (k-3)(n-k+2)]\ldots[z - 1(n-2)]$$

of degree $k/2$. The k-dimensional Radon transform on X is then inverted by the formula

$$c(f + f \circ A) = P_k(L)((\widehat{f})^\vee), \quad f \in \mathcal{E}(X)$$

where c is the constant in (37).

In particular, for f even, $f = $ constant $\Leftrightarrow \widehat{f}$ is constant.

(iii) For $k = n - 1$, Ξ is identified with $\mathbf{S}^n/\mathbf{Z}_2$ and the mapping $f \to \widehat{f}$ is surjective.

Proof. We first prove (ii) in a similar way as in the noncompact case. The Riemannian structure in (3) is now replaced by

$$ds^2 = dr^2 + \sin^2 r \, d\sigma^2 \, ;$$

the Laplace–Beltrami operator is now given by

(64) $$L = \frac{\partial^2}{\partial r^2} + (n - 1)\cot r \frac{\partial}{\partial r} + (\sin r)^{-2} L_S$$

instead of (33) and

$$(\widehat{f})^\vee(x) = \Omega_k \int_0^\pi (M^r f)(x) \sin^{k-1} r \, dr \, .$$

For a fixed x we put $F(r) = (M^r f)(x)$. The analog of Lemma 1.9 now reads as follows.

Lemma 1.18. *Let m be an integer, $0 < m < n = \dim X$. Then*

$$\int_0^\pi \sin^m r L_r F \, dr$$

$$= (n - m - 1) \left[m \int_0^\pi \sin^m r F(r) \, dr - (m - 1) \int_0^\pi \sin^{m-2} r F(r) \, dr \right].$$

If $m = 1$, the term $(m - 1) \int_0^\pi \sin^{m-2} r F(r) \, dr$ should be replaced by $F(o) + F(\pi)$.

Since (40) is still valid the lemma implies

$$[L_x - m(n - m - 1)] \int_0^\pi \sin^m r (M^r f)(x) \, dr$$

$$= -(n - m - 1)(m - 1) \int_0^\pi \sin^{m-2} r (M^r f)(x) \, dr \, ,$$

and the desired inversion formula follows by iteration since

$$F(0) + F(\pi) = f(x) + f(Ax).$$

In the case when k is even, Part (i) follows from (ii). Next suppose $k = n - 1$, n even. For each ξ there are exactly two points x and Ax at maximum distance, namely $\frac{\pi}{2}$, from ξ and we write

$$\widehat{f}(x) = \widehat{f}(Ax) = \widehat{f}(\xi).$$

We have then

(65) $$\widehat{f}(x) = \Omega_n(M^{\frac{\pi}{2}} f)(x).$$

Next we recall some well-known facts about spherical harmonics. We have

(66) $$L^2(X) = \sum_0^\infty \mathcal{H}_s,$$

where the space $\mathcal{H}_s$ consist of the restrictions to X of the homogeneous harmonic polynomials on $\mathbf{R}^{n+1}$ of degree s.

(a) $Lh_s = -s(s + n - 1)h_s$ $(h_s \in \mathcal{H}_s)$ for each $s \geq 0$. This is immediate from the decomposition

$$L_{n+1} = \frac{\partial^2}{\partial r^2} + \frac{n}{r} \frac{\partial}{\partial r} + \frac{1}{r^2} L$$

of the Laplacian L_{n+1} of $\mathbf{R}^{n+1}$ (cf. (33)). Thus the spaces $\mathcal{H}_s$ are precisely the eigenspaces of L.

(b) Each $\mathcal{H}_s$ contains a function $(\not\equiv 0)$ which is invariant under the group K of rotations around the vertical axis (the x_{n+1}-axis in $\mathbf{R}^{n+1}$). This function φ_s is nonzero at the North Pole o and is uniquely determined by the condition $\varphi_s(o) = 1$. This is easily seen since by (64) φ_s satisfies the ordinary differential equation

$$\frac{d^2\varphi_s}{dr^2} + (n-1) \cot r \frac{d\varphi_s}{dr} = -s(s+n-1)\varphi_s, \quad \varphi_s'(o) = 0.$$

It follows that $\mathcal{H}_s$ is irreducible under the orthogonal group $\mathbf{O}(n+1)$.

(c) Since the mean-value operator $M^{\pi/2}$ commutes with the action of $\mathbf{O}(n+1)$ it acts as a scalar c_s on the irreducible space $\mathcal{H}_s$. Since we have

$$M^{\pi/2}\varphi_s = c_s\varphi_s, \quad \varphi_s(o) = 1,$$

we obtain

(67) $$c_s = \varphi_s\left(\tfrac{\pi}{2}\right).$$

Lemma 1.19. *The scalar $\varphi_s(\pi/2)$ is zero if and only if s is odd.*

Proof. Let H_s be the K-invariant homogeneous harmonic polynomial whose restriction to X equals φ_s. Then H_s is a polynomial in $x_1^2 + \cdots + x_n^2$ and x_{n+1} so if the degree s is odd, x_{n+1} occurs in each term whence $\varphi_s(\pi/2) = H_s(1,0,\ldots 0,0) = 0$. If s is even, say $s = 2d$, we write

$$H_s = a_0(x_1^2 + \cdots + x_n^2)^d + a_1 x_{n+1}^2(x_1^2 + \cdots + x_n^2)^{d-1} + \cdots + a_d x_{n+1}^{2d}.$$

Using $L_{n+1} = L_n + \partial^2/\partial x_{n+1}^2$ and formula (64) in Ch. I the equation $L_{n+1}H_s \equiv 0$ gives the recursion formula

$$a_i(2d - 2i)(2d - 2i + n - 2) + a_{i+1}(2i + 2)(2i + 1) = 0$$

$(0 \le i < d)$. Hence $H_s(1,0\ldots 0)$, which equals a_0, is $\ne 0$; Q.E.D.

Now each $f \in \mathcal{E}(X)$ has a uniformly convergent expansion

$$f = \sum_0^\infty h_s \quad (h_s \in \mathcal{H}_s)$$

and by (65)

$$\widehat{f} = \Omega_n \sum_0^\infty c_s h_s.$$

If $\widehat{f} = 0$ then by Lemma 1.19, $h_s = 0$ for s even so f is skew. Conversely $\widehat{f} = 0$ if f is skew so Part (i) is proved for the case $k = n - 1$, n even.

If k is odd, $0 < k < n - 1$, the proof just carried out shows that $\widehat{f}(\xi)=0$ for all $\xi \in \Xi$ implies that f has integral 0 over every $(k + 1)$-dimensional sphere with radius 1 and center 0. Since $k + 1$ is even and $< n$ we conclude by (ii) that $f + f \circ A = 0$ so Part (i) is proved in general.

For (iii) $(k = n - 1)$, $\widehat{f}$ in (65) coincides with the map $\widehat{f} \circ j$ in (97) given later so the surjectivity in Theorem 2.2 implies (iii) here. This completes the proof of Theorem 1.17. For $k < n - 1$ the range of the map $f \to \widehat{f}$ is determined by Gonzalez [1994] and Kakehi [1993] as the kernel of a certain fourth degree operator. The analog for the complex projective space was proved by Kakehi [1992]. See also Gonzalez and Kakehi [2003] for a wider perspective.

Example. (Minkowski's theorem.)

Let $\Omega \subset \mathbf{R}^3$ be a compact convex body with smooth boundary and 0 in its interior. For a unit vector ω let

$$\langle x, \omega \rangle = H(\omega)$$

be *the* supporting plane with $H(\omega) > 0$, perpendicular to ω. The sum

$$B(\omega) = H(\omega) + H(-\omega)$$

is called the **width (Breite)** of Ω in the direction ω.

We consider also the projection of Ω onto the plane through 0 perpendicular to ω. The arc length $U(\omega)$ of the boundary curve $C(\omega)$ is called the **circumference (Umfang)** in the direction ω. Using spherical harmonics Minkowski [1911] proved the following result.

Theorem 1.20. *The width $B(\omega)$ is constant (in ω) if and only if the circumference $U(\omega)$ is a constant.*

His method actually gives the following more general result which combined with Theorem 1.17 (i) implies Theorem 1.20 immediately.

Theorem 1.21. *With Ω, $B(\omega)$ and $U(\omega)$ as above we have*

$$(68) \qquad U(\omega) = \frac{1}{2} \int_{E_\omega} B(\sigma)\, ds(\sigma),$$

where E_ω is the equator of $\mathbf{S}^2(0)$ perpendicular to ω and ds the arc-element on E_ω.

Proof. Let ω_0 be the vector $(0, 0, 1)$ and consider the boundary curve $C(\omega_0)$ in the $x_1 x_2$-plane. The vertical tangent planes to Ω intersect the $x_1 x_2$-plane in the lines

$$(69) \qquad \langle x, \omega_\varphi \rangle = H(\omega_\varphi), \qquad \omega_\varphi = (\cos\varphi, \sin\varphi).$$

Thus $C(\omega_0)$ is the envelope of these lines so its parametric equation is obtained by combining (69) with its derivative with respect to φ. Thus, putting $h(\varphi) = H(\omega_\varphi)$ we get the parametric representation of $C(\omega_0)$:

$$x = h(\varphi)\cos\varphi - \frac{\partial h}{\partial \varphi}\sin\varphi, \quad y = h(\varphi)\sin\varphi + \frac{\partial h}{\partial \varphi}\cos\varphi.$$

Thus

$$\frac{dx}{d\varphi} = -\left(h + \frac{\partial^2 h}{\partial \varphi^2}\right)\sin\varphi, \quad \frac{dy}{d\varphi} = \left(h + \frac{\partial^2 h}{\partial \varphi^2}\right)\cos\varphi.$$

Since these cannot simultaneously vanish the arc length of $C(\omega_0)$ is given by

$$U(\omega_0) = \int_0^{2\pi} \left(h(\varphi) + \frac{\partial^2 h}{\partial \varphi^2}\right) d\varphi.$$

The second term integrates to 0 and

$$\int_0^{2\pi} h(\varphi)\, d\varphi = \int_0^{2\pi} H(\omega_\varphi)\, d\varphi = \int_0^{2\pi} H(-\omega_\varphi)\, d\varphi,$$

so

$$U(\omega_0) \;=\; \frac{1}{2} \int_0^{2\pi} B(\omega_\varphi)\, d\varphi \,,$$

which proves the theorem.

We shall now extend Theorem 1.17 to arbitrary k, $1 \leq k < n$, with special form for the X-ray transform where $k = 1$. The method is similar to that for $\mathbf{H}^n$ but the geometry is a bit more complicated.

For clarity we first work out the case $k = 1$, $n = 2$ so $X = \mathbf{S}^2 \subset \mathbf{R}^3$ with center the origin 0 and $o = (0,0,1)$ the North Pole. Let $x \in X$, $\xi \in \Xi$ with $d(x,\xi) = p$. Let $x_0 \in \xi$ with $d(x,x_0) = p$. Thus the geodesic (xx_0) is perpendicular to ξ at x_0 and the two-plane through ξ is perpendicular to the plane E_0 through 0, x, and x_0.

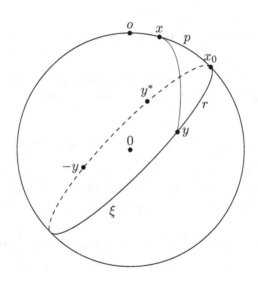

FIGURE III.3.

We put $r = d(x_0, y)$, $q = d(x,y)$ and select $g \in G = \mathbf{O}(3)$ such that $g \cdot o = x$. Then, as in (41),

(70)
$$(\widehat{f})^\vee_p(x) = \int_\xi (M^q f)(x)\, dm(y)\,.$$

The sphere $S_r(x_0)$ in ξ consists of two points y and y^* (symmetric to y with respect to the plane E_0). From (70) we get

$$(\widehat{f})^\vee_p(x) = 2 \int_0^\pi (M^q f)(x)\, dr\,.$$

With K_x the isotropy group of x, $(M^q f)(x)$ is the average of f over the K_x-orbit of y. Since f is even it also equals the average of f over the K_x-orbit of $-y$. These orbits are circles on $\mathbf{S}^2$ with centers x and $-x$. Thus we can limit r to the interval $(0, \frac{\pi}{2})$ and write

$$(71) \qquad (\widehat{f})_p^{\vee}(x) = 4 \int_0^{\frac{\pi}{2}} (M^q f)(x)\, dr\,.$$

By spherical trigonometry for the right-angled triangle (xx_0y) we have

$$(72) \qquad \cos q = \cos p \cos r\,.$$

Putting

$$(73) \qquad F(\cos q) = (M^q f)(x), \quad \widehat{F}(\cos p) = (\widehat{f})_p^{\vee}(x),$$

and $v = \cos p$, $u = v \cos r = \cos q$ we obtain

$$(74) \qquad \widehat{F}(v) = 4 \int_0^v F(u)(v^2 - u^2)^{-\frac{1}{2}}\, du\,,$$

which as in Theorem 2.6, Ch. I inverts to

$$(75) \qquad F(u) = \frac{1}{2\pi} \frac{d}{du} \int_0^u (u^2 - v^2)^{-\frac{1}{2}} v \widehat{F}(v)\, dv\,.$$

Figure II,2 (Ch. II, §4) illustrates the geometric meaning of $(\widehat{f})_p^{\vee}(x)$. As remarked before (41) the set $\{gkg^{-1} \cdot \xi : k \in K\}$ constitutes the set of great circles with distance p from x so

$$(76) \qquad (\widehat{f})_p^{\vee}(x) = \int_K \widehat{f}(gkg^{-1} \cdot \xi)\, dk\,.$$

We observe now that the left hand side is smooth in p. By the G-invariance of the operations $\widehat{}$ and $\vee$ it suffices to show this for $x = o$, where

$$(77) \qquad (\widehat{f})_p^{\vee}(o) = \int_K \widehat{f}(k \cdot \xi_p)\, dk\,, \qquad \xi_p = \xi\,.$$

Perturbing ξ_p to $\xi_{p+\Delta p}$ with the plane of $\xi_{p+\Delta p}$ remaining perpendicular to E_0 we see that (77) extends to a smooth function in p and by its geometric interpretation is even in p. Thus

$$\frac{d}{dp}(\widehat{f})_p^{\vee}(o) = 0 \qquad \text{for } p = 0\,.$$

Putting $u = 1$ in (75) we obtain

(78) $$f(x) = \frac{1}{2\pi}\left[\frac{d}{du}\int_0^u (\widehat{f})^{\vee}_{\cos^{-1} v}(x)v(u^2 - v^2)^{-\frac{1}{2}}\,dv\right]_{u=1}.$$

More generally we consider $X = \mathbf{S}^n$ and Ξ the set of k-dimensional totally geodesic submanifolds of X. The members of Ξ are intersections of X with $(k+1)$-dimensional planes in $\mathbf{R}^{n+1}$ through the origin. Again we take f even. Let $x \in X$, $\xi \in \Xi$ with $d(x,\xi) = p$. We have $X = \mathbf{O}(n+1)/\mathbf{O}(n)$ where $\mathbf{O}(n) = K$ is the isotropy subgroup of the North Pole $o = (0,\ldots,0,1)$. Then as in (41)

(79) $$(\widehat{f})^{\vee}_p(x) = \int_\xi (M^{d(x,y)}f)(x)\,dm(y)$$

and

(80) $$(M^{d(x,y)}f)(x) = \int_K f(gkg^{-1}\cdot y)\,dk.$$

Since f is even and gkg^{-1} a linear transformation, $(M^{d(x,y)}f)(x)$ is even in y.

Again we choose $x_0 \in \xi$ minimizing $d(x,y)$ for $y \in \xi$, so $d(x,x_0) = p$ and the geodesic arc (xx_0) is perpendicular to the $(k+1)$-plane containing ξ. Let $r = d(x_0,x)$. Then $q = d(x,y)$ is constant for y on each sphere $S_r(x_0) \subset \xi$ (cf. (72)). Thus expressing (79) in geodesic polar coordinates in ξ with center x_0 we obtain

(81) $$(\widehat{f})^{\vee}_p(x) = \Omega_k \int_0^\pi (M^q f)(x)\sin^{k-1} r\,dr.$$

Figure III,4 shows the $(k-1)$-spheres $S_r(x_0)$ and $S_{\pi-r}(x_0)$ in the k-sphere ξ. Note that by (80) $(M^q f)(x)$ is the average of f on the K_x-orbit of y, K_x being the isotropy subgroup of x. By the evenness of f this will also equal the average $M^{d(-x,-y)}(-x)$ of f over the K_x orbit of $-y$. Also $\sin r = \sin(\pi - r)$ so in (81) we can restrict r to the interval $(0, \frac{\pi}{2})$. Thus

$$(\widehat{f})^{\vee}_p(x) = 2\Omega_k \int_0^{\frac{\pi}{2}} (M^q f)(x)\sin^{k-1} r\,dr.$$

Using again the substitution (73) we obtain

$$v^{k-1}\widehat{F}(v) = 2\Omega_k \int_0^v F(u)(v^2 - u^2)^{\frac{k}{2}-1}\,du,$$

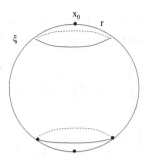

FIGURE III.4.

which is inverted as in Theorem 2.6, Ch. I by

(82) $$F(u) = \frac{c}{2} u \left(\frac{d}{d(u^2)} \right)^k \int_0^u (u^2 - v^2)^{\frac{k}{2}-1} v^k \widehat{F}(v) \, dv ,$$

c being a constant. Since $F(1) = f(x)$ this proves the following analog of Theorem 1.11.

Theorem 1.22. *The k-dimensional totally geodesic Radon transform $f \to \widehat{f}$ on $\mathbf{S}^n$ is for f symmetric inverted by*

$$f(x) = \frac{c}{2} \left[\left(\frac{d}{d(u^2)} \right)^k \int_0^u (\widehat{f})^{\vee}_{\cos^{-1}(v)}(x) v^k (u^2 - v^2)^{\frac{k}{2}-1} \, dv \right]_{u=1} ,$$

where

$$c^{-1} = (k-1)! \Omega_{k+1}/2^{k+1} .$$

For the case $k = 1$ we put Theorem 1.22 in more explicit form. Given $x \in \mathbf{S}^n$ let E_x be the corresponding "equator" that is the set of points at distance $\pi/2$ from x. For later purpose we state the result without the evenness assumption for f.

Theorem 1.23. *The X-ray transform on $\mathbf{S}^n$ is inverted by the formula*

$$\tfrac{1}{2}(f(x) + f(-x)) = \int_{E_x} f(\omega) \, d\omega - \frac{1}{2\pi} \int_0^{\frac{\pi}{2}} \frac{d}{dp} ((\widehat{f})^{\vee}_p(x)) \frac{dp}{\sin p}$$

for each $f \in \mathcal{E}(\mathbf{S}^n)$. Here $d\omega$ is the normalized measure on the equator E_x.

Remark. Because of the integration over E_x this is not an exact inversion. However, if f is even and $\widehat{f} \equiv 0$, then the theorem implies $M^{\frac{\pi}{2}} f = f$, which, together with Theorem 1.1, Ch. VI, implies $f \equiv 0$. See also Corollary 1.24 below.

Proof. Replacing $f(x)$ with $\frac{1}{2}(f(x) + f(-x))$ has no effect on $\widehat{f}$, so with $\widehat{F}(\cos p) = (\widehat{f})^\vee_p(x)$, we have for the right hand side of (82)

$$\frac{1}{2\pi} \left\{ \frac{d}{du} \int_0^u (u^2 - v^2)^{-\frac{1}{2}} v \widehat{F}(v)\, dv \right\}_{u=1}$$

$$= -\frac{1}{2\pi} \left\{ \frac{d}{du} \int_0^u \frac{d}{dv} (u^2 - v^2)^{\frac{1}{2}} \widehat{F}(v)\, dv \right\}_{u=1},$$

which by integration by parts becomes

$$-\frac{1}{2\pi} \left\{ \frac{d}{du} \left[-u\widehat{F}(0) - \int_0^u (u^2 - v^2)^{1/2} \frac{d}{dv} \widehat{F}(v)\, dv \right] \right\}_{u=1}$$

$$= \frac{1}{2\pi} \widehat{F}(0) + \frac{1}{2\pi} \int_0^1 (1 - v^2)^{-\frac{1}{2}} \frac{d\widehat{F}}{dv}\, dv$$

$$= \frac{1}{2\pi} (\widehat{f})^\vee_{\frac{\pi}{2}}(x) - \frac{1}{2\pi} \int_0^{\frac{\pi}{2}} \frac{d}{dp} (\widehat{f})^\vee_p(x) \frac{dp}{\sin p}.$$

The first term is $(2\pi)^{-1}$ times the average of the integrals of f over geodesics at distance $\pi/2$ from x which thus lie in E_x. It represents a rotation-invariant functional on E_x so taking $f \equiv 1$ the term becomes the first term on the right in the formula.

Corollary 1.24. *Suppose* $f \in \mathcal{E}(\mathbf{S}^n)$ *has support in the ball* $B = \{x \in \mathbf{S}^n : d(o, x) < \frac{\pi}{4}\}$. *Then*

$$f(x) = -\frac{1}{\pi} \int_0^{\frac{\pi}{2}} \frac{d}{dp} ((\widehat{f})^\vee_p(x)) \frac{dp}{\sin p}, \quad x \in B.$$

In fact, if $x \in B$ *then* $f(-x) = 0$ *and if* $y \in E_x$ *then*

$$d(o, y) \geq d(x, y) - d(o, x) \geq \frac{\pi}{2} - \frac{\pi}{4}$$

so $f(y) = 0$.

We shall now discuss the analog for $\mathbf{S}^n$ of the support theorem (Theorem 1.6) relative to the X-ray transform $f \to \widehat{f}$.

Theorem 1.25 (The Support Theorem on $\mathbf{S}^n$). *Let C be a closed spherical cap on $\mathbf{S}^n$, C' the cap on $\mathbf{S}^n$ symmetric to C with respect to the origin $0 \in \mathbf{R}^{n+1}$. Let $f \in C(\mathbf{S}^n)$ be symmetric and assume*

$$\widehat{f}(\gamma) = 0$$

for every geodesic γ which does not enter the "arctic zones" C and C'. (See Fig. III.4.)

(i) *If $n \geq 3$ then $f \equiv 0$ outside $C \cup C'$.*

(ii) *If $n = 2$ the same conclusion holds if all derivatives of f vanish on the equator.*

Proof. (i)   Given a point $x \in \mathbf{S}^n$ outside $C \cup C'$ we can (say by the Hahn-Banach theorem) find a 3-dimensional subspace ξ of $\mathbf{R}^{n+1}$ which contains $\mathbf{R}x$ but does not intersect $C \cup C'$. Then $\xi \cap \mathbf{S}^n$ is a 2-sphere and f has integral 0 over each great circle on it. By Theorem 1.17, $f \equiv 0$ on $\xi \cap \mathbf{S}^n$, so $f(x) = 0$.

(ii)   Here we consider the lower hemisphere $\mathbf{S}^2_-$ of the unit sphere and its tangent plane π at the South Pole S. The central projection μ from the origin is a bijection of $\mathbf{S}^2_-$ onto π which intertwines the two Radon transforms as follows: If γ is a (half) great circle on $\mathbf{S}^2_-$ and ℓ the line $\mu(\gamma)$ in π we have (Fig. III.5)

$$(83) \qquad \cos d(S, \gamma)\widehat{f}(\gamma) = 2 \int_\ell (f \circ \mu^{-1})(x)(1 + |x|^2)^{-1}\, dm(x).$$

The proof follows by elementary geometry: Let on Fig. III.5, $x = \mu(s)$, φ and θ the lengths of the arcs SM, Ms. The plane $o'So''$ is perpendicular to ℓ and intersects the semi-great circle γ in M. If $q = |So''|, p = |o''x|$ we have for $f \in C(\mathbf{S}^2)$ symmetric,

$$\widehat{f}(\gamma) = 2 \int_\gamma f(s)d\theta = 2 \int_\ell (f \circ \mu^{-1})(x)\frac{d\theta}{dp}\, dp.$$

Now

$$\tan \varphi = q, \quad \tan \theta = \frac{p}{(1 + q^2)^{1/2}}, \quad |x|^2 = p^2 + q^2.$$

so

$$\frac{dp}{d\theta} = (1 + q^2)^{1/2}(1 + \tan^2 \theta) = (1 + |x|^2)/(1 + q^2)^{1/2}.$$

Thus

$$\frac{dp}{d\theta} = (1 + |x|^2) \cos \varphi$$

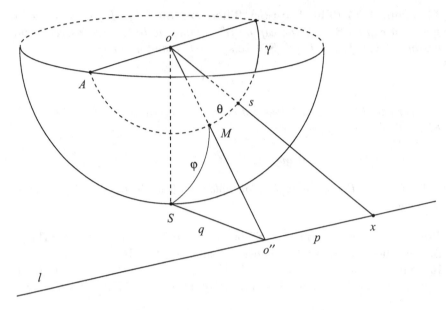

FIGURE III.5.

and since $\varphi = d(S, \gamma)$ this proves (83). Considering the triangle $o'xS$ we obtain

$$(84) \qquad\qquad |x| = \tan d(S, s).$$

Thus the vanishing of all derivatives of f on the equator implies rapid decrease of $f \circ \mu^{-1}$ at ∞.

Let C have spherical radius β. If $\varphi > \beta$ we have by assumption, $\widehat{f}(\gamma) = 0$ so by (83) and Theorem 2.6 in Chapter I,

$$(f \circ \mu^{-1})(x) = 0 \quad \text{for } |x| > \tan \beta,$$

whence by (84),

$$f(s) = 0 \quad \text{for } d(S, s) > \beta.$$

Remark. Because of the example in Remark 2.9 in Chapter I the vanishing condition in (ii) cannot be dropped.

There is a generalization of formula (83) to d-dimensional totally geodesic submanifolds of $\mathbf{S}^n$ as well as of $\mathbf{H}^n$ (Kurusa [1992], [1994], Berenstein-Casadio Tarabusi [1993]). This makes it possible to transfer the range characterizations of the d-plane Radon transform in $\mathbf{R}^n$ (Chapter I, §6) to the d-dimensional totally geodesic Radon transform in $\mathbf{H}^n$. A very explicit range description is given by Ishikawa [1997]. In addition to the above references see also Berenstein-Casadio Tarabusi-Kurusa [1997].

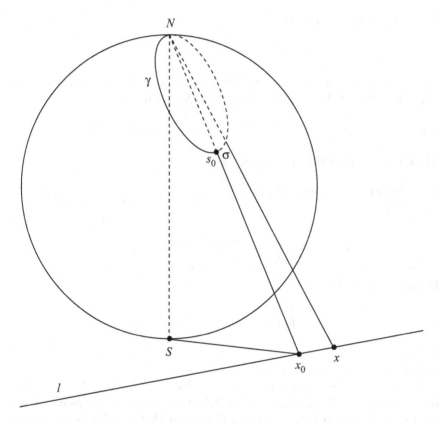

FIGURE III.6.

D.   The Spherical Slice Transform

We shall now briefly consider a variation on the Funk transform and consider integrations over circles on $\mathbf{S}^2$ passing through the North Pole. This Radon transform is given by $f \rightarrow \widehat{f}$ where f is a function on $\mathbf{S}^2$,

$$(85) \qquad\qquad \widehat{f}(\gamma) = \int_{\gamma} f(s)\, dm(s),$$

γ being a circle on $\mathbf{S}^2$ passing through N and dm the arc-element on γ.

It is easy to study this transform by relating it to the X-ray transform on $\mathbf{R}^2$ by means of stereographic projection from N.

We consider a two-sphere $\mathbf{S}^2$ of *diameter* 1, lying on top of its tangent plane $\mathbf{R}^2$ at the South Pole. Let $\nu : \mathbf{S}^2 - N \rightarrow \mathbf{R}^2$ be the stereographic projection. The image $\nu(\gamma)$ is a line $\ell \subset \mathbf{R}^2$. (See Fig. III.6.) The plane through the diameter NS perpendicular to ℓ intersects γ in s_0 and ℓ in x_0. Then Ns_0 is a diameter in γ, and in the right angle triangle NSx_0, the line Ss_0 is perpendicular to Nx_0. Thus, d denoting the Euclidean distance in

$\mathbf{R}^3$, and $q = d(S, x_0)$, we have

$$(86) \qquad d(N, s_0) = (1 + q^2)^{-1/2}, \quad d(s_0, x_0) = q^2(1 + q^2)^{-1/2}.$$

Let σ denote the circular arc on γ for which $\nu(\sigma)$ is the segment (x_0, x) on ℓ. If θ is the angle between the lines Nx_0, Nx then

$$(87) \qquad \sigma = (2\theta) \cdot \tfrac{1}{2}(1 + q^2)^{-1/2}, \quad d(x_0, x) = \tan\theta(1 + q^2)^{1/2}.$$

Thus, $dm(x)$ being the arc-element on ℓ,

$$\begin{aligned}
\frac{dm(x)}{d\sigma} &= \frac{dm(x)}{d\theta} \cdot \frac{d\theta}{d\sigma} = (1 + q^2)^{1/2} \cdot (1 + \tan^2\theta)(1 + q^2)^{1/2} \\
&= (1 + q^2)\left(1 + \frac{d(x_0, x)^2}{1 + q^2}\right) = 1 + |x|^2.
\end{aligned}$$

Hence we have

$$(88) \qquad \widehat{f}(\gamma) = \int_\ell (f \circ \nu^{-1})(x)(1 + |x|^2)^{-1}\, dm(x),$$

a formula quite similar to (83).

If f lies on $C^1(\mathbf{S}^2)$ and vanishes at N then $f \circ \nu^{-1} = 0(x^{-1})$ at ∞. Also of $f \in \mathcal{E}(\mathbf{S}^2)$ and all its derivatives vanish at N then $f \circ \nu^{-1} \in \mathcal{S}(\mathbf{R}^2)$. As in the case of Theorem 1.25 (ii) we can thus conclude the following corollaries of Theorem 3.1, Chapter I and Theorem 2.6, Chapter I.

Corollary 1.26. *The transform $f \to \widehat{f}$ is one-to-one on the space $C_0^1(\mathbf{S}^2)$ of C^1-functions vanishing at N.*

In fact, $(f \circ \nu^{-1})(x)/(1 + |x|^2) = 0(|x|^{-3})$ so Theorem 3.1, Chapter I applies.

Corollary 1.27. *Let B be a spherical cap on $\mathbf{S}^2$ centered at N. Let $f \in C^\infty(\mathbf{S}^2)$ have all its derivatives vanish at N. If*

$$\widehat{f}(\gamma) = 0 \text{ for all } \gamma \text{ through } N, \quad \gamma \subset B$$

then $f \equiv 0$ on B.

In fact $(f \circ \nu^{-1})(x) = 0(|x|^{-k})$ for each $k \geq 0$. The assumption on $\widehat{f}$ implies that $(f \circ \nu^{-1})(x)(1 + |x|^2)^{-1}$ has line integral 0 for all lines outside $\nu(B)$ so by Theorem 2.6, Ch. I, $f \circ \nu^{-1} \equiv 0$ outside $\nu(B)$.

Remark. In Cor. 1.27 the condition of the vanishing of all derivatives at N cannot be dropped. This is clear from Remark 2.9 in Chapter I where the rapid decrease at ∞ was essential for the conclusion of Theorem 2.6.

§2   Compact Two-Point Homogeneous Spaces. Applications

We shall now extend the inversion formula in Theorem 1.17 to compact two-point homogeneous spaces X of dimension $n > 1$. By virtue of Wang's classification [1952] these are also the compact symmetric spaces of rank one (see Matsumoto [1971] and Szabo [1991] for more direct proofs), so their geometry can be described very explicitly. Here we shall use some geometric and group theoretic properties of these spaces ((i)–(vii) below) and refer to Helgason ([1959], p. 278, [1965a], §5–6 or [DS], Ch. VII, §10) for their proofs.

Let U denote the group $I(X)$ of isometries of X. Fix an origin $o \in X$ and let K denote the isotropy subgroup U_o. Let $\mathfrak{k}$ and $\mathfrak{u}$ be the Lie algebras of K and U, respectively. Then $\mathfrak{u}$ is semisimple. Let $\mathfrak{p}$ be the orthogonal complement of $\mathfrak{k}$ in $\mathfrak{u}$ with respect to the Killing form B of $\mathfrak{u}$. Changing the distance function on X by a constant factor we may assume that the differential of the mapping $u \to u \cdot o$ of U onto X gives an isometry of $\mathfrak{p}$ (with the metric of $-B$) onto the tangent space X_o. This is the canonical metric X which we shall use.

Let L denote the diameter of X, that is the maximal distance between any two points. If $x \in X$ let A_x denote the set of points in X of distance L from x. By the two-point homogeneity the isotropy subgroup U_x acts transitively on A_x; thus $A_x \subset X$ is a submanifold, **the antipodal manifold** associated to x. With X_x denoting the tangent space to X at x, Exp_x denotes the map $X_x \to X$ which maps lines through the origin in X_x to geodesics through x in X. Thus the curve $t \to \mathrm{Exp}_x(tA)$ has tangent vector A for $t = 0$.

(i)   *Each A_x is a totally geodesic submanifold of X; with the Riemannian structure induced by that of X it is another two-point homogeneous space.*

(ii)   *Let Ξ denote the set of all antipodal manifolds in X; since U acts transitively on Ξ, the set Ξ has a natural manifold structure. Then the mapping $j : x \to A_x$ is a one-to-one diffeomorphism; also $x \in A_y$ if and only if $y \in A_x$.*

(iii)   *Each geodesic in X has period $2L$. If $x \in X$ the mapping $\mathrm{Exp}_x : X_x \to X$ gives a diffeomorphism of the ball $B_L(0)$ onto the open set $X - A_x$.*

Fix a vector $H \in \mathfrak{p}$ of length L (i.e., $L^2 = -B(H, H)$). For $Z \in \mathfrak{p}$ let T_Z denote the linear transformation $Y \to [Z, [Z, Y]]$ of $\mathfrak{p}$, $[\,,\,]$ denoting the Lie bracket in $\mathfrak{u}$. For simplicity, we now write Exp instead of Exp_o. A point $Y \in \mathfrak{p}$ is said to be **conjugate** to o if the differential $d\mathrm{Exp}$ is singular at Y.

The line $\mathfrak{a} = \mathbf{R}H$ is a maximal abelian subspace of $\mathfrak{p}$. The eigenvalues of T_H are 0, $\alpha(H)^2$ and possibly $(\alpha(H)/2)^2$, where $\pm\alpha$ (and possibly $\pm\alpha/2$)

are the roots of $\mathfrak{u}$ with respect to $\mathfrak{a}$. Let

(89) $$\mathfrak{p} = \mathfrak{a} + \mathfrak{p}_\alpha + \mathfrak{p}_{\alpha/2}$$

be the corresponding decomposition of $\mathfrak{p}$ into eigenspaces; the dimensions $q = \dim(\mathfrak{p}_\alpha)$, $p = \dim(\mathfrak{p}_{\alpha/2})$ are called the **multiplicities** of α and $\alpha/2$, respectively.

(iv) *Suppose H is conjugate to o. Then $\mathrm{Exp}(\mathfrak{a}+\mathfrak{p}_\alpha)$, with the Riemannian structure induced by that of X, is a sphere, totally geodesic in X, having o and $\mathrm{Exp}H$ as antipodal points and having curvature $\pi^2 L^2$. Moreover,*

$$A_{\mathrm{Exp}H} = \mathrm{Exp}(\mathfrak{p}_{\alpha/2}).$$

(v) *If H is not conjugate to o then $\mathfrak{p}_{\alpha/2} = 0$ and*

$$A_{\mathrm{Exp}H} = \mathrm{Exp}\,\mathfrak{p}_\alpha.$$

(vi) *The differential at Y of Exp is given by*

$$d\mathrm{Exp}_Y = d\tau(\exp Y) \circ \sum_0^\infty \frac{T_Y^k}{(2k+1)!},$$

where for $u \in U$, $\tau(u)$ is the isometry $x \to u \cdot x$.

(vii) *In analogy with (33) the Laplace-Beltrami operator L on X has the expression*

$$L = \frac{\partial^2}{\partial r^2} + \frac{1}{A(r)} A'(r) \frac{\partial}{\partial r} + L_{S_r},$$

where L_{S_r} is the Laplace-Beltrami operator on $S_r(o)$ and $A(r)$ its area.

(viii) *The spherical mean-value operator M^r commutes with the Laplace-Beltrami operator.*

Lemma 2.1. *The surface area $A(r)$ $(0 < r < L)$ is given by*

$$A(r) = \Omega_n \lambda^{-p}(2\lambda)^{-q} \sin^p(\lambda r) \sin^q(2\lambda r),$$

where p and q are the multiplicities above and $\lambda = |\alpha(H)|/2L$.

Proof. Because of (iii) and (vi) the surface area of $S_r(o)$ is given by

$$A(r) = \int_{|Y|=r} \det\left(\sum_0^\infty \frac{T_Y^k}{(2k+1)!}\right) d\omega_r(Y),$$

where $d\omega_r$ is the surface on the sphere $|Y| = r$ in $\mathfrak{p}$. Because of the two-point homogeneity the integrand depends on r only, so it suffices to evaluate it for $Y = H_r = \frac{r}{L}H$. Since the nonzero eigenvalues of T_{H_r} are $\alpha(H_r)^2$ with multiplicity q and $(\alpha(H_r)/2)^2$ with multiplicity p, a trivial computation gives the lemma.

We consider now Problems A, B and C from Chapter II, §2 for the homogeneous spaces X and Ξ, which are acted on transitively by the same group U. Fix an element $\xi_o \in \Xi$ passing through the origin $o \in X$. If $\xi_o = A_{o'}$, then an element $u \in U$ leaves ξ_o invariant if and only if it lies in the isotropy subgroup $K' = U_{o'}$; we have the identifications

$$X = U/K, \quad \Xi = U/K',$$

and $x \in X$ and $\xi \in \Xi$ are incident if and only if $x \in \xi$.

On Ξ we now choose a Riemannian structure such that the diffeomorphism $j : x \to A_x$ from (ii) is an isometry. Let L and Λ denote the Laplacians on X and Ξ, respectively. With $\check{x}$ and $\widehat{\xi}$ defined as in Ch. II, §1, we have

$$\widehat{\xi} = \xi, \quad \check{x} = \{j(y) : y \in j(x)\};$$

the first relation amounts to the incidence description above and the second is a consequence of the property $x \in A_y \Leftrightarrow y \in A_x$ listed under (ii).

The sets $\check{x}$ and $\widehat{\xi}$ will be given the measures $d\mu$ and dm, respectively, induced by the Riemannian structures of Ξ and X. The Radon transform and its dual are then given by

$$\widehat{f}(\xi) = \int_\xi f(x)\, dm(x), \quad \check{\varphi}(x) = \int_{\check{x}} \varphi(\xi)\, d\mu(\xi).$$

However,

$$\check{\varphi}(x) = \int_{\check{x}} \varphi(\xi)\, d\mu(\xi) = \int_{y \in j(x)} \varphi(j(y))\, d\mu(j(y)) = \int_{j(x)} (\varphi \circ j)(y)\, dm(y),$$

so

(90) $$\check{\varphi} = (\varphi \circ j)\widehat{} \circ j.$$

Because of this correspondence between the transforms $f \to \widehat{f}$, $\varphi \to \check{\varphi}$ it suffices to consider the first one. Let $\mathbf{D}(X)$ denote the algebra of differential operators on X, invariant under U. It can be shown that $\mathbf{D}(X)$ is generated by L. Similarly $\mathbf{D}(\Xi)$ is generated by Λ.

Theorem 2.2 (The Inversion of the Antipodal map). *(i)   The mapping $f \to \widehat{f}$ is a linear one-to-one mapping of $\mathcal{E}(X)$ onto $\mathcal{E}(\Xi)$ and*

$$(Lf)\widehat{} = \Lambda \widehat{f}.$$

(ii)   Except for the case when X is an even-dimensional elliptic space

$$f = P(L)((\widehat{f})^\vee), \quad f \in \mathcal{E}(X),$$

where P is a polynomial, independent of f, explicitly given below, (93)–(96). In all cases

$$\text{degree } P = \tfrac{1}{2} \text{ dimension of the antipodal manifold}.$$

Proof. We first prove (ii). Let dk be the Haar measure on K such that $\int dk = 1$ and let Ω_X denote the total measure of an antipodal manifold in X. Then $\mu(\breve{o}) = m(A_o) = \Omega_X$ and if $u \in U$,

$$\breve{\varphi}(u \cdot o) = \Omega_X \int_K \varphi(uk \cdot \xi_o) \, dk \,.$$

Hence

$$(\widehat{f})^\vee (u \cdot o) = \Omega_X \int_K \left(\int_{\xi_o} f(uk \cdot y) \, dm(y) \right) dk = \Omega_X \int_{\xi_o} (M^r f)(u \cdot o) \, dm(y) \,,$$

where r is the distance $d(o, y)$ in the space X between o and y. If $d(o, y) < L$ there is a unique geodesic in X of length $d(o, y)$ joining o to y and since ξ_0 is totally geodesic, $d(o, y)$ is also the distance in ξ_0 between o and y. Thus using geodesic polar coordinates in ξ_0 in the last integral we obtain

$$(91) \qquad (\widehat{f})^\vee (x) = \Omega_X \int_0^L (M^r f)(x) A_1(r) \, dr \,,$$

where $A_1(r)$ is the area of a sphere of radius r in ξ_0. By Lemma 2.1 we have

$$(92) \qquad A_1(r) = C_1 \sin^{p_1}(\lambda_1 r) \sin^{q_1}(2\lambda_1 r) \,,$$

where C_1 and λ_1 are constants and p_1, q_1 are the multiplicities for the antipodal manifold. In order to prove (ii) on the basis of (91) we need the following complete list of the compact symmetric spaces of rank one and their corresponding antipodal manifolds (see Nagano [1952], p. 52):

X		A_0
Spheres	$\mathbf{S}^n (n = 1, 2, \ldots)$	point
Real projective spaces	$\mathbf{P}^n(\mathbf{R})(n = 2, 3, \ldots)$	$\mathbf{P}^{n-1}(\mathbf{R})$
Complex projective spaces	$\mathbf{P}^n(\mathbf{C})(n = 4, 6, \ldots)$	$\mathbf{P}^{n-2}(\mathbf{C})$
Quaternion projective spaces	$\mathbf{P}^n(\mathbf{H})(n = 8, 12, \ldots)$	$\mathbf{P}^{n-4}(\mathbf{H})$
Cayley plane	$\mathbf{P}^{16}(\mathbf{Cay})$	$\mathbf{S}^8$

We list here the multiplicities p and q for the compact rank-one symmetry space (Cartan [1927], Araki [1962], or [DS], p. 532).

$X = \mathbf{P}^n(\mathbf{R})$

$p = 0, q = n - 1, \lambda = \pi/4L$

$A(r) = \Omega_n(2\lambda)^{-n+1}\sin^{n-1}(2\lambda r)$

$A_1(r) = \Omega_{n-1}(2\lambda)^{-n+2}\sin^{n-2}(2\lambda r)$

$X = \mathbf{P}^n(\mathbf{C})$

$p = n - 2, q = 1, \lambda = \pi/2L$

$A(r) = \frac{1}{2}\Omega_n(2\lambda)^{-n+1}\sin^{n-2}(\lambda r)\sin(2\lambda r)$

$A_1(r) = \frac{1}{2}\Omega_{n-2}\lambda^{-n+3}\sin^{(n-4)}(\lambda r)\sin 2(\lambda r)$

$X = \mathbf{P}^n(\mathbf{H})$

$p = n - 4, q = 3, \lambda = \pi/2L$

$A(r) = \frac{1}{8}\Omega_n\lambda^{-n+1}\sin^{(n-4)}(\lambda r)\sin^3(2\lambda r)$

$A_1(r) = \frac{1}{8}\Omega_{n-4}\lambda^{-n+5}\sin^{(n-8)}(\lambda r)\sin^3(2\lambda r)$

$X = \mathbf{P}^{16}(\mathbf{Cay})$

$p = 8, q = 7, \lambda = \pi/2L$

$A(r) = (1/2^7)\Omega_{16}\lambda^{-15}\sin^8(\lambda r)\sin^7(2\lambda r)$

$A_1(r) = (1/2^7)\lambda^{-7}\Omega_8\sin^7(2\lambda r)$

Here the superscripts denote the real dimension. For the lowest dimensions, note that

$$\mathbf{P}^1(\mathbf{R}) = \mathbf{S}^1, \quad \mathbf{P}^2(\mathbf{C}) = \mathbf{S}^2, \quad \mathbf{P}^4(\mathbf{H}) = \mathbf{S}^4.$$

For the case $\mathbf{S}^n$, (ii) is trivial and the case $X = \mathbf{P}^n(\mathbf{R})$ was already done in Theorem 1.17. The remaining cases are done by classification starting with (91). The mean-value operator M^r still commutes with the Laplacian L,

$$M^r L = LM^r,$$

and this implies

$$L_x((M^r f)(x)) = L_r((M^r f)(x)),$$

where L_r is the radial part of L. Because of (vii) above and Lemma 2.1 it is given by

$$L_r = \frac{\partial^2}{\partial r^2} + \lambda\{p\cot(\lambda r) + 2q\cot(2\lambda r)\}\frac{\partial}{\partial r}.$$

For each of the two-point homogeneous spaces we prove the analog of Lemma 1.18. Then by the pattern of the proof of Theorem 1.17, part (ii) of Theorem 2.2 can be proved.

The polynomial P is explicitly given in the list below. Note that for $\mathbf{P}^n(\mathbf{R})$ the metric is normalized by means of the Killing form so it differs from that of Theorem 1.17 by a nontrivial constant.

The polynomial P is now given as follows:

For $X = \mathbf{P}^n(\mathbf{R})$, n odd

$$
(93) \quad P(L) = c \left(L - \frac{(n-2)1}{2n}\right) \left(L - \frac{(n-4)3}{2n}\right) \cdots \left(L - \frac{1(n-2)}{2n}\right)
$$

$$
c = \tfrac{1}{4}(-4\pi^2 n)^{\frac{1}{2}(n-1)} .
$$

For $X = \mathbf{P}^n(\mathbf{C})$, $n = 4, 6, 8, \ldots$

$$
(94) \quad P(L) = c \left(L - \frac{(n-2)2}{2(n+2)}\right) \left(L - \frac{(n-4)4}{2(n+2)}\right) \cdots \left(L - \frac{2(n-2)}{2(n+2)}\right)
$$

$$
c = (-8\pi^2(n+2))^{1-\frac{n}{2}} .
$$

For $X = \mathbf{P}^n(\mathbf{H})$, $n = 8, 12, \ldots$

$$
(95) \quad P(L) = c \left(L - \frac{(n-2)4}{2(n+8)}\right) \left(L - \frac{(n-4)6}{2(n+8)}\right) \cdots \left(L - \frac{4(n-2)}{2(n+8)}\right)
$$

$$
c = \tfrac{1}{2}[-4\pi^2(n+8)]^{2-n/2} .
$$

For $X = \mathbf{P}^{16}(\mathbf{Cay})$

$$
(96) \quad P(L) = c \left(L - \tfrac{14}{9}\right)^2 \left(L - \tfrac{15}{9}\right)^2 , \quad c = 3^6 \pi^{-8} 2^{-13} .
$$

Since the original computation of P (Helgason [1965a]) was so complicated we give now a unified proof of these formulas, using the method of Rouvière [2001] for the noncompact case. The compact case has some complications because of the antipodal set; as a result it is best to leave out the case $\mathbf{P}^n(\mathbf{R})$ which is no harm since this case was done already in Theorem 1.17.

The simplification results from rewriting (91) in the style of (60), Ch. I. We have

$$
(\widehat{f})^\vee(o) = \Omega_X \int_0^L \frac{1}{A(r)} \int_{S_r(o)} f(\omega) \, d\omega \, A_1(r) \, dr ,
$$

where $d\omega$ is the induced measure on $S_r(o)$. Writing $\sigma(x) = \Omega_X A_1(r)/A(r)$ for $r = d(o, x)$ we have if dx is the Riemannian measure on X,

$$
(\widehat{f})^\vee(o) = \int_X f(x)\sigma(x) \, dx .
$$

Since $\widehat{}$ and $\vee$ commute with the action of U we have

$$(\widehat{f})^\vee(u \cdot o) = \int_X f(u \cdot x)\sigma(x)\,dx$$

$$= \int_U f(h \cdot o)\sigma(u^{-1}h \cdot o)\,dh = \int_U f(h \cdot o)\sigma(h^{-1}u \cdot o)\,dh\,,$$

where dh is a suitable Haar measure on U and $\sigma(h \cdot o) = \sigma(h^{-1} \cdot o)$. Thus

$$(\widehat{f})^\vee = f \times \sigma$$

as in Theorem 2.7, Ch.II. Here σ is an integrable function on X and in accordance with Corollary 2.8, Ch.II we shall find a U-invariant differential operator D on X (a polynomial in the Laplacian) such that $D\sigma = \delta$.

Let $g_a(r) = \sin^a(\lambda r)$ and consider the radial part L_r of L in the form

$$L_r f = A^{-1}(Af')'\,,$$

and note that

$$A(r) = \Omega_n \lambda^{-n+1} g_{n-1}(r) \cos^q(\lambda r)\,.$$

We have

$$g_a(r) = g_{a-2}(r) - g_{a-2}(r)\cos^2(\lambda r)$$
$$g_a'(r) = \lambda a\, g_{a-1}(r) \cos(\lambda r)\,,$$

so

$$\left(L_r + \lambda^2 a(a+n+q-1)\right) g_a = \lambda^2 a(a+n-2)\, g_{a-2}\,.$$

Lemma 2.3. *Define G_a on X by*

$$G_a(x) = g_a(d(o, x))\,.$$

If $a+n \geq 2$, G_a is a locally integrable function on X which as a distribution satisfies

$$(L + \lambda^2 a(a+n+q-1))\, G_a = \begin{cases} \lambda^2 a(a+n-2)G_{a-2} & \text{if } a+n > 2, \\ \lambda^{-n+2}\Omega_n\, a\, \delta & \text{if } a+n = 2. \end{cases}$$

Proof. The functional G_a is defined by

$$G_a(f) = \int_X g_a(d(o, x))f(x)\,dx\,, \quad f \in \mathcal{D}(X)\,,$$

and since $g_a(d(o, x))$ is radial, G_a is K-invariant and we can take f radial and write $f(x) = f(d(o, x))$. Then

$$G_a(f) = \int_0^L g_a(r)f(r)A(r)\, dr$$

$$= \Omega_n\, \lambda^{-n+1} \int_0^L g_{a+n-1}(r)\cos^q(\lambda r)f(r)\, dr\,.$$

$$(LG_a)(f) = G_a(Lf) = \int_0^L g_a(r)(L_rf)(r)A(r)\, dr$$

$$= \int_0^L g_a(r)(Af')'(r)\, dr$$

$$= \left[A(r)f'(r)g_a(r)\right]_0^L - \int_0^L A(r)f'(r)g_a'(r)\, dr\,.$$

The boundary term vanishes both for $r = 0$ and $r = L$. Thus the expression reduces to

$$-\int_0^L f'(r)A(r)g_a'(r)\, dr$$

$$= -\left[f(r)A(r)g_a'(r)\right]_0^L + \int_0^L f(r)(Ag_a')'(r)\, dr\,.$$

If $a + n > 2$ the boundary term vanishes at $r = 0$ and $r = L$ and we have

$$(LG_a)(f) = \int_0^L f(r)(L_r g_a)(r)A(r)\, dr\,,$$

so the formula for $L_r\, g_a$ gives the lemma in this case.

If $a + n = 2$ the boundary term at $r = 0$ contributes

$$f(o) \lim_{r \to 0} A(r)\, g_a'(r) = f(o)\,\lambda\, a\, \Omega_n \lambda^{-n+1}\,,$$

while the contribution at $r = L$ is 0 because of the formula for g_a' and the relation $\lambda = \pi/2L$. This proves the second part of the lemma.

We shall now use the lemma on the formula

$$(\widehat{f})^{\vee} = f \times \sigma\,,$$

where $\sigma(r)$ is the radial function

$$\sigma(r) = \Omega_X \frac{A_1(r)}{A(r)} = \Omega_X \frac{\Omega_{n-2\ell}}{\Omega_n} \lambda^{2\ell} \sin^{-2\ell}(\lambda r),$$

$$\sigma(r) = C \sin^{-2\ell}(\lambda r),$$

where C is a constant dealt with later and $\ell = 1, 2, 4$ in the three respective cases $\mathbf{P}^n(\mathbf{C})$, $\mathbf{P}^n(\mathbf{H})$ and $\mathbf{P}^{16}(\mathbf{Cay})$ for which $q = 1, 3$ and 7. The antipodal manifolds have dimensions $2k = n - 2\ell$ in the three cases. We apply the lemma k-times to $\sigma = C\, G_{-2\ell}$, and end up with a multiple of δ. In fact if we put $L_a = L + \lambda_a^2(a + n + q - 1)$ and

$$P_k(L) = L_{-2\ell}L_{-2(\ell-2)} \ldots L_{2-n}$$

we deduce from the lemma

$$P_k(L)\, \sigma = c\, \delta \qquad c = \text{constant}.$$

We thus recover the formulas (93)–(96) up to the constant c. The constant c is recovered by using the formula for $f \equiv 1$, noting the formula

$$\Omega_X = \int_0^L A_1(r)\, dr\,,$$

and the formula for L,

$$L^2 = p\left(\frac{\pi^2}{2}\right) + 2q\,\pi^2$$

obtained in [GGA], p. 169.

The formula (93) for $\mathbf{P}^n(\mathbf{R})$ comes from Theorem 1.17 if we note that the Killing form metric on $\mathbf{S}^n$ is obtained by multiplying the usual Riemannian metric (with curvature $+1$) by $2n$. The new Laplacian is then $1/2n$ times the Laplacian in Theorem 1.17. This verifies (93).

That $f \to \widehat{f}$ is injective follows from (ii) except for the case $X = \mathbf{P}^n(\mathbf{R})$, n even. But in this exceptional case, the injectivity follows from Theorem 1.17.

For the surjectivity we use once more the fact that the mean-value operator M^r commutes with the Laplacian (property (viii)). We have

(97) $$\widehat{f}(j(x)) = c(M^L f)(x),$$

where c is a constant. Thus by (90),

$$(\widehat{f})^{\vee}(x) = (\widehat{f} \circ j)\widehat{\,}(j(x)) = cM^L(\widehat{f} \circ j)(x),$$

so

(98) $(\widehat{f})^{\vee} = c^2 M^L M^L f$.

Thus if X is not an even-dimensional projective space f is a constant multiple of $M^L P(L) M^L f$ which by (97) shows $f \to \widehat{f}$ surjective. For the remaining case $\mathbf{P}^n(\mathbf{R})$, n even, we use the expansion of $f \in \mathcal{E}(\mathbf{P}^n(\mathbf{R}))$ in spherical harmonics

$$f = \sum_{k,m} a_{km} S_{km} \quad (k \text{ even}) .$$

Here $k \in \mathbf{Z}^+$, and S_{km} $(1 \leq m \leq d(k))$ is an orthonormal basis of the space of spherical harmonics of degree k. Here the coefficients a_{km} are rapidly decreasing in k. On the other hand, by (65) and (67),

(99) $\widehat{f} = \Omega_n M^{\frac{\pi}{2}} f = \Omega_n \sum_{k,m} a_{km} \varphi_k \left(\frac{\pi}{2}\right) S_{km} \quad (k \text{ even}) .$

The spherical function φ_k is given by

$$\varphi_k(s) = \frac{\Omega_{n-1}}{\Omega_n} \int\limits_0^{\pi} (\cos\theta + i\sin\theta\cos\varphi)^k \sin^{n-2}\varphi \, d\varphi$$

so $\varphi_{2k}\left(\frac{\pi}{2}\right) \sim k^{-\frac{n-1}{2}}$. Thus by Lemma 1.19 every series $\sum_{k,m} b_{k,m} S_{2k,m}$ with $b_{2k,m}$ rapidly decreasing in k can be put in the form (99). This verifies the surjectivity of the map $f \to \widehat{f}$.

It remains to prove $(Lf)\widehat{} = \Lambda\widehat{f}$. For this we use (90), (vii), and (97). By the definition of Λ we have

$$(\Lambda\varphi)(j(x)) = L(\varphi \circ j)(x), \qquad x \in X, \varphi \in \mathcal{E}(X) .$$

Thus

$$(\Lambda\widehat{f})(j(x)) = (L(\widehat{f} \circ j))(x) = cL(M^L f)(x) = cM^L(Lf)(x) = (Lf)\widehat{}(j(x)) .$$

This finishes our proof of Theorem 2.2.

Corollary 2.4. *Let X be a compact two-point homogeneous space and suppose f satisfies*

$$\int\limits_{\gamma} f(x) \, ds(x) = 0$$

for each (closed) geodesic γ in X, ds being the element of arc-length. Then

(i)   *If X is a sphere, f is skew.*

(ii)   *If X is not a sphere, $f \equiv 0$.*

Taking a convolution with f we may assume f smooth. Part (i) is already contained in Theorem 1.17. For Part (ii) we use the classification; for $X = \mathbf{P}^{16}(\mathbf{Cay})$ the antipodal manifolds are totally geodesic spheres so using Part (i) we conclude that $\widehat{f} \equiv 0$ so by Theorem 2.2, $f \equiv 0$. For the remaining cases $\mathbf{P}^n(\mathbf{C})$ $(n = 4, 6, \ldots)$ and $\mathbf{P}^n(\mathbf{H})$, $(n = 8, 12, \ldots)$ (ii) follows similarly by induction as the initial antipodal manifolds, $\mathbf{P}^2(\mathbf{C})$ and $\mathbf{P}^4(\mathbf{H})$, are totally geodesic spheres.

Corollary 2.5. *Let B be a bounded open set in $\mathbf{R}^{n+1}$, symmetric and star-shaped with respect to 0, bounded by a hypersurface. Assume for a fixed k $(1 \leq k < n)$*

$$(100) \qquad\qquad Area\,(B \cap P) = \ constant$$

for all $(k+1)$-planes P through 0. Then B is an open ball.

In fact, we know from Theorem 1.17 that if f is a symmetric function on $X = \mathbf{S}^n$ with $\widehat{f}(\mathbf{S}^n \cap P)$ constant (for all P) then f is a constant. We apply this to the function

$$f(\theta) = \rho(\theta)^{k+1} \quad \theta \in \mathbf{S}^n$$

if $\rho(\theta)$ is the distance from the origin to each of the two points of intersection of the boundary of B with the line through 0 and θ; f is well defined since B is symmetric. If $\theta = (\theta_1, \ldots, \theta_k)$ runs through the k-sphere $\mathbf{S}^n \cap P$ then the point

$$x = \theta r \quad (0 \leq r < \rho(\theta))$$

runs through the set $B \cap P$ and

$$Area\,(B \cap P) = \int\limits_{\mathbf{S}^n \cap P} d\omega(\theta) \int\limits_0^{\rho(\theta)} r^k \, dr\,.$$

It follows that Area $(B \cap P)$ is a constant multiple of $\widehat{f}(\mathbf{S}^n \cap P)$ so (100) implies that f is constant. This proves the corollary.

§3   Noncompact Two-Point Homogeneous Spaces

Theorem 2.2 has an analog for noncompact two-point homogeneous spaces which we shall now describe. By Tits' classification [1955], p. 183, of homogeneous manifolds L/H for which L acts transitively on the tangents to L/H it is known, in principle, what the noncompact two-point homogeneous spaces are. As in the compact case they turn out to be symmetric. A direct proof of this fact was given by Nagano [1959] and Helgason [1959]. The theory of symmetric spaces then implies that the noncompact two-point homogeneous spaces are the Euclidean spaces and the noncompact spaces $X = G/K$, where G is a connected semisimple Lie group with finite center and real rank one and K a maximal compact subgroup.

Let $\mathfrak{g} = \mathfrak{k} + \mathfrak{p}$ be the direct decomposition of the Lie algebra of G into the Lie algebra $\mathfrak{k}$ of K and its orthogonal complement $\mathfrak{p}$ (with respect to the Killing form of $\mathfrak{g}$). Fix a 1-dimensional subspace $\mathfrak{a} \subset \mathfrak{p}$ and let

$$(101) \qquad\qquad \mathfrak{p} = \mathfrak{a} + \mathfrak{p}_\alpha + \mathfrak{p}_{\alpha/2}$$

be the decomposition of $\mathfrak{p}$ into eigenspaces of T_H (in analogy with (89)). Let ξ_o denote the totally geodesic submanifold $\mathrm{Exp}(\mathfrak{p}_{\alpha/2})$; in the case $\mathfrak{p}_{\alpha/2} = 0$ we put $\xi_o = \mathrm{Exp}(\mathfrak{p}_\alpha)$. By the classification and duality for symmetric spaces we have the following complete list of the spaces G/K. In the list the superscript denotes the real dimension; for the lowest dimensions note that

$$\mathbf{H}^1(\mathbf{R}) = \mathbf{R}, \quad \mathbf{H}^2(\mathbf{C}) = \mathbf{H}^2(\mathbf{R}), \quad \mathbf{H}^4(\mathbf{H}) = \mathbf{H}^4(\mathbf{R}).$$

X		ξ_0
Real hyperbolic spaces	$\mathbf{H}^n(\mathbf{R})(n = 2, 3, \ldots)$,	$\mathbf{H}^{n-1}(\mathbf{R})$
Complex hyperbolic spaces	$\mathbf{H}^n(\mathbf{C})(n = 4, 6, \ldots)$,	$\mathbf{H}^{n-2}(\mathbf{C})$
Quaternion hyperbolic spaces	$\mathbf{H}^n(\mathbf{H})(n = 8, 12, \ldots)$,	$\mathbf{H}^{n-4}(\mathbf{H})$
Cayley hyperbolic spaces	$\mathbf{H}^{16}(\mathbf{Cay})$,	$\mathbf{H}^8(\mathbf{R})$

Let Ξ denote the set of submanifolds $g \cdot \xi_0$ of X as g runs through G; Ξ is given the canonical differentiable structure of a homogeneous space. Each $\xi \in \Xi$ has a measure m induced by the Riemannian structure of X and the Radon transform on X is defined by

$$\widehat{f}(\xi) = \int_\xi f(x)\, dm(x), \quad f \in C_c(X).$$

The dual transform $\varphi \to \check{\varphi}$ is defined by

$$\check{\varphi}(x) = \int_{\xi \ni x} \varphi(\xi)\, d\mu(\xi), \quad \varphi \in C(\Xi),$$

where μ is the invariant average on the set of ξ passing through x. Let L denote the Laplace-Beltrami operator on X, Riemannian structure being that given by the Killing form of $\mathfrak{g}$.

Theorem 3.1. *The Radon transform $f \to \widehat{f}$ is a one-to-one mapping of $\mathcal{D}(X)$ into $\mathcal{D}(\Xi)$ and, except for the case $X = \mathbf{H}^n(\mathbf{R})$, n even, is inverted by the formula*

$$f = Q(L)((\widehat{f})^\vee).$$

Here Q is given by

$X = \mathbf{H}^n(\mathbf{R}), n$ *odd*:
$$Q(L) = \gamma \left(L + \tfrac{(n-2)1}{2n}\right) \left(L + \tfrac{(n-4)3}{2n}\right) \cdots \left(L + \tfrac{1(n-2)}{2n}\right).$$

$X = \mathbf{H}^n(\mathbf{C}):$
$$Q(L) = \gamma \left(L + \tfrac{(n-2)2}{2(n+2)}\right) \left(L + \tfrac{(n-4)4}{2(n+2)}\right) \cdots \left(L + \tfrac{2(n-2)}{2(n+2)}\right).$$

$X = \mathbf{H}^n(\mathbf{H}):$
$$Q(L) = \gamma \left(L + \tfrac{(n-2)4}{2(n+8)}\right) \left(L + \tfrac{(n-4)6}{2(n+8)}\right) \cdots \left(L + \tfrac{4(n-2)}{2(n+8)}\right).$$

$X = \mathbf{H}^{16}(\mathbf{Cay}):$
$$Q(L) = \gamma \left(L + \tfrac{14}{9}\right)^2 \left(L + \tfrac{15}{9}\right)^2.$$

The constants γ are obtained from the constants c in (93)–(96) by multiplication by the factor Ω_X which is the volume of the antipodal manifold in the compact space corresponding to X. This factor is explicitly determined for each X in [GGA], Chapter I, §4.

§4   Support Theorems Relative to Horocycles

In our noncompact two-point homogeneous space X fix a geodesic $\gamma(t)$, $\gamma(0) = o$. Since X is symmetric the geodesic symmetries are isometries and $a_t = s_{\frac{t}{2}} s_{\gamma(0)}$ is a one parameter group of "transvections" of X, $t \in \mathbf{R}$. Here s_p is the geodesic symmetry with respect to p and $s_t = s_{\gamma(t)}$. Let

$$N = \{g \in G \,|\, a_t^{-1} g a_t \to 0 \text{ as } t \to +\infty\}.$$

This generalizes the group N in Ch. II, §4. The orbits of N and its conjugates are the **horocycles** in X and again they are permuted transitively by G. For the hyperbolic space it is clear what is meant by the interior of a horocycle. For our X the **interior** of the horocycle $N \cdot o$ is defined as

$$\bigcup_{t>0} N a_t \cdot o.$$

The point o was arbitrary so the interior of an arbitrary horocycle is well defined. (Notion introduced in Faraut [1982]).

A function $f \in \mathcal{E}(X)$ will be said to be exponentially decreasing if $f(x) e^{m\, d(0,x)}$, $x \in X$ is bounded for each $m > 0$.

Theorem 4.1. *Let X be a symmetric space of the noncompact type and of rank one. Let ξ be a horocycle in X and $f \in \mathcal{E}(X)$ an exponentially decreasing function whose X-ray transform $\widehat{f}$ satisfies*

(102)
$$\widehat{f}(\gamma) = 0 \text{ whenever } \gamma \cap \xi = \emptyset.$$

Then

$$f(x) = 0 \text{ for } x \notin \text{ interior of } \xi.$$

Proof. We start with the case when X is the hyperbolic plane with the Riemannian structure

(103) $$ds^2 = \frac{dx^2 + dy^2}{y^2}, \quad y > 0, o = (0,1).$$

The geodesics are the semi-circles

$$\gamma_{u,r} : x = u + r\cos\theta, \, y = r\sin\theta, \quad 0 < \theta < \pi,$$

and the half-lines $x = $ const., $y > 0$. Since the horocycles are permuted transitively by G we can take ξ as the line $y = 1$ and $o = (0,1)$. On $\gamma_{u,r}$ we have $ds = (\sin\theta)^{-1} d\theta$ so the X-ray transform becomes

$$\widehat{f}(\gamma_{u,r}) = \int_0^\pi f(u + r\cos\theta, \, r\sin\theta)(\sin\theta)^{-1} \, d\theta.$$

Thus our assumption (102) amounts to

(104) $$\int_{\gamma_{u,r}} \frac{f(x,y)}{y} \, d\omega = 0, \quad r < 1,$$

where $d\omega$ is the Euclidean arc element. For points $(x,y) \in \gamma_{u,r}$ we have by (103)

$$d(o,(x,y)) \geq d((x,1),(x,y)) = d(o,(0,y)) = \int_y^1 \frac{dt}{t} = -\log y$$

so $e^{d(0,(x,y))} \geq \frac{1}{y}$. Thus the decay of f implies that $f(x,y)/y$ extends (by reflection) to a smooth function on $\mathbf{R}^2$ by $F(x,y) = f(x,|y|)/|y|$. By (104) we have

(105) $$\int_{S_r(x)} F(s) \, d\omega(s) = 0,$$

for x on the x-axis and $r < 1$. With $B_r(x)$ as the ball with boundary $S_r(x)$ we have, since $d\omega$ is the Euclidean arc element,

(106) $$\int_{B_r(x)} F(u,v) \, du \, dv = 0,$$

whence

(107) $$\int_{B_r(0)} (\partial_1 F)(x + u, v) \, du \, dv = 0,$$

where $\partial_1 = \partial/\partial u$. Using the divergence theorem on the vector field $F(x + u, v)\partial/\partial u$ we get from (107) since the unit normal is $\left(\frac{u}{r}, \frac{v}{r}\right)$ and our vector field is horizontal,

$$(108) \qquad \int_{S_r(0)} F(x + u, v)u \, d\omega(u, v) = 0.$$

Now (105) implies trivially

$$(109) \qquad \int_{S_r(0)} F(x + u, v)x \, d\omega(u, v) = 0$$

and these formulas imply

$$\int_{S_r(x)} F(s)s_1 \, d\omega(s) = 0, \qquad s = (s_1, s_2).$$

Iterating the implication (105) $\Rightarrow$ (109) we obtain

$$\int_{S_r(x)} F(s)P(s_1) \, d\omega(s) = 0,$$

where P is an arbitrary polynomial. Since $F(s_1, s_2) = F(s_1, -s_2)$ and since the polynomials $P(s_1)$ form a separating algebra on $\gamma_{x,r}$ we obtain $F \equiv 0$ on $S_r(x)$. Thus $f \equiv 0$ on the strip $0 < y < 1$. This proves the theorem for $X = \mathbf{H}^2$.

In the general case we can take ξ as the horocycle $N \cdot o$ as above and assume that (102) holds. Let θ be the involution of G with fixed points set K. If X_α is a root vector in the Lie algebra of N, that is an eigenvector of $\mathrm{ad}H$ with H in the Lie algebra of $\{a_t, t \in \mathbf{R}\}$ then $[X_\alpha, \theta X_\alpha]$ spans the Lie algebra of A. If $G_\alpha \subset G$ is the analytic subgroup with Lie algebra $\mathbf{R}H + \mathbf{R}X_\alpha + \mathbf{R}\theta X_\alpha$ then $G_\alpha \cdot o$ is a totally geodesic submanifold of X isometric to $\mathbf{H}^2$ (up to a factor). Also the horocycle $\exp(tX_\alpha) \cdot o$ is the intersection $(G_\alpha \cdot o) \cap (N \cdot o)$. By the first part of the proof, $f(a_{-t} \cdot o) = 0$ for $t > 0$. Using this on the translated function $x \to f(n \cdot x)$, n being a fixed element of N, we get $f(na_{-t} \cdot o)$ for $t > 0$ and this concludes the proof.

We shall now write out explicitly the horocycle Radon transform (108) in Chapter II for the hyperbolic space $\mathbf{H}^n$. In §1 we wrote down two models of $\mathbf{H}^n$, the unit ball model (1) §1 and the quadric model Q^+. Now it will be useful to use a third one, namely the upper half space

$$(x_1, \ldots, x_n) \in \mathbf{R}^n, \qquad x_n > 0$$

with the metric

(110)
$$ds^2 = \frac{dx_1^2 + \cdots + dx_n^2}{x_n^2},$$

which is isometric to Q_-^+ under the map

$$(x_1, \ldots, x_n) = (y_{n+1} - y_n)^{-1}(y_1, \ldots, y_n, 1), \qquad y \in Q_-^+,$$

(cf. e.g. [**GGA**], Ch. I, Ex. C). In the metric (110) the geodesics are the circular arcs perpendicular to the plane $x_n = 0$; among these are the half lines perpendicular to $x_n = 0$. The horocycles perpendicular to these last geodesics are the planes $x_n = $ const. The other horocycles are the Euclidean $(n-1)$-spheres tangential to the boundary.

Let $\xi \subset \mathbf{H}^n$ be a horocycle in the half space model. It is a Euclidean sphere with center (x', r) (where $x' = (x_1, \ldots x_{n-1})$) and radius r. We consider the intersection of ξ with the $x_{n-1}x_n$ plane. It is the circle γ : $x_{n-1} = r \sin \theta$, $x_n = r(1 - \cos \theta)$ where θ is the angle measured from the point of contact of ξ with $x_n = 0$. The plane $x_n = r(1-\cos\theta)$ intersects ξ in an $(n-2)$-sphere whose points are $x' + r \sin \theta \omega'$ where $\omega' = (\omega_1, \ldots, \omega_{n-1})$ is a point on the unit sphere S_{n-2} in $\mathbf{R}^{n-1}$. Let $d\omega'$ be the surface element on S_{n-2}.

Proposition 4.2. *Let f be exponentially decreasing on $\mathbf{H}^n$. Then in the notation above,*
(111)

$$\widehat{f}(\xi) = \int_0^\pi \int_{S_{n-2}} f(x' + r\sin\theta\omega', r(1 - \cos\theta))\, d\omega' \left(\frac{\sin\theta}{1 - \cos\theta}\right)^{n-2} \frac{d\theta}{1 - \cos\theta}.$$

Proof. Since N acts on $\mathbf{H}^n$ by translation and since $f \to \widehat{f}$ commutes with the N-action we may assume $x' = 0$.

The plane $\pi_\theta : x_n = r(1 - \cos\theta)$ has the non-Euclidean metric

$$\frac{dx_1^2 + \cdots + dx_{n-1}^2}{r^2(1 - \cos\theta)^2}$$

and the intersection $\pi_\theta \cap \xi$ is an $(n-2)$-sphere with induced metric

$$\frac{r^2 \sin^2\theta(d\omega')^2}{r^2(1 - \cos\theta)^2},$$

where $(d\omega')^2$ is the metric on the $(n-2)$-dimensional unit sphere in $\mathbf{R}^{n-1}$. The non-Euclidean volume element on $\xi \cap \pi_\theta$ is thus

$$\left(\frac{\sin\theta}{1 - \cos\theta}\right)^{n-2} d\omega'.$$

The non-Euclidean arc element on γ is by (110) equal to $d\theta/(1 - \cos\theta)$. Putting these facts together (111) follows by integrating over ξ by slices $\xi \cap \pi_\theta$.

Theorem 4.3. *Let $\xi_0 \subset \mathbf{H}^n$ be a fixed horocycle. Let f be exponentially decreasing and assume*

$$\widehat{f}(\xi) = 0$$

for each horocycle ξ lying outside ξ_0. Then

$$f(x) = 0 \quad \text{for} \quad x \quad \text{outside} \quad \xi_0.$$

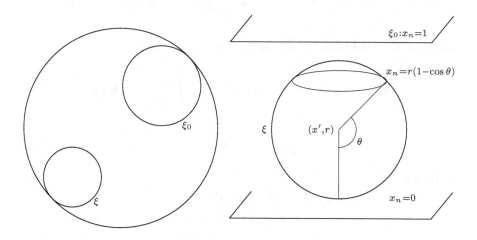

FIGURE III.7.   FIGURE III.8.
Ball Model   Half-space Model

Proof. In the half space model we may take ξ_0 as the plane $x_n = 1$. Assuming $\widehat{f}(\xi) = 0$ we take the Fourier transform in the x' variable of the right hand side of (111), in other words integrate it against $e^{-i\langle x', \eta' \rangle}$ where $\eta' \in \mathbf{R}^{n-1}$.

Then

$$\int_0^\pi \int_{S_{n-2}} \widetilde{f}(\eta', r(1 - \cos\theta))e^{-ir\sin\theta\langle\eta',\omega'\rangle}\, d\omega' \left(\frac{\sin\theta}{1 - \cos\theta}\right)^{n-2} \frac{d\theta}{1 - \cos\theta} = 0.$$

By rotational invariance the ω' integral only depends on the norm $|\eta'|r\sin\theta$ so we write

$$J(r\sin\theta|\eta'|) = \int_{S_{n-2}} e^{-ir\sin\theta\langle\eta',\omega'\rangle}\,d\omega'.$$

and thus

$$\int_0^\pi \tilde{f}(\eta', r(1-\cos\theta))J(r\sin\theta|\eta'|)\left(\frac{\sin\theta}{1-\cos\theta}\right)^{n-2}\frac{d\theta}{1-\cos\theta} = 0.$$

Here we substitute $u = r(1-\cos\theta)$ and obtain

$$(112)\qquad \int_0^{2r} \tilde{f}(\eta', u)J((2ur-u^2)^{1/2}|\eta'|)\frac{r}{u^{n-1}}(2ur-u^2)^{\frac{1}{2}(n-3)}\,du = 0.$$

Since the distance from the origin $(0,1)$ to (x', u) satisfies

$$d((0,1),(x',u)) \geq d((0,1),(0,u)) = \int_u^1 \frac{dx_n}{x_n} = -\log u$$

so

$$e^{d((0,1),(x',u))} \geq \frac{1}{u},$$

and since

$$\tilde{f}(\eta', u) = \int_{\mathbf{R}^{n-1}} f(x', u)e^{-i\langle x',\eta'\rangle}\,dx',$$

we see from the exponential decrease of f, that the function $u \to \tilde{f}(\eta', u)/u^{n-1}$ is continuous down to $u = 0$.

The case $n = 3$. In this simplest case (112) takes the form

$$(113)\qquad \int_0^{2r} \tilde{f}(\eta', u)u^{-2}J((2ur-u^2)^{1/2}|\eta'|)\,du = 0.$$

We need here standard result for Volterra integral equation (cf. Yosida [1960]).

Proposition 4.4. *Let $a < b$ and $f \in C[a,b]$ and $K(s,t)$ of class C^1 on $[a,b] \times [a,b]$. Then the integral equation*

$$(114)\qquad \varphi(s) + \int_a^s K(s,t)\varphi(t)\,dt = f(s)$$

has a unique continuous solution $\varphi(t)$. In particular, if $f \equiv 0$ then $\varphi \equiv 0$.

Corollary 4.5. *Assume $K(s, s) \neq 0$ for $s \in [a, b]$. Then the equation*

$$(115) \qquad \int_a^s K(s, t)\psi(t)\, dt = 0 \quad implies \ \psi \equiv 0 \,.$$

This follows from Prop. 4.4 by differentiation. Using Cor. 4.5 on (113) we deduce $\tilde{f}(\eta', u) = 0$ for $u \leq 2r$ with $2r \leq 1$ proving Theorem 4.3 for $n = 3$.

The case $n = 2$. Here (112) leads to the generalized Abel integral equation $(0 < \alpha < 1)$.

$$(116) \qquad \int_a^s \frac{G(s, t)}{(s - t)^\alpha} \varphi(t)\, dt = f(s) \,.$$

Theorem 4.6. *With f continuous, G of class C^1 and $G(s, s) \neq 0$ for all $s \in [a, b]$, equation (116) has a unique continuous solution φ. In particular, $f \equiv 0 \Rightarrow \varphi \equiv 0$.*

This is proved by integrating the equation against $1/(x - s)^{1-\alpha}$ whereby the statement is reduced to Cor. 4.5 (cf. Yosida, *loc.cit.* and Ch. I, §2 in this book).

This proves Theorem 4.3 for $n = 2$.

The general case. Here the parity of n makes a difference. For n odd we just use the following lemma.

Lemma 4.7. *Assume $\varphi = C^1([a, b])$ and that $K(s, t)$ has all derivatives with respect to s up to order $m - 2$ equal to 0 on the diagonal (s, s). Assume the $(m - 1)^{th}$ order derivative is nowhere 0 on the diagonal. Then (115) still holds.*

In fact, by repeated differentiation of (115) one can show that (114) holds with a kernel

$$\frac{K^{(m)}(s, t)}{\{K_s^{(m-1)}(s, t)\}_{t=s}}$$

and $f \equiv 0$.

This lemma proves Theorem 4.3 for n odd. For n even we write (112) in the general form

$$(117) \qquad \int_0^s F(u)H((su - u^2)^{1/2})(su - u^2)^{\frac{1}{2}(n-3)}\, du = 0 \quad n \ \text{even} \ \geq 2 \,,$$

where $H(0) \neq 0$.

Theorem 4.8. *Assume $F \in C([0,1])$ satisfies (117) for $0 \leq s \leq 1$ and $H \in C^\infty$ arbitrary with $H(0) \neq 0$. Then $F \equiv 0$ on $[0,1]$.*

Proof. We proceed by induction on n, the case $n = 2$ being covered by Theorem 4.6. We assume the theorem holds for n and any function H satisfying $H(0) \neq 0$. We consider (117) with n replaced by $n + 2$ and take d/ds. The result is with $H_1(x) = H'(x)x + (n-1)H(x)$,

$$\int_0^s F(u)uH_1((su - u^2)^{\frac{1}{2}})(su - u^2)^{\frac{1}{2}(n-3)}\,du = 0.$$

Since $H_1(0) \neq 0$ we conclude $F \equiv 0$ by induction. This finishes the proof of Theorem 4.3.

Remark 4.9. *Theorem 4.8 can be viewed as a variation of the uniqueness part for the Abel integral equation (116). The method would also solve the equation (117) with a function on the right hand side.*

There is a counterpart to Theorem 4.1 for the horocycle transform relative to balls (X still of rank one). We say a ball B in X is **enclosed** by a horocycle ξ if B is contained in the interior of ξ. We say B is **exterior** to ξ if it is disjoint from the interior of ξ. Consider now the horocycle transform

$$\widehat{f}(\xi) = \int_\xi f(x)\,dm(x)$$

dm being the measure on ξ induced by the Riemannian structure of X.

Theorem 4.10. *Let f be exponentially decreasing. The following conditions are equivalent for any ball $B \subset X$.*

(i) $\widehat{f}(\xi) = 0$ whenever B is enclosed by ξ.

(ii) $\widehat{f}(\xi) = 0$ whenever B is exterior to ξ.

(iii) $f(x) = 0$ for $x \notin B$.

The proof is omitted. It is more technical (see Helgason [2005] or [1973], which also contains a higher rank version) although for $\mathbf{H}^3$ an elementary proof can be given (Lax–Phillips [1979], Theorem 3.13) and also for $\mathbf{H}^2$ (Helgason [1983a], Theorem 4.1).

Exercises and Further Results

1. Radon's Method for Lines in $\mathbf{R}^2$.

Consider for f in $\mathcal{S}(\mathbf{R}^2)$ the integral

(i)
$$\int\limits_{|x|\geq q} f(x)(|x|^2 - q^2)^{-\frac{1}{2}}\, dx$$

and calculate it in two ways: Let ω be a unit vector and $\omega^\perp$ the vector ω rotated $+\frac{\pi}{2}$. The exterior $|x| \geq q$ is swept out by half-tangents to $|x| = q$ in two ways, namely

$$x = q\omega + s\omega^\perp, \quad s \in \mathbf{R}^+, \quad x = q\omega + s\omega^\perp, \quad s \in -\mathbf{R}^+.$$

With $\omega = (\cos\varphi, \sin\varphi)$ and $\partial(x_1, x_2)/\partial(s, \varphi) = s$ the integral (1) equals

$$\int\limits_0^{2\pi} d\varphi \int\limits_0^\infty f(q\omega + s\omega^\perp)\, ds \quad \text{and} \quad \int\limits_0^{2\pi} d\varphi \int\limits_{-\infty}^0 f(q\omega + s\omega^\perp)\, ds$$

as well as their average

$$\frac{1}{2}\int\limits_0^{2\pi} \widehat{f}(\omega, q)\, d\varphi = \pi(\widehat{f})^\vee_q(0),$$

in the notation (6)§1. But in polar coordinates (i) evaluates to

$$2\pi \int\limits_q^\infty (M^r f)(0) r (r^2 - q^2)^{-\frac{1}{2}}\, dr .$$

This yields equation (8) for $k = 1$, $\Omega_k = 2$, which proves Theorem 1.4(iii) for $n = 2$.

Remarkably, the identical formula holds for all dimensions n (Theorem 1.4).

2. Alternative Inversion on $\mathbf{S}^n$. (Rubin [1998b]).

For n even, $xy =$ inner product,

$$f = \frac{1}{2(2\pi)^n} P(L) \int\limits_{S^n} \widehat{f}(y) \log\frac{1}{|xy|}\, dy + \frac{(n-1)!}{2(2\pi)^n} \int\limits_{S^n} \widehat{f}(y)\, dy,$$

where

$$P(L) = \prod\limits_{k=1}^{n/2} (-L + 2(k-1)(n+1-2k)).$$

3.   Spheres Through 0. (Cormack–Quinto [1980], Quinto [1982]).

Given $f \in C(\mathbf{R}^n)$ let

$$(Sf)(y) = (M^{|y|/2}f)(y/2),$$

the mean-value of f over a sphere with center $y/2$ passing through 0.

(i)   For $\omega \in \mathbf{S}^{n-1}$, $p \in \mathbf{R}$ put

$$\varphi(\omega, p) = f(p\omega)|p|^{n-2}.$$

Then

$$(Sf)(y) = \left(\frac{2}{|y|}\right)^{n-2} \check{\varphi}(y).$$

(ii)   Let E denote the set of restrictions of entire functions on $\mathbf{C}^n$ to $\mathbf{R}^n$. Then $f \to Sf$ is a bijection of E onto E.

(iii)   The map $S : L^2(\mathbf{R}^n) \to L^2(\mathbf{R}^n)$ is injective.

(iv)   If B is a ball with center O then S maps $L^2(B)$ into itself and the kernel is the closure of the span of the functions

$$f(x) = |x|^{2-n+k}Y_\ell(x/|x|),$$

where $(n-4)/2 < k < \ell$ and $\ell - k$ even and Y_ℓ is a homogeneous spherical harmonic of degree ℓ.

(v)   Let

$$(mf)(x) = \frac{2^{n-1}}{\Omega_n}|x|^{2-2n}f(x/|x|^2).$$

Then if $x \neq 0$, $f \in C(\mathbf{R}^n)$,

$$(Sf)(x) = |x|^{1-n}(mf)\hat{}(x/|x|, 1/|x|).$$

(vi)   If $f \in C^\infty(\mathbf{R}^n)$ and $(Sf)(y) = 0$ for $|y| \leq \mathcal{A}$ then $f(y) = 0$ for $|y| \leq \mathcal{A}$.

Bibliographical Notes

As mentioned earlier, it was shown by Funk [1916] that a function f on the two-sphere, symmetric with respect to the center, can be determined by the integrals of f over the great circles. Minkowski's Theorem 1.20 played a role in this involving Lemma 1.19.

The Radon transform on Euclidean, hyperbolic, and elliptic spaces corresponding to k-dimensional totally geodesic submanifolds was defined in the author's paper [1959]. Here and in [1990] and [2006] are proved the inversion formulas in Theorems 1.4,1.5, 1.9, 1.11, 1.12, 1.22, and 1.23. Semyanisty [1961] then gave a treatment for k odd by analytic continuation of the Riesz potential and Rubin [1998b] gave a formula for k odd in terms of the logarithmic potential. The alternative version for $\mathbf{H}^2$ was obtained by Berenstein and Casadio Tarabusi in their paper [1991] which also deals with the case of $\mathbf{H}^k$ in $\mathbf{H}^n$ (where the regularization is more complex). Still another interesting variation of Theorem 1.11 (for $k = 1, n = 2$) is given by Lissianoi and Ponomarev [1997]. By calculating the dual transform $\check{\varphi}_p(z)$ they derive from formula (30) in Chapter II an inversion formula which has a formal analogy to (47) in Chapter II. The underlying reason may be that to each geodesic γ in $\mathbf{H}^2$ one can associate a pair of horocycles tangential to $|z| = 1$ at the endpoints of γ having the same distance from o as γ. For another inversion for $\mathbf{H}^n$ see Gindikin [1995].

The support theorem (Theorem 1.6) was proved by the author ([1964], [1980b]) and its consequence, Cor. 1.7, pointed out in [1980d]. Interesting generalizations are contained in Boman [1991], Boman and Quinto [1987], [1993]. For the case of $\mathbf{S}^{n-1}$ see Quinto [1983] and in the stronger form of Theorem 1.25, Kurusa [1994]. The variation (85) of the Funk transform was considered by Abouelaz and Daher [1993] at least for K-invariant functions. The theory of the Radon transform for antipodal manifolds in compact two-point homogeneous spaces (Theorem 2.2) is from Helgason [1965a]. A substantial simplification of the computation was given by Rouvière [2001]. A simplified proof of Cor. 2.4 is given by Klein, Thorbergsson and Verhóczki [2008]. R. Michel has in [1972] and [1973] used Theorem 2.2 in showing that on the compact rank-one spaces $X \neq \mathbf{S}^n$ no deformation of the metric is possible with the geodesics remaining closed and of same length. For $\mathbf{S}^2$ such a deformation is possible (Guillemin [1976]). See also Besse [1978], Goldschmidt [1990], Gasqui and Goldschmidt [2004], Estezet [1988].

The support theorems (Theorems 4.1 and 4.9) relative to the X-ray transform and the horocycle transform are from Helgason [1987] and [2005]. On the other hand, Theorem 4.3 is an extension of Theorem 3.14 in Lax–Phillips [1979] where the case $n = 3$ is handled. Only Corollary 4.5 is needed for this case. A local version is proved in Lax–Philips [1981].

The range question has been settled in different ways by Bernstein-Casadio Tarabusi [1993], same authors with Kurusa [1997], Kakehi [1993], and Ishikawa [1997].

CHAPTER IV

THE X-RAY TRANSFORM ON A SYMMETRIC SPACE

The X-ray transform which we studied for $\mathbf{R}^n$ of course makes sense for an arbitrary complete Riemannian manifold X. For a continuous function f on X, the X-ray transform $\widehat{f}$ is defined by

$$\widehat{f}(\gamma) = \int_\gamma f(x)\, ds(x)\,,$$

γ being a complete geodesic and ds the element of arc length. One can then pose the question of injectivity (keeping in mind the negative result for $\mathbf{S}^n$) and, if the geodesics are reasonably behaved, hope for an inversion formula. In this chapter we study these questions for a symmetric space where the pattern of geodesics is well understood, even globally.

For a general Riemannian manifold little seems known about the X-ray transform. For a symmetric space $X \neq \mathbf{S}^n$ we shall see that the X-ray transform is injective and has an inversion formula. First we deal with the compact case.

§1 Compact Symmetric Spaces. Injectivity and Local Inversion. Support Theorem

First we prove the injectivity of the X-ray transform. For $X = \mathbf{S}^n$ see Ch. III, Cor. 2.4.

Theorem 1.1. *Let $X = U/K$ be a compact symmetric space $(X \neq \mathbf{S}^n)$ and $f \to \widehat{f}$ the X-ray transform where $f \in \mathcal{E}(X)$ and*

$$\widehat{f}(\xi) = \int_\xi f(x)\, ds(x)\,,$$

ξ being a closed geodesic. Then $\widehat{f} \equiv 0$ implies $f \equiv 0$.

Proof. For X two-point homogeneous (i.e., of rank one) we know this from Cor. 2.4 in Ch. III. If X has rank ℓ higher than one each $x \in X$ lies on a totally geodesic torus T of dimension 2 (even ℓ).

We view T as $\mathbf{R}^2/\mathbf{Z}^2$ and recall that the geodesics are viewed without orientation. Separating out the closed geodesics which correspond to horizontal and vertical lines the closed geodesics are parametrized as follows:

S. Helgason, *Integral Geometry and Radon Transforms*,
DOI 10.1007/978-1-4419-6055-9_4, © Springer Science+Business Media, LLC 2010

$$\xi = \xi(x_0, y_0, p, q) : (x(t), y(t)) = (x_0 + pt, y_0 + qt) \quad 0 \le t \le 1$$
where p and q are relatively prime integers and $p > 0$, $q \not\le 0$.

$$\xi' = \xi(x_0, y_0, 1, 0) : (x(t), y(t)) = (x_0 + t, y_0), \ 0 \le t \le 1.$$

$$\xi'' = \xi(x_0, y_0, 0, 1) : (x(t), y(t)) = (x_0, y_0 + t), \ 0 \le t \le 1.$$

Then

$$\widehat{f}(\xi) = (p^2 + q^2)^{\frac{1}{2}} \int_0^1 f(x_0 + tp, y_0 + tq)\, dt$$

and this holds also for $\widehat{f}(\xi')$ and $\widehat{f}(\xi'')$ by putting $(p, q) = (1, 0)$ and $(p, q) = (0, 1)$, respectively.

We can expand f into an absolutely convergent Fourier series,

$$f(x, y) = \sum_{j \neq 0, k \neq 0} a_{jk} e_{j,k}(x, y) + \sum_k a_{0,k} e_{0,k}(x, y) + \sum_{j \neq 0} a_{j,0} e_{j,0}(x, y),$$

where $e_{j,k}(x, y) = e^{2\pi i(jx + ky)}$. Then

$$\widehat{e_{j,k}}(\xi) = (p^2 + q^2)^{1/2} e_{j,k}(x_o, y_o) \int_0^1 e^{2\pi i(jp + kq)t}\, dt,$$

which vanishes unless $jp + kq = 0$. Given $(j \neq 0, k \neq 0)$ there is just one pair p, q, $p > 0$, $q \neq 0$, $(p, q) = 1$ for which $jp + kq = 0$. Put $c_{j,k}(p, q) = (p^2 + q^2)^{1/2}$ for this pair p, q. We thus have for $j \neq 0, k \neq 0$

$$\widehat{e}_{j,k}(\xi) = c_{j,k}(p, q) \ \delta_{jp+kq,0} \ e_{j,k}(x_0, y_0),$$
$$\widehat{e}_{j,k}(\xi') = \delta_{j,0} \ e_{j,k}(x_0, y_0), \ \widehat{e}_{j,k}(\xi'') = \delta_{k,0} \ e_{j,k}(x_0, y_0)$$

and

$$\widehat{e}_{0,k}(\xi) = (p^2 + q^2)^{\frac{1}{2}} \delta_{k,0} \ e_{0,k}(x_0, y_0),$$
$$\widehat{e}_{0,k}(\xi') = e_{0,k}(x_0, y_0), \ \widehat{e}_{0,k}(\xi'') = \delta_{k,0} \ e_{0,k}(x_0, y_0),$$
$$\widehat{e}_{j,0}(\xi) = (p^2 + q^2)^{\frac{1}{2}} \delta_{j,0} \ e_{j,0}(x_0, y_0),$$
$$\widehat{e}_{j,0}(\xi') = \delta_{j,0} \ e_{j,0}(x_0, y_0), \ \widehat{e}_{j,0}(\xi'') = e_{j,0}(x_0.y_0).$$

Assume now f satisfies

$$\widehat{f}(\xi) = 0, \ \widehat{f}(\xi') = 0, \ \widehat{f}(\xi'') = 0$$

for all ξ, ξ' and ξ''. Then by the expansion for f,

$$\sum_{j\neq 0, k\neq 0} a_{j,k} c_{j,k}(p,q) \delta_{jp+kq,0} e_{j,k}(x_0, y_0) + a_{0,0}(p^2 + q^2)^{\frac{1}{2}} = 0,$$

$$\sum_{j\neq 0, k\neq 0} a_{j,k} \delta_{j,0} e_{j,k}(x_0, y_0) + \sum_k a_{0,k} e_{0,k}(x_0, y_0) + 0 = 0,$$

$$\sum_{j\neq 0, k\neq 0} a_{j,k} \delta_{k,0} e_{j,k}(x_0, y_0) + \sum_k a_{0,k} \delta_{k,0} e_{0,k}(x_0, y_0) + \sum_{j\neq 0} a_{j,0} e_{j,0}(x_0, y_0) = 0.$$

Viewing these as Fourier series in (x_0, y_0) we deduce for fixed $p > 0$, $q \neq 0$, $(p, q) = 1$,

$$a_{j,k} c_{j,k}(p,q) \delta_{jp+kq,0} = 0, \ j \neq 0, k \neq 0, a_{0,0} = 0$$
$$a_{0,k} = 0, \ a_{j,0} = 0 \ (j \neq 0).$$

Now choosing p and q such that $jp + kq = 0$ we deduce that all a_{jk} must vanish so $f \equiv 0$.

We shall now deal with the inversion problem, using and generalizing the results from the sphere in Ch. III. Let U/K be an irreducible compact simply connected symmetric space, U being a compact semisimple Lie group with K as the fixed point group of an involutive automorphism σ of U. Let $\mathfrak{u}$ denote the Lie algebra of U, $\mathfrak{u} = \mathfrak{k} + \mathfrak{p}$ its decompositions into the eigenspaces of $d\sigma$. Let $\mathfrak{a} \subset \mathfrak{p}$ be a maximal abelian subspace of $\mathfrak{p}$ and $\mathfrak{p} = \bigoplus_\alpha \mathfrak{p}_\alpha$ the decomposition of $\mathfrak{p}$ into the joint eigenspaces of $(\mathrm{ad}H)^2$, $(H \in \mathfrak{a})$

$$\mathfrak{p}_\alpha = \left\{ Z \in \mathfrak{p} : (\mathrm{ad}H)^2 Z = \alpha(H)^2 Z \quad H \in \mathfrak{a} \right\}.$$

The linear forms $\alpha \neq 0$ are the **restricted roots**. These have a natural ordering. Let $\bar{\delta}$ denote the highest, let $\mathfrak{a}_\delta$ be the one-dimensional orthocomplement of the null space of $\bar{\delta}$ in $\mathfrak{a}$. Let $m(\bar{\delta})$ denote the multiplicity of $\bar{\delta}$, that is the dimension of $\mathfrak{p}_\delta$. Our reduction of the X-ray transform to the sphere is based on the following conjugacy result. (Helgason [1966] or [DS], Ch. VII). The maximal sectional curvature on X equals $|\bar{\delta}|^2$ in the metric defined by the negative of the Killing form. We normalize the metric such that the maximal curvature is 1. In addition to Theorem 1.2 we use Lemma 1.3 and some further results from the above sources.

Theorem 1.2 (The Transitivity of Minimal Geodesics).

(i)   *The shortest closed geodesics in X have length 2π and they are permuted transitively by U.*

(ii)   *X has totally geodesic spheres of curvature 1. Their maximum dimension is $1 + m(\bar{\delta})$. All such spheres $\mathbf{S}^{1+m(\bar{\delta})}$ are conjugate under U.*

Let $\mathfrak{a}_\delta$ and $\mathfrak{p}_\delta$ be as above. Then the submanifold $X_\delta = \mathrm{Exp}(\mathfrak{a}_\delta + \mathfrak{p}_\delta)$ is a sphere, totally geodesic in X of dimension $1 + m(\bar{\delta})$ and curvature 1. Let $A(\bar{\delta})$ be the first point on the line $\mathfrak{a}_\delta$ such that $\mathrm{Exp}A(\bar{\delta})$ is the antipodal point to o in X_δ. By [DS], VII, §11, the subgroup $S \subset K$ fixing both o and $\mathrm{Exp}A(\bar{\delta})$ is quite large. In fact, we have the following result. For the definition of Ad see Ch. VIII, §1.

Lemma 1.3. *The restriction of* $\mathrm{Ad}_U(S)$ *to the tangent space* $(X_\delta)_o$ *contains* $\mathbf{SO}((X_\delta)_o)$.

Definition. For $x \in X$ the *midpoint locus* A_x associated to x is the set of midpoints $m(\xi)$ of all the closed minimal geodesics ξ starting at x. Let $e_1(\xi), e_2(\xi)$ denote the midpoints of the arcs of ξ which join x to $m(\xi)$. Let E_x denote the set of these $e(\xi)$ as ξ varies. We call E_x the *equator* associated to x.

Lemma 1.4. A_o *and* E_o *are* K-*orbits and* A_o *is totally geodesic in* X.

Proof. That A_o is a K-orbit is clear from Theorem 1.2 (i). If $x \in A_o$ the geodesic symmetry s_x fixes o and thus permutes the closed minimal geodesics through o. Thus s_x maps A_o into itself so by [DS], VII, Cor. 10.5 A_0 is totally geodesic in X. Let e_1 and e_2 be two points in E_o lying on minimal geodesics γ_1, γ_2 starting at o. To show them conjugate under K we may assume by Theorem 1.2 that both γ_1 and γ_2 have midpoints equal to $\mathrm{Exp}A_{\bar{\delta}}$. But then γ_1 and γ_2 are K-conjugate, hence S-conjugate. Thus we can assume e_1 and e_2 lie on the same minimal geodesic through o and $\mathrm{Exp}A(\bar{\delta})$. But then, by Lemma 1.3, they are S-conjugate.

Definition. The **Funk transform** for $X = U/K$ is the map $f \to \widehat{f}$ where,

$$(1) \qquad \widehat{f}(\xi) = \int_\xi f(x)\,dm(x)\,,$$

ξ being a closed geodesic in X of minimal length and dm the arc element on ξ.

Let Ξ denote the set of closed minimal geodesics. By Theorem 1.2 we have the pair of homogeneous spaces

$$(2) \qquad X = U/K\,, \quad \Xi = U/H\,,$$

where the group H is the stabilizer of some specific $\xi_o \in \Xi$. We have then the corresponding dual transform

$$(3) \qquad \check{\varphi}(gK) = \int_K \varphi(gk \cdot \xi_o)\,.$$

We shall now study the transform (1) using the theory in Chapter III for the sphere. Given $f \in \mathcal{E}(X)$ we consider its restriction $f|X_\delta$ to the sphere X_δ. For $0 \le p \le \frac{\pi}{2}$ we fix a geodesic $\xi_p \subset X_\delta$ at a distance p from o. We consider X_δ and Ξ_δ (the space of geodesics in X_δ) as in Ch. II, §4, A. Let f_* denote the Funk transform $(f|X_\delta)\widehat{}$ (as a function on the set of great circles on X_δ) and let φ_p^* denote the corresponding dual transform for the pair X_δ and Ξ_δ. This means that for $x \in X_\delta$, $\varphi_p^*(x)$ is the average of φ over the set of closed geodesics in X_δ which have distance p from x. For $x = o$, the group S permutes these transitively, so

$$(4) \qquad \varphi_p^*(o) = \int_S \varphi(s \cdot \xi_p)\, ds.$$

Note that this is (for given p) independent of the choice of ξ_p. Then Theorem 1.24 in Ch. III implies

$$(5) \qquad \tfrac{1}{2}(f(o) + f(\mathrm{Exp}A(\bar{\delta}))) = \int_{E_o'} f(\omega')\, d\omega' - \frac{1}{2\pi} \int_0^{\frac{\pi}{2}} \frac{d}{dp}((f_*)_p^*(o))\frac{dp}{\sin p},$$

where E_o' is the equator in X_δ corresponding to o and $d\omega'$ is the averaging measure. In view of Lemma 1.3, $E_o' = S \cdot \mathrm{Exp}\left(\tfrac{1}{2}A(\bar{\delta})\right)$. We shall now use (5) on the averaged functions

$$f^\natural(x) = \int_K f(k \cdot x)\, dk, \qquad \varphi^\natural(\xi) = \int_K \varphi(k \cdot \xi)\, dk.$$

Since $A_o = K \cdot \mathrm{Exp}A(\bar{\delta})$ the left hand side of (5) becomes

$$\tfrac{1}{2}\left(f(o) + \int_{A_o} f(\omega)\, d\omega\right),$$

where $d\omega$ stands for the average on A_o. The first term on the right is now replaced by

$$\int_K \int_{E_o'} f(k \cdot \omega')\, d\omega' = \int_K dk \int_S f\left(ks \cdot \mathrm{Exp}(\tfrac{1}{2}A(\bar{\delta}))\right) ds$$

$$= \int_K f\left(k \cdot \mathrm{Exp}(\tfrac{1}{2}A(\bar{\delta}))\right) dk = \int_{E_o} f(\omega)\, d\omega,$$

where $d\omega$ is the average. Recall that E_0 is contained in $S_{\pi/2}(o)$. For the second term on the right in (5) we use (4) so

$$
\begin{aligned}
((f^\natural)_*)^*_p(o) &= \int_S (f^\natural)_*(s \cdot \xi_p)\, ds = \int_S (f^\natural)\widehat{\;}(s \cdot \xi_p)\, ds \\
&= \int_K dk \int_S (\widehat{f})^\natural(s \cdot \xi_p)\, ds = \int_S \left(\int_K \widehat{f}(ks \cdot \xi_p)\, dk \right) ds \\
&= \int_K \widehat{f}(k \cdot \xi_p)\, dk = (\widehat{f})^\vee_p(o),
\end{aligned}
$$

where $\check{\varphi}_p$ denotes the dual transform (3) with $\xi_o = \xi_p$. This proves

$$
(6) \qquad \tfrac{1}{2}\left(f(o) + \int_{A_o} f(\omega)\, d\omega \right) = \int_{E_0} f(\omega)\, d\omega - \frac{1}{2\pi} \int_0^{\frac{\pi}{2}} \frac{d}{dp}((\widehat{f})^\vee_p(o)) \frac{dp}{\sin p}.
$$

Because of Theorem 1.2 and Lemma 1.3 the orbit $\Xi_p = K \cdot \xi_p$ consists of all closed minimal geodesics in X each lying in some totally geodesic sphere $S^{1+m(\delta)}$ through o having distance p from o. Thus

$$
(7) \qquad\qquad (\widehat{f})^\vee_p(o) = \int_{\Xi_p} \widehat{f}(\xi)\, d\omega^o_p,
$$

where ω^o_p is the normalized K-invariant measure on this orbit.

The set $\Xi_p(x)$ is defined similarly for each $x \in X$. If $u \in U$ is such that $u \cdot o = x$ then $\Xi_p(x) = u \cdot \Xi_p$, $A_x = u \cdot A_o$ and $u \cdot E_o = E_x$. Thus (6) implies the following result.

Theorem 1.5. *The Funk transform $f \to \widehat{f}$ on the space $X = U/K$ satisfies the following identity*

$$
(8) \qquad \tfrac{1}{2}\left(f(x) + \int_{A_x} f(\omega)\, d\omega \right)
$$

$$
= \int_{E_x} f(\omega)\, d\omega - \frac{1}{2\pi} \int_0^{\frac{\pi}{2}} \frac{d}{dp}\left(\int_{\Xi_p(x)} \widehat{f}(\xi)\, d\omega_p(\xi) \right) \frac{dp}{\sin p}.
$$

This is not quite an inversion formula since it includes the integrals over A_x and E_x. However, if we restrict the support of f in advance we can invert $f \to \widehat{f}$.

Corollary 1.6. *Let $f \in \mathcal{D}(X)$ have support in $B_{\pi/2}(o)$ ($\frac{\pi}{2}$-cap around North Pole). Then for $x \in B_{\frac{\pi}{2}}(o)$ we have*

$$f(x) = 2 \int\limits_{E_x} f(\omega)\, d\omega - \frac{1}{\pi} \int\limits_0^{\frac{\pi}{2}} \frac{d}{dp} \Big(\int\limits_{\Xi_p(x)} \widehat{f}(\xi)\, d\omega_p(\xi) \Big) \frac{dp}{\sin p}.$$

In fact, we claim

(9) $$A_x \cap B_{\frac{\pi}{2}}(o) = \emptyset \quad \text{if } x \in B_{\frac{\pi}{2}}(o)$$

so the left hand side of (5) is just $\frac{1}{2}f(x)$. To see (9) let $g \in U$ be such that $g \cdot o = x$. Then $gA_o = A_x$ and

$$\begin{aligned} d(o, gk \cdot \mathrm{Exp}(\bar{\delta})) &= d(g^{-1} \cdot o, k \cdot \mathrm{Exp}A(\bar{\delta})) \\ &\geq d(o, k \cdot \mathrm{Exp}A(\bar{\delta})) - d(o, g^{-1} \cdot o) \\ &\geq \pi - \frac{\pi}{2} = \frac{\pi}{2} \end{aligned}$$

so (9) follows.

Shrinking the support further we get a genuine inversion formula.

Corollary 1.7. *Let $f \in \mathcal{D}(X)$ have support in $B_{\pi/4}(o)$. Then if $x \in B_{\pi/4}(o)$,*

$$f(x) = -\frac{1}{\pi} \int\limits_0^{\frac{\pi}{2}} \frac{d}{dp} \Big(\int\limits_{\Xi_p(x)} \widehat{f}(\xi)\, d\omega_p(x) \Big) \frac{dp}{\sin p}.$$

Here we must show

$$B_{\pi/4}(o) \cap E_x = \emptyset \quad \text{if } x \in B_{\pi/4}(o)$$

or better still that

(10) $$B_{\pi/4}(o) \cap S_{\pi/2}(x) = \emptyset \quad \text{if } x \in B_{\pi/4}(o).$$

However, $z \in S_{\pi/2}(x)$ implies

$$d(o, z) \geq d(x, z) - d(o, x) \geq \frac{\pi}{2} - \frac{\pi}{4} = \frac{\pi}{4}$$

proving (10).

While Corollary 1.7 only gives a restricted inversion formula, injectivity holds in general.

Theorem 1.8. *The Funk transform is injective if X is not a sphere.*

Proof. (Klein, Thorbergsson and Verhóczki [2009].)

Assume $f \in \mathcal{E}(X)$ real-valued such that $\widehat{f} \equiv 0$ but $f \not\equiv 0$. Then f is not constant. By Sard's theorem f has a *regular value* b in its range. This means that $df_x \neq 0$ for all $x \in X$ for which $f(x) = b$; it implies that $N = f^{-1}(b)$ is a submanifold of X of codimension 1. We may assume $o \in N$.

For each of the maximally curved spheres $S = S^{1+m(\delta)}$ in Theorem 1.2 the great circles are minimal length geodesics in X. Thus $(f|S)^{\wedge} = 0$ so by Funk's theorem f is odd on each such S. Let $q \in A_0$ and fix a sphere S as above through o and q. Since $f(o) = b$ we have $f(q) = -b$. The midpoint locus A_q goes through o and $kA_q = A_{kq}$ $(k \in K)$. Now kq is the antipodal point to o in the sphere kS so $f(kq) = -f(o) = -b$. On the other hand, each point in A_{kq} is antipodal to kq in some other sphere S^1 so on A_{kq} f has value $-f(kq) = b$. Hence $kA_q = A_{kq} \subset N$ so for the tangent spaces at o we have $Ad(k)(A_q)_o \subset N_o$. Thus the span of the spaces $Ad(k)(A_q)_o$ $(k \in K)$ is a proper non zero K-invariant subspace of X_o, contradicting the irreducibility of X.

We shall now prove a support theorem for X of rank one. As suggested by Ch. III, Theorem 1.26 we assume rapid decay towards the equator E_o.

Theorem 1.9 (Support theorem). *Suppose X has rank one. Let $0 < \delta < \frac{\pi}{2}$. Suppose $f \in \mathcal{E}(B_{\frac{\pi}{2}}(o))$ satisfies*

(i) $\widehat{f}(\xi) = 0$ *for* $d(o, \xi) > \delta$.

(ii) *For each* $m > 0$

$$f(x) \cos d(o, x)^{-m} \text{ is bounded.}$$

Then

(11)
$$f(x) = 0 \quad \text{for} \quad d(o, x) > \delta.$$

Proof. The restriction $f|X_\delta$ can by the decay assumption (ii) be extended to a symmetric function on the sphere X_δ. By the support theorem for the sphere, (11) holds for all $x \in X_\delta$ with $d(o, x) > \delta$. By the K-invariance of our assumptions (11) holds for all $x \in X_\delta$ $(k \in K)$. Since each $x \in X$ lies on a closed geodesic through o, $X = \bigcup_{k \in K} k \cdot X_\delta$ so the result follows.

§2 Noncompact Symmetric Spaces. Global Inversion and General Support Theorem

We pass now to the X-ray transform for a symmetric space $X = G/K$ of the noncompact type (G connected semisimple with finite center, K maximal compact).

With d denoting the distance in X and $o \in X$ some fixed point we now define two subspaces of $C(X)$. Let

$$F(X) = \{f \in C(X) : \sup_x d(o,x)^k |f(x)| < \infty \text{ for each } k \geq 0\}.$$

$$\mathfrak{F}(X) = \{f \in C(X) : \sup_x e^{kd(0,x)} |f(x)| < \infty \text{ for each } k \geq 0\}.$$

Because of the triangle inequality these spaces do not depend on the choice of o. We can informally refer to $F(X)$ as the space of continuous **rapidly decreasing functions** and to $\mathfrak{F}(X)$ as the space of continuous **exponentially decreasing functions** . We shall now prove the analog of the support theorem (Theorem 2.6, Ch. I, Theorem 1.6, Ch. III) for the X-ray transform on a symmetric space of the noncompact type. This general analog turns out to be a direct corollary of the Euclidean case and the hyperbolic case, already done.

Corollary 2.1 (General Support Theorem). *Let X be a symmetric space of the noncompact type, B any ball in X.*

(i) If a function $f \in \mathfrak{F}(X)$ satisfies

$$(12) \qquad \widehat{f}(\xi) = 0 \text{ whenever } \xi \cap B = \emptyset, \quad \xi \text{ a geodesic,}$$

then

$$(13) \qquad f(x) = 0 \quad \text{for } x \notin B.$$

In particular, the X-ray transform is one-to-one on $\mathfrak{F}(X)$

(ii) If X has rank greater than one statement (i) holds with $\mathfrak{F}(X)$ replaced by $F(X)$.

Proof. Let o be the center of B, r its radius, and let γ be an arbitrary geodesic in X through o.

Assume first X has rank greater than one. By a standard conjugacy theorem for symmetric spaces, γ lies in a 2-dimensional, flat, totally geodesic submanifold of X. Using Theorem 2.6, Ch. I on this Euclidean plane we deduce $f(x) = 0$ if $x \in \gamma, d(o,x) > r$. Since γ is arbitrary (13) follows.

Next suppose X has rank one. Identifying $\mathfrak{p}$ with the tangent space X_o let $\mathfrak{a}$ be the tangent line to γ. We can then consider the eigenspace decomposition (89) in Ch. III. If $\mathfrak{b} \subset \mathfrak{p}_\alpha$ is a line through the origin then $S = \mathrm{Exp}(\mathfrak{a} + \mathfrak{b})$ is a totally geodesic submanifold of X (cf. (iv) in the beginning of §2). Being 2-dimensional and not flat, S is necessarily a hyperbolic space. From Theorem 1.6, Ch. III we therefore conclude $f(x) = 0$ for $x \in \gamma, d(o,x) > r$. Again (13) follows since γ is arbitrary.

We shall now prove an inversion formula for the X-ray transform on $Y = G/K$, somewhat simpler than that of Theorem 1.5 since now there is no midpoint locus A_x and no equator. Since the method parallels that of Theorem 1.5 we shall assume $Y = G/K$ to be irreducible. We also normalize the metric such that the maximal negative curvature is -1. A geodesic in Y which lies in a totally geodesic hyperbolic space of curvature -1 will be called a **flexed geodesic**. The duality between the symmetric spaces $Y = G/K$ and $X = U/K$ (see e.g. [DS], Chapter V) shows that their tangent spaces correspond under multiplication by i inside the joint complexification of $\mathfrak{u}$ and $\mathfrak{g}$ (the Lie algebra of G). This correspondence commutes with the adjoint action of K. Lie triple systems are mapped into Lie triple systems and sectional curvatures are mapped into their negatives. Thus Theorem 1.2 implies the following result.

Theorem 2.2. (i)   The space $Y = G/K$ has hyperbolic totally geodesic submanifolds of curvature -1. Their maximum dimension is $1 + m(\bar{\delta})$ and these spaces $\mathbf{H}^{1+m(\bar{\delta})}$ are all conjugate under G.

(ii)   The flexed geodesics in Y are permuted transitively by G.

We consider now the hyperbolic analog of X_δ, say Y_δ of curvature -1, dimension $1 + m(\bar{\delta})$, passing through $o = eK$. As proved in Ch. III, Theorem 1.12,

$$f(x) = -\frac{1}{\pi} \int\limits_{o}^{\infty} \frac{d}{dp}\left((\hat{f})_p^\vee(x)\right) \frac{dp}{\sinh p}, \quad x \in Y_\delta .$$

For $p \geq 0$ let $\Xi_p(x)$ denote the set of all flexed geodesic ξ in Y, each lying in a totally geodesic $\mathbf{H}^{1+m(\bar{\delta})}$ passing through x with $d(x, \xi) = p$. Let ω_p denote the normalized measure on $\Xi_p(x)$, invariant under the isotropy group at x. The proof of Theorem 1.2 then gives the following result.

Theorem 2.3 (Inversion). Let $f \in C_c^\infty(Y)$. Then

$$f(x) = -\frac{1}{\pi} \int\limits_{o}^{\infty} \frac{d}{dp}\left(\int\limits_{\Xi_p(x)} \hat{f}(\xi) \, d\omega_p(\xi) \right) \frac{dp}{\sinh p} .$$

Another and more elementary proof had been given by Rouvière [2004, 2006]. His proof relied on reduction to $\mathbf{H}^2$ only.

§3 Maximal Tori and Minimal Spheres in Compact Symmetric Spaces

Let $\mathfrak{u}$ be a compact semisimple Lie algebra, θ an involutive automorphism of $\mathfrak{u}$ with fixed point algebra $\mathfrak{k}$. Let U be the simply connected Lie group with Lie algebra $\mathfrak{u}$ and $\mathrm{Int}(\mathfrak{u})$ the adjoint group of u. Then θ extends to an involutive automorphism of U and $\mathrm{Int}(\mathfrak{u})$. We denote these extensions also by θ and let K and K_θ denote the respective fixed point groups under θ. The symmetric space $X_\theta = \mathrm{Int}(\mathfrak{u})/K_\theta$ is called the **adjoint space** of $(\mathfrak{u}, \theta)$ ([DS], p. 327), and is covered by $X = U/K$, this latter space being simply connected since K is automatically connected.

The flat totally geodesic submanifolds of X_θ of maximal dimension are permuted transitively by $\mathrm{Int}(\mathfrak{u})$ according to a classical theorem of Cartan. Let E_θ be one such manifold passing through the origin eK_θ in X_θ and let H_θ be the subgroup of $\mathrm{Int}(\mathfrak{u})$ preserving E_θ. We then have the pairs of homogeneous spaces

$$(14) \qquad X_\theta = \mathrm{Int}(\mathfrak{u})/K_\theta \,, \quad \Xi_\theta = \mathrm{Int}(\mathfrak{u})/H_\theta \,.$$

The corresponding Radon transform $f \to \hat{f}$ from $C(X_\theta)$ to $C(\Xi_\theta)$ amounts to

$$(15) \qquad \widehat{f}(E) = \int_E f(x)\, dm(x)\,, \quad E \in \Xi_\theta \,,$$

E being any flat totally geodesic submanifold of X_θ of maximal dimension and dm the volume element. If X_θ has rank one, E is a geodesic and we are in the situation of Corollary 2.4 in Ch. III. The transform (15) is often called the **flat Radon transform**.

Concerning the injectivity of this Radon transform on X_θ see Grinberg [1992].

The sectional curvatures of the space X lie in an interval $[0, \kappa]$. The space X contains totally geodesic spheres of curvature κ and all such spheres S of maximal dimension are conjugate under U (Theorem 1.2). Fix one such sphere S_0 through the origin eK and let H be the subgroup of U preserving S_0. The we have another double fibration

$$X = U/K \,, \quad \Xi = U/H$$

and the accompanying Radon transform

$$\widehat{f}(S) = \int_S f(x)\, d\sigma(x)\,.$$

$S \in \Xi$ being arbitrary and $d\sigma$ being the volume element on S.

It is proved by Grinberg [1994a] that injectivity holds in many cases, although the general question is not fully settled.

Exercises and Further Results

1.   Some Non-Euclidean Identities.

(i)   Let D be the non-Euclidean disk (Ch. II, §4), γ a geodesic in D, Γ the Euclidean chord joining the endpoints of γ.

Then the distances from the origin o satisfy

$$d(o, \Gamma) = 2d(o, \gamma).$$

Here d corresponds to

$$ds^2 = \frac{|dz|^2}{(1 - |z|^2)^2}.$$

(ii)   Let z, w be any points in D. Let γ be the geodesic joining them and ξ the horocycle joining them. Let $r = d_\gamma(z, w)$ and $\rho = d_\xi(z, w)$ their distances in γ and ξ, respectively. Then (Helgason [1994b], Ch. II, Ex. **C**1),

$$\rho = \sinh r.$$

(iii)   In D let γ_0 be the geodesic from -1 to $+1$ and $z \in D$. Then the distance $d(z, \gamma_0)$ is given by

$$\sinh 2d(z, \gamma_o) = \frac{2|\operatorname{Im} z|}{1 - |z|^2}$$

(Beardon [1983], §7.20).

(iv)   Let ξ_0 denote the horocycle $N \cdot o$ in D and γ a geodesic passing through $+1$ and a point z on ξ_0. Let $t = d(o, \gamma)$ and $r = d(o, z)$. Then

$$\sinh 2t = 2 \sinh r.$$

(*Hint:* Write $z = n_x \cdot o$, and $d(o, \gamma) = d(n_{-x} \cdot o, \gamma_0)$. Then use Part (iii) and $\sinh r = |x|$.)

(v)   In the upper half plane with metric

$$ds^2 = \frac{dx^2 + dy^2}{4y^2}$$

the non-Euclidean circle $S_r(i)$ equals the Euclidean circle $S_{\mathrm{sh}2r}(i\mathrm{ch}2r)$.

(vi)   In the non-Euclidean disk with diameter γ a circular arc $\widetilde{\gamma}$ connecting the endpoints of γ has its points at constant distance from γ.

(vii)   In Figure IV.1 ξ is perpendicular to γ.

$P = i \,\mathrm{th}\, t$

$Q = x(r, t)$

$\xi = $ geodesic

r and t non-Euclidean distances

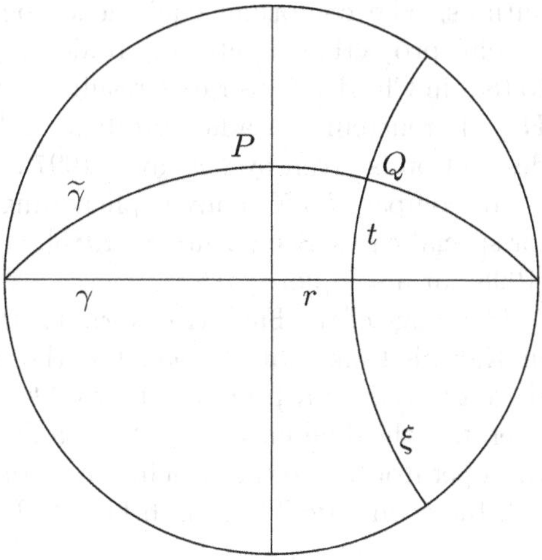

FIGURE IV.1.

In the notation of Fig. IV.1 show that the mapping $(r, t \to x(r, t))$, where

$$x(r, t) = \frac{i \,\mathrm{ch}\, r \,\mathrm{th}\, t + \mathrm{sh}\, r}{i \,\mathrm{sh}\, r \,\mathrm{th}\, t + \mathrm{ch}\, r}$$

is a diffeomorphism of $\mathbf{R} \times \mathbf{R}$ onto D. Show that the arc of $\tilde{\gamma}$ from P to Q has length $(\mathrm{ch}\, 2t)r$ and that

$$\int_D f(x) \, dx = \int_{\mathbf{R}} \int_{\mathbf{R}} f(x(r, t)) \mathrm{ch}\, 2t \, dt \, dr \,,$$

dx denoting the non-Euclidean measure on D. (See [GGA], pp. 77, 555.)

Bibliographical Notes

The inversion formula in Theorem 2.3 was proved by Rouvière [2004, 2006]. He has another equivalent formula without the sinh.

The compact analogs (Theorem 1.5, Cor. 1.6–1.7) were given by the author [2007]. The support theorems Cor. 2.1 and Theorem 1.9 are from Helgason [1980d], [2007]. The injectivity of the X-ray transform on the torus which enters into Theorem 1.1 was proved by Michel [1977], Strichartz [1992] and Gindikin (oral communication). Theorem 1.8 is due to Klein, Thorbergsson and Verhóczki [2009].

Range questions for the transforms here have been investigated by many authors. The conformal equivalence between $\mathbf{R}^n$ and $\mathbf{H}^n$ (via a stereographic projection) relates the Radon transforms on the two spaces similar to (83) in Ch. III. Thus range results for $\mathbf{R}^n$ (Ch. I §6) can be transferred to $\mathbf{H}^n$. (Berenstein, Casadio–Tarabusi and Kurusa [1997]). Another explicit description is given by Ishikawa [1997].

For compact U/K many explicit range descriptions are known but only for special cases. See Grinberg [1985], Gonzalez [1994], and Kakehi [1992]–[1998] for a sample.

The range of the Funk transform for the n-sphere is determined precisely in Kakehi–Tsukamoto [1993]. For the 3-sphere the space of geodesics is parametrized by a product of two 2-spheres. This has an ultrahyperbolic operator (the difference of the two Laplacians) and the range is the kernel of this operator. Since the Ásgeirsson theorem holds for this product (Ch. VI, §2), the geometric Theorem 6.11, Ch. I, holds here too.

CHAPTER V

ORBITAL INTEGRALS AND THE WAVE OPERATOR FOR ISOTROPIC LORENTZ SPACES

In Chapter II, §3 we discussed the problem of determining a function on a homogeneous space by means of its integrals over generalized spheres. We shall now solve this problem for the **isotropic Lorentz spaces** (Theorem 4.1 below). As we shall presently explain these spaces are the Lorentzian analogs of the two-point homogeneous spaces considered in Chapter III.

§1   Isotropic Spaces

Let X be a manifold. A **pseudo-Riemannian structure** of signature (p, q) is a smooth assignment $y \to g_y$ where $y \in X$ and g_y is a symmetric non-degenerate bilinear form on $X_y \times X_y$ of signature (p, q). This means that for a suitable basis $Y_1, \ldots, Y_{p+q}$ of X_y we have

$$g_y(Y) = y_1^2 + \cdots + y_p^2 - y_{p+1}^2 - \cdots - y_{p+q}^2$$

if $Y = \sum_1^{p+q} y_i Y_i$. If $q = 0$ we speak of a **Riemannian** structure and if $p = 1$ we speak of a **Lorentzian** structure. Connected manifolds X with such structures g are called pseudo-Riemannian (respectively Riemannian, Lorentzian) manifolds.

A manifold X with a pseudo-Riemannian structure g has a differential operator of particular interest, the so-called Laplace-Beltrami operator. Let $(x_1, \ldots, x_{p+q})$ be a coordinate system on an open subset U of X. We define the functions g_{ij}, g^{ij}, and $\bar{g}$ on U by

$$g_{ij} = g\left(\frac{\partial}{\partial x_i}, \frac{\partial}{\partial x_j}\right), \quad \sum_j g_{ij} g^{jk} = \delta_{ik}, \quad \bar{g} = |\det(g_{ij})|.$$

The *Laplace-Beltrami operator* L is defined on U by

$$Lf = \frac{1}{\sqrt{\bar{g}}} \left(\sum_k \frac{\partial}{\partial x_k} \left(\sum_i g^{ik} \sqrt{\bar{g}} \frac{\partial f}{\partial x_i} \right) \right)$$

for $f \in \mathcal{C}^\infty(U)$. It is well known that this expression is invariant under coordinate changes so L is a differential operator on X.

An **isometry** of a pseudo-Riemannian manifold X is a diffeomorphism preserving g. It is easy to prove that L is **invariant** under each isometry φ, that is $L(f \circ \varphi) = (Lf) \circ \varphi$ for each $f \in \mathcal{E}(X)$. Let $I(X)$ denote the group of all isometries of X. For $y \in X$ let $I(X)_y$ denote the subgroup of $I(X)$

S. Helgason, *Integral Geometry and Radon Transforms*,
DOI 10.1007/978-1-4419-6055-9_5, © Springer Science+Business Media, LLC 2010

fixing y (the isotropy subgroup at y) and let H_y denote the group of linear transformations of the tangent space X_y induced by the action of $I(X)_y$. For each $a \in \mathbf{R}$ let $\sum_a(y)$ denote the "sphere"

$$(1) \qquad \sum_a(y) = \{Z \in X_y : g_y(Z, Z) = a, \quad Z \neq 0\}.$$

Definition. The pseudo-Riemannian manifold X is called *isotropic* if for each $a \in \mathbf{R}$ and each $y \in X$ the group H_y acts transitively on $\sum_a(y)$.

Proposition 1.1. *An isotropic pseudo-Riemannian manifold X is homogeneous; that is, $I(X)$ acts transitively on X.*

Proof. The pseudo-Riemannian structure on X gives an affine connection preserved by each isometry $g \in I(X)$. Any two points $y, z \in X$ can be joined by a curve consisting of finitely many geodesic segments $\gamma_i (1 \leq i \leq p)$. Let g_i be an isometry fixing the midpoint of γ_i and reversing the tangents to γ_i at this point. The product $g_p \cdots g_1$ maps y to z, whence the homogeneity of X.

A.   The Riemannian Case

The following simple result shows that the isotropic spaces are natural generalizations of the spaces considered in the last chapter.

Proposition 1.2. *A Riemannian manifold X is isotropic if and only if it is two-point homogeneous.*

Proof. If X is two-point homogeneous and $y \in X$ the isotropy subgroup $I(X)_y$ at y is transitive on each sphere $S_r(y)$ in X with center y so X is clearly isotropic. On the other hand if X is isotropic it is homogeneous (Prop. 1.1) hence complete; thus by standard Riemannian geometry any two points in X can be joined by means of a geodesic. Now the isotropy of X implies that for each $y \in X, r > 0$, the group $I(X)_y$ is transitive on the sphere $S_r(y)$, whence the two-point homogeneity.

B.   The General Pseudo-Riemannian Case

Let X be a manifold with pseudo-Riemannian structure g and curvature tensor R. Let $y \in X$ and $S \subset X_y$ a 2-dimensional subspace on which g_y is nondegenerate. The **curvature** of X along the section S spanned by Z and Y is defined by

$$K(S) = -\frac{g_p(R_p(Z, Y)Z, Y)}{g_p(Z, Z)g_p(Y, Y) - g_p(Z, Y)^2}$$

The denominator is in fact $\neq 0$ and the expression is independent of the choice of Z and Y.

We shall now construct isotropic pseudo-Riemannian manifolds of signature (p, q) and constant curvature. Consider the space $\mathbf{R}^{p+q+1}$ with the flat pseudo-Riemannian structure

$$B_e(Y) = y_1^2 + \cdots + y_p^2 - y_{p+1}^2 - \cdots - y_{p+q}^2 + e\, y_{p+q+1}^2\,, \quad (e = \pm 1).$$

Let Q_e denote the quadric in $\mathbf{R}^{p+q+1}$ given by

$$B_e(Y) = e.$$

The orthogonal group $\mathbf{O}(B_e)$ $(= \mathbf{O}(p, q+1)$ or $\mathbf{O}(p+1, q))$ acts transitively on Q_e; the isotropy subgroup at $o = (0, \ldots, 0, 1)$ is identified with $\mathbf{O}(p, q)$.

Theorem 1.3. *(i)   The restriction of B_e to the tangent spaces to Q_e gives a pseudo-Riemannian structure g_e on Q_e of signature (p, q).*

(ii)   We have

(2) $$Q_{-1} \cong \mathbf{O}(p, q+1)/\mathbf{O}(p, q) \qquad \text{(diffeomorphism)}$$

and the pseudo-Riemannian structure g_{-1} on Q_{-1} has constant curvature -1.

(iii)   We have

(3) $$Q_{+1} = \mathbf{O}(p+1, q)/\mathbf{O}(p, q) \qquad \text{(diffeomorphism)}$$

and the pseudo-Riemannian structure g_{+1} on Q_{+1} has constant curvature $+1$.

(iv)   The flat space $\mathbf{R}^{p+q}$ with the quadratic form $g_o(Y) = \sum_1^p y_i^2 - \sum_{p+1}^{p+q} y_j^2$ and the spaces

$$\mathbf{O}(p, q+1)/\mathbf{O}(p, q), \quad \mathbf{O}(p+1, q)/\mathbf{O}(p, q)$$

are all isotropic and (up to a constant factor on the pseudo-Riemannian structure) exhaust the class of pseudo-Riemannian manifolds of constant curvature and signature (p, q) except for local isometry.

Proof. If s_o denotes the linear transformation

$$(y_1, \ldots, y_{p+q}, y_{p+q+1}) \to (-y_1, \ldots, -y_{p+q}, y_{p+q+1})$$

then the mapping $\sigma : g \to s_o g s_o$ is an involutive automorphism of $\mathbf{O}(p, q+1)$ whose differential $d\sigma$ has fixed point set $\mathfrak{o}(p, q)$ (the Lie algebra of $\mathbf{O}(p, q)$). The (-1)-eigenspace of $d\sigma$, say $\mathfrak{m}$, is spanned by the vectors

(4) $$Y_i \;=\; E_{i, p+q+1} + E_{p+q+1, i} \qquad (1 \le i \le p),$$
(5) $$Y_j \;=\; E_{j, p+q+1} - E_{p+q+1, j} \qquad (p+1 \le j \le p+q).$$

Here E_{ij} denotes a square matrix with entry 1 where the ith row and the jth column meet, all other entries being 0.

The mapping $\psi : g\mathbf{O}(p,q) \to g \cdot o$ has a differential $d\psi$ which maps $\mathfrak{m}$ bijectively onto the tangent plane $y_{p+q+1} = 1$ to Q_{-1} at o and $d\psi(X) = X \cdot o$ ($X \in \mathfrak{m}$). Thus

$$d\psi(Y_k) = (\delta_{1k}, \ldots, \delta_{p+q+1,k}), \quad (1 \le k \le p+q).$$

Thus

$$B_{-1}(d\psi(Y_k)) = 1 \quad \text{if} \quad 1 \le k \le p \text{ and } -1 \text{ if } p+1 \le k \le p+q,$$

proving (i). Next, since the space (2) is symmetric its curvature tensor satisfies

$$R_o(X,Y)(Z) = [[X,Y],Z],$$

where $[\, ,\,]$ is the Lie bracket. A simple computation then shows for $k \ne \ell$

$$K(\mathbf{R}Y_k + \mathbf{R}Y_\ell) = -1 \quad (1 \le k, \ell \le p+q)$$

and this implies (ii). Part (iii) is proved in the same way. For (iv) we first verify that the spaces listed are isotropic. Since the isotropy action of $\mathbf{O}(p, q+1)_o = \mathbf{O}(p,q)$ on $\mathfrak{m}$ is the ordinary action of $\mathbf{O}(p,q)$ on $\mathbf{R}^{p+q}$ it suffices to verify that $\mathbf{R}^{p+q}$ with the quadratic form g_o is isotropic. But we know $\mathbf{O}(p,q)$ is transitive on $g_e = +1$ and on $g_e = -1$ so it remains to show $\mathbf{O}(p,q)$ transitive on the cone $\{Y \ne 0 : g_e(Y) = 0\}$. By rotation in $\mathbf{R}^p$ and in $\mathbf{R}^q$ it suffices to verify the statement for $p = q = 1$. But for this case it is obvious. The uniqueness in (iv) follows from the general fact that a symmetric space is determined locally by its pseudo-Riemannian structure and curvature tensor at a point (see e.g. [DS], pp. 200–201). This finishes the proof.

The spaces (2) and (3) are the pseudo-Riemannian analogs of the spaces $\mathbf{O}(p,1)/\mathbf{O}(p)$, $\mathbf{O}(p+1)/\mathbf{O}(p)$ from Ch. III, §1. But the other two-point homogeneous spaces listed in Ch. III, §2–§3 have similar pseudo-Riemannian analogs (indefinite elliptic and hyperbolic spaces over $\mathbf{C}$, $\mathbf{H}$ and $\mathbf{Cay}$). As proved by Wolf [1967], p. 384, each non-flat isotropic pseudo-Riemannian manifold is locally isometric to one of these models.

We shall later need a lemma about the connectivity of the groups $\mathbf{O}(p,q)$. Let $I_{p,q}$ denote the diagonal matrix (d_{ij}) with

$$d_{ii} = 1 \quad (1 \le i \le p), \, d_{jj} = -1 \quad (p+1 \le j \le p+q),$$

so a matrix g with transpose ${}^t g$ belongs to $\mathbf{O}(p,q)$ if and only if

(6) $$\qquad {}^t g I_{p,q} g = I_{p,q}.$$

If $y \in \mathbf{R}^{p+q}$ let

$$y^T = (y_1, \ldots, y_p, 0 \ldots 0), \quad y^S = (0, \ldots, 0, y_{p+1}, \ldots, y_{p+q})$$

and for $g \in \mathbf{O}(p, q)$ let g_T and g_S denote the matrices

$$
\begin{aligned}
(g_T)_{ij} &= g_{ij} & (1 \le i, j \le p), \\
(g_S)_{k\ell} &= g_{k\ell} & (p+1 \le k, \ell \le p+q)
\end{aligned}
$$

If $g_1, \ldots, g_{p+q}$ denote the column vectors of the matrix g then (3.6) means for the scalar products

$$
\begin{aligned}
g_i^T \cdot g_i^T - g_i^S \cdot g_i^S &= 1, & 1 \le i \le p, \\
g_j^T \cdot g_j^T - g_j^S \cdot g_j^S &= -1, & p+1 \le j \le p+q, \\
g_j^T \cdot g_k^T &= g_j^S \cdot g_k^S, & j \ne k.
\end{aligned}
$$

Lemma 1.4. *We have for each $g \in \mathbf{O}(p, q)$*

$$
|\det(g_T)| \ge 1, \quad |\det(g_S)| \ge 1.
$$

The components of $\mathbf{O}(p, q)$ are obtained by

(7)	$\det g_T \ge 1$	,	$\det g_S \ge 1$;	(identity component)
(8)	$\det g_T \le -1$	,	$\det g_S \ge 1$;	
(9)	$\det g_T \ge -1$	,	$\det g_S \le -1$,	
(10)	$\det g_T \le -1$	,	$\det g_S \le -1$.	

Thus $\mathbf{O}(p, q)$ has four components if $p \ge 1, q \ge 1$, two components if p or $q = 0$.

Proof. Consider the Gram determinant

$$
\det \begin{pmatrix}
g_1^T \cdot g_1^T & g_1^T \cdot g_2^T & \cdots & g_1^T \cdot g_p^T \\
g_2^T \cdot g_1^T & & & \\
\vdots & & & \\
g_p^T \cdot g_1^T & \cdots & & g_p^T \cdot g_p^T
\end{pmatrix},
$$

which equals $(\det g_T)^2$. Using the relations above it can also be written

$$
\det \begin{pmatrix}
1 + g_1^S \cdot g_1^S & g_1^S \cdot g_2^S & \cdots & g_1^S \cdot g_p^S \\
g_2^S \cdot g_1^S & & \cdots & \\
\vdots & & & \\
g_p^S \cdot g_1^S & & & 1 + g_p^S \cdot g_p^S
\end{pmatrix},
$$

which equals 1 plus a sum of lower order Gram determinants each of which is still positive. Thus $(\det g_T)^2 \ge 1$ and similarly $(\det g_S)^2 \ge 1$. Assuming now $p \ge 1, q \ge 1$ consider the decomposition of $\mathbf{O}(p, q)$ into the four pieces (7), (8), (9), (10). Each of these is $\ne \emptyset$ because (8) is obtained from (7) by multiplication by $I_{1,p+q-1}$ etc. On the other hand, since the functions $g \to$

$\det(g_T)$, $g \to \det(g_S)$ are continuous on $\mathbf{O}(p,q)$ the four pieces above belong to different components of $\mathbf{O}(p,q)$. But by Chevalley [1946], p. 201, $\mathbf{O}(p,q)$ is homeomorphic to the product of $\mathbf{O}(p,q) \cap \mathbf{U}(p+q)$ with a Euclidean space. Since $\mathbf{O}(p,q) \cap \mathbf{U}(p+q) = \mathbf{O}(p,q) \cap \mathbf{O}(p+q)$ is homeomorphic to $\mathbf{O}(p) \times \mathbf{O}(q)$ it just remains to remark that $\mathbf{O}(n)$ has two components.

C.   The Lorentzian Case

The isotropic Lorentzian manifolds are more restricted than one might at first think on the basis of the Riemannian case. In fact there is a theorem of Lichnerowicz and Walker [1945] (see Wolf [1967], Ch. 12) which implies that an isotropic Lorentzian manifold has constant curvature. Thus we can deduce the following result from Theorem 1.3.

Theorem 1.5. *Let X be an isotropic Lorentzian manifold (signature $(1,q)$, $q \geq 1$). Then X has constant curvature so (after a multiplication of the Lorentzian structure by a positive constant) X is locally isometric to one of the following:*

$$\mathbf{R}^{1+q}(\textit{flat, signature } (1,q)),$$
$$Q_{-1} = \mathbf{O}(1, q+1)/\mathbf{O}(1,q): \ y_1^2 - y_2^2 - \cdots - y_{q+2}^2 = -1,$$
$$Q_{+1} = \mathbf{O}(2, q)/\mathbf{O}(1,q): \ y_1^2 - y_2^2 - \cdots - y_{q+1}^2 + y_{q+2}^2 = 1,$$

the Lorentzian structure being induced by $y_1^2 - y_2^2 - \cdots \mp y_{q+2}^2$.

§2   Orbital Integrals

The orbital integrals for isotropic Lorentzian manifolds are analogs to the spherical averaging operator M^r considered in Ch. I, §2, and Ch. III, §1. We start with some geometric preparation.

For manifolds X with a Lorentzian structure g we adopt the following customary terminology: If $y \in X$ the cone

$$C_y = \{Y \in X_y : g_y(Y,Y) = 0\}$$

is called the **null cone** (or the **light cone**) in X_y with vertex y. A nonzero vector $Y \in X_y$ is said to be **timelike, isotropic** or **spacelike** if $g_y(Y,Y)$ is positive, 0, or negative, respectively. Similar designations apply to geodesics according to the type of their tangent vectors.

While the geodesics in $\mathbf{R}^{1+q}$ are just the straight lines, the geodesics in Q_{-1} and Q_{+1} can be found by the method of Ch. III, §1.

Proposition 2.1. *The geodesics in the Lorentzian quadrics Q_{-1} and Q_{+1} have the following properties:*

(i)   The geodesics are the nonempty intersections of the quadrics with two-planes in $\mathbf{R}^{2+q}$ through the origin.

(ii)   For Q_{-1} the spacelike geodesics are closed, for Q_{+1} the timelike geodesics are closed.

(iii)   The isotropic geodesics are certain straight lines in $\mathbf{R}^{2+q}$.

Proof. Part (i) follows by the symmetry considerations in Ch. III, §1. For Part (ii) consider the intersection of Q_{-1} with the two-plane

$$y_1 = y_4 = \cdots = y_{q+2} = 0.$$

The intersection is the circle $y_2 = \cos t$, $y_3 = \sin t$ whose tangent vector $(0, -\sin t, \cos t, 0, \ldots, 0)$ is clearly spacelike. Since $\mathbf{O}(1, q+1)$ permutes the spacelike geodesics transitively the first statement in (ii) follows. For Q_{+1} we intersect similarly with the two-plane

$$y_2 = \cdots = y_{q+1} = 0.$$

For (iii) we note that the two-plane $\mathbf{R}(1, 0, \ldots, 0, 1) + \mathbf{R}(0, 1, \ldots, 0)$ intersects Q_{-1} in a pair of straight lines

$$y_1 = t, y_2 \pm 1, y_3 = \cdots = y_{q+1} = 0, y_{q+2} = t$$

which clearly are isotropic. The transitivity of $\mathbf{O}(1, q + 1)$ on the set of isotropic geodesics then implies that each of these is a straight line. The argument for Q_{+1} is similar.

Lemma 2.2. *The quadrics Q_{-1} and Q_{+1} $(q \geq 1)$ are connected.*

Proof. The q-sphere being connected, the point $(y_1, \ldots, y_{q+2})$ on $Q_{\mp 1}$ can be moved continuously on $Q_{\mp 1}$ to the point

$$(y_1, (y_2^2 + \cdots + y_{q+1}^2)^{1/2}, 0, \ldots, 0, y_{q+2})$$

so the statement follows from the fact that the hyperboloids $y_1^2 - y_1^2 \mp y_3^2 = \mp 1$ are connected.

Lemma 2.3. *The identity components of $\mathbf{O}(1, q+1)$ and $\mathbf{O}(2, q)$ act transitively on Q_{-1} and Q_{+1}, respectively, and the isotropy subgroups are connected.*

Proof. The first statement comes from the general fact (see e.g. [DS], pp. 121–124) that when a separable Lie group acts transitively on a connected manifold then so does its identity component. For the isotropy groups we use the description (7) of the identity component. This shows quickly that

$$\mathbf{O}_o(1, q + 1) \cap \mathbf{O}(1, q) = \mathbf{O}_o(1, q),$$
$$\mathbf{O}_o(2, q) \cap \mathbf{O}(1, q) = \mathbf{O}_o(1, q)$$

the subscript o denoting identity component. Thus we have

$$Q_{-1} = \mathbf{O}_o(1, q+1)/\mathbf{O}_o(1, q),$$
$$Q_{+1} = \mathbf{O}_o(2, q)/\mathbf{O}_o(1, q),$$

proving the lemma.

We now write the spaces in Theorem 1.5 in the form $X = G/H$ where $H = \mathbf{O}_o(1, q)$ and G is either $G^0 = \mathbf{R}^{1+q} \cdot \mathbf{O}_o(1, q)$ (semi-direct product) $G^- = \mathbf{O}_o(1, q+1)$ or $G^+ = \mathbf{O}_o(2, q)$. Let o denote the origin $\{H\}$ in X, that is

$$o = (0, \ldots, 0) \qquad \text{if } X = \mathbf{R}^{1+q}$$
$$o = (0, \ldots, 0, 1) \qquad \text{if } X = Q_{-1} \text{ or } Q_{+1}.$$

In the cases $X = Q_{-1}, X = Q_{+1}$ the tangent space X_o is the hyperplane $\{y_1, \ldots, y_{q+1}, 1\} \subset \mathbf{R}^{2+q}$.

The timelike vectors at o fill up the "interior" C_o^o of the cone C_o. The set C_o^o consists of two components. The component which contains the timelike vector

$$v_o = (-1, 0, \ldots, 0)$$

will be called the **solid retrograde cone** in X_o. It will be denoted by D_o. The component of the hyperboloid $g_o(Y, Y) = r^2$ which lies in D_o will be denoted $S_r(o)$. If y is any other point of X we define $C_y, D_y, S_r(y) \subset X_y$ by

$$C_y = g \cdot C_o, \quad D_y = g \cdot D_o, \quad S_r(y) = g \cdot S_r(o)$$

if $g \in G$ is chosen such that $g \cdot o = y$. This is a valid definition because the connectedness of H implies that $h \cdot D_o \subset D_o$. We also define

$$B_r(y) = \{Y \in D_y : 0 < g_y(Y, Y) < r^2\}.$$

If Exp denotes the exponential mapping of X_y into X, mapping rays through 0 onto geodesics through y we put

$$\mathbf{D}_y = \operatorname{Exp} D_y, \quad \mathbf{C}_y = \operatorname{Exp} C_y$$
$$\mathbf{S}_r(y) = \operatorname{Exp} S_r(y), \quad \mathbf{B}_r(y) = \operatorname{Exp} B_r(y).$$

Again $\mathbf{C}_y$ and $\mathbf{D}_y$ are respectively called the **light cone** and **solid retrograde cone** in X with vertex y. For the spaces $X = Q_+$ we always assume $r < \pi$ in order that Exp will be one-to-one on $B_r(y)$ in view of Prop. 2.1(ii).

Figure V.1 illustrates the situation for Q_{-1} in the case $q = 1$. Then Q_{-1} is the hyperboloid

$$y_1^2 - y_2^2 - y_3^2 = -1$$

and the y_1-axis is vertical. The origin o is

$$o = (0, 0, 1)$$

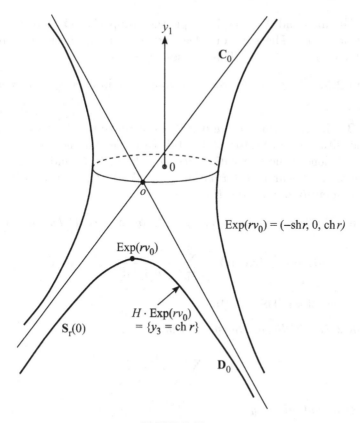

FIGURE V.1.

and the vector $v_o = (-1, 0, 0)$ lies in the tangent space

$$(Q_{-1})_o = \{y : y_3 = 1\}$$

pointing downward. The mapping $\psi : gH \to g \cdot o$ has differential $d\psi : \mathfrak{m} \to (Q_{-1})_o$ and

$$d\psi(E_{13} + E_{31}) = -v_o$$

in the notation of (4). The geodesic tangent to v_o at o is

$$t \to \mathrm{Exp}(tv_o) = \exp(-t(E_{13} + E_{31})) \cdot o = (-\sinh t, 0, \cosh t)$$

and this is the section of Q_{-1} with the plane $y_2 = 0$. Note that since H preserves each plane $y_3 = \mathrm{const.}$, the "sphere" $\mathbf{S}_r(o)$ is the plane section $y_3 = \cosh r, y_1 < 0$ with Q_{-1}.

Lemma 2.4. *The negative of the Lorentzian structure on $X = G/H$ induces on each $\mathbf{S}_r(y)$ a Riemannian structure of constant negative curvature $(q > 1)$.*

Proof. The manifold X being isotropic the group $H = \mathbf{O}_o(1, q)$ acts transitively on $\mathbf{S}_r(o)$. The subgroup leaving fixed the geodesic from o with tangent vector v_o is $\mathbf{O}_o(q)$. This implies the lemma.

Lemma 2.5. *The timelike geodesics from y intersect $\mathbf{S}_r(y)$ under a right angle.*

Proof. By the group invariance it suffices to prove this for $y = o$ and the geodesic with tangent vector v_o. For this case the statement is obvious.

Let $\tau(g)$ denote the translation $xH \to gxH$ on G/H and for $Y \in \mathfrak{m}$ let T_Y denote the linear transformation $Z \to [Y, [Y, Z]]$ of $\mathfrak{m}$ into itself. As usual, we identify $\mathfrak{m}$ with $(G/H)_o$.

Lemma 2.6. *The exponential mapping* $\mathrm{Exp} : \mathfrak{m} \to G/H$ *has differential*

$$d\mathrm{Exp}_Y = d\tau(\exp Y) \circ \sum_0^\infty \frac{T_Y^n}{(2n+1)!} \qquad (Y \in \mathfrak{m}).$$

For the proof see [DS], p. 215.

Lemma 2.7. *The linear transformation*

$$A_Y = \sum_0^\infty \frac{T_Y^n}{(2n+1)!}$$

has determinant given by

$$\det A_Y = \left\{ \frac{\sinh(g(Y,Y))^{1/2}}{(g(Y,Y))^{1/2}} \right\}^q \quad for\ Q_{-1}$$

$$\det A_Y = \left\{ \frac{\sin(g(Y,Y))^{1/2}}{(g(Y,Y))^{1/2}} \right\}^q \quad for\ Q_{+1}$$

for Y timelike.

Proof. Consider the case of Q_{-1}. Since $\det(A_Y)$ is invariant under H it suffices to verify this for $Y = cY_1$ in (4), where $c \in \mathbf{R}$. We have $c^2 = g(Y, Y)$ and $T_{Y_1}(Y_j) = Y_j$ $(2 \le j \le q+1)$. Thus T_Y has the eigenvalue 0 and $g(Y, Y)$; the latter is a q-tuple eigenvalue. This implies the formula for the determinant. The case Q_{+1} is treated in the same way.

From this lemma and the description of the geodesics in Prop. 2.1 we can now conclude the following result.

Proposition 2.8. *(i)   The mapping* $\mathrm{Exp} : \mathfrak{m} \to Q_{-1}$ *is a diffeomorphism of D_o onto $\mathbf{D}_o$.*

(ii)   The mapping $\mathrm{Exp} : \mathfrak{m} \to Q_{+1}$ *gives a diffeomorphism of $B_\pi(o)$ onto $\mathbf{B}_\pi(o)$.*

Let dh denote a bi-invariant measure on the unimodular group H. Let $u \in \mathcal{D}(X)$, $y \in X$ and $r > 0$. Select $g \in G$ such that $g \cdot o = y$ and select $x \in \mathbf{S}_r(o)$. Consider the integral

$$\int_H u(gh \cdot x) \, dh \,.$$

Since the subgroup $K \subset H$ leaving x fixed is compact it is easy to see that the set

$$C_{g,x} = \{h \in H : gh \cdot x \in \text{ support } (u)\}$$

is compact; thus the integral above converges. By the bi-invariance of dh it is independent of the choice of g (satisfying $g \cdot o = y$) and of the choice of $x \in \mathbf{S}_r(o)$. In analogy with the Riemannian case (Ch. III, §1) we thus define the operator M^r (**the orbital integral**) by

$$(11) \qquad\qquad (M^r u)(y) = \int_H u(gh \cdot x) \, dh \,.$$

If g and x run through suitable compact neighborhoods, the sets $C_{g,x}$ are enclosed in a fixed compact subset of H so $(M^r u)(y)$ depends smoothly on both r and y. It is also clear from (11) that the operator M^r is invariant under the action of G: if $m \in G$ and $\tau(m)$ denotes the transformation $nH \to mnH$ of G/H onto itself then

$$M^r(u \circ \tau(m)) = (M^r u) \circ \tau(m) \,.$$

If dk denotes the normalized Haar measure on K we have by standard invariant integration

$$\int_H u(h \cdot x) \, dh = \int_{H/K} d\dot{h} \int_K u(hk \cdot x) \, dk = \int_{H/K} u(h \cdot x) \, d\dot{h} \,,$$

where $d\dot{h}$ is an H-invariant measure on H/K. But if $d\mathbf{w}^r$ is the volume element on $\mathbf{S}_r(o)$ (cf. Lemma 2.4) we have by the uniqueness of H-invariant measures on the space $H/K \approx \mathbf{S}_r(o)$ that

$$(12) \qquad\qquad \int_H u(h \cdot x) \, dh = \frac{1}{A(r)} \int_{\mathbf{S}_r(o)} u(z) \, d\mathbf{w}^r(z) \,,$$

where $A(r)$ is a positive scalar. But since g is an isometry we deduce from (12) that

$$(M^r u)(y) = \frac{1}{A(r)} \int_{\mathbf{S}_r(y)} u(z) \, d\mathbf{w}^r(z) \,.$$

Now we have to determine $A(r)$.

Lemma 2.9. *For a suitable fixed normalization of the Haar measure dh on H we have*

$$A(r) = r^q, \quad (\sinh r)^q, \quad (\sin r)^q$$

for the cases

$$\mathbf{R}^{1+q}, \quad \mathbf{O}(1, q+1)/\mathbf{O}(1, q), \quad \mathbf{O}(2, q)/\mathbf{O}(1, q),$$

respectively.

Proof. The relations above show that $dh = A(r)^{-1}\, d\mathbf{w}^r\, dk$. The mapping $\mathrm{Exp} : D_o \to \mathbf{D}_o$ preserves length on the geodesics through o and maps $S_r(o)$ onto $\mathbf{S}_r(o)$. Thus if $z \in S_r(o)$ and Z denotes the vector from 0 to z in X_o the ratio of the volume of elements of $\mathbf{S}_r(o)$ and $S_r(o)$ at z is given by $\det(d\mathrm{Exp}_Z)$. Because of Lemmas 2.6–2.7 this equals

$$1, \left(\frac{\sinh r}{r}\right)^q, \left(\frac{\sin r}{r}\right)^q$$

for the three respective cases. But the volume element $d\omega^r$ on $S_r(o)$ equals $r^q d\omega^1$. Thus we can write in the three respective cases

$$dh = \frac{r^q}{A(r)}\, d\omega^1\, dk, \quad \frac{\sinh^q r}{A(r)}\, d\omega^1\, dk, \quad \frac{\sin^q r}{A(r)}\, d\omega^1\, dk.$$

But we can once and for all normalize dh by $dh = d\omega^1\, dk$ and for this choice our formulas for $A(r)$ hold.

Let $\square$ denote the **wave operator** on $X = G/H$, that is the Laplace-Beltrami operator for the Lorentzian structure g.

Lemma 2.10. *Let $y \in X$. On the solid retrograde cone $\mathbf{D}_y$, the wave operator $\square$ can be written*

$$\square = \frac{\partial^2}{\partial r^2} + \frac{1}{A(r)}\frac{dA}{dr}\frac{\partial}{\partial r} - L_{\mathbf{S}_r(y)},$$

where $L_{\mathbf{S}_r(y)}$ is the Laplace-Beltrami operator on $\mathbf{S}_r(y)$.

Proof. We can take $y = o$. If $(\theta_1, \ldots, \theta_q)$ are coordinates on the "sphere" $S_1(o)$ in the flat space X_o then $(r\theta_1, \ldots, r\theta_q)$ are coordinates on $S_r(o)$. The Lorentzian structure on D_o is therefore given by

$$dr^2 - r^2\, d\theta^2,$$

where $d\theta^2$ is the Riemannian structure of $S_1(o)$. Since A_Y in Lemma 2.7 is a diagonal matrix with eigenvalues 1 and $r^{-1}A(r)^{1/q}$ (q-times) it follows from Lemma 2.6 that the image $\mathbf{S}_r(o) = \mathrm{Exp}(S_r(o))$ has Riemannian structure

$r^2 \, d\theta^2$, $\sinh^2 r \, d\theta^2$ and $\sin^2 r \, d\theta^2$ in the cases $\mathbf{R}^{1+q}$, Q_{-1} and Q_{+1}, respectively. By the perpendicularity in Lemma 2.5 it follows that the Lorentzian structure on $\mathbf{D}_o$ is given by

$$dr^2 - r^2 \, d\theta^2 \,, \quad dr^2 - \sinh^2 r \, d\theta^2 \,, \quad dr^2 - \sin^2 r \, d\theta^2$$

in the three respective cases. Now the lemma follows immediately.

The operator M^r is of course the Lorentzian analog to the spherical mean value operator for isotropic Riemannian manifolds. We shall now prove that in analogy to the Riemannian case (cf. (3.41), Ch. III) the operator M^r commutes with the wave operator $\square$.

Theorem 2.11. *For each of the isotropic Lorentz spaces* $X = G^-/H$, G^+/H *or* G^0/H *the wave operator* $\square$ *and the orbital integral* M^r *commute:*

$$\square M^r u = M^r \square u \quad \text{for } u \in \mathcal{D}(X).$$

(For G^+/H we assume $r < \pi$.)

Given a function u on G/H we define the function $\tilde{u}$ on G by $\tilde{u}(g) = u(g \cdot o)$.

Lemma 2.12. *There exists a differential operator* $\widetilde{\square}$ *on G invariant under all left and all right translations such that*

$$\widetilde{\square}\tilde{u} = (\square u)^{\sim} \quad \text{for } u \in \mathcal{D}(X).$$

Proof. We consider first the case $X = G^-/H$. The bilinear form

$$K(Y, Z) = \tfrac{1}{2} \operatorname{Tr}(YZ)$$

on the Lie algebra $\mathfrak{o}(1, q+1)$ of G^- is nondegenerate; in fact K is nondegenerate on the complexification $\mathfrak{o}(q+2, \mathbf{C})$ consisting of all complex skew symmetric matrices of order $q+2$. A simple computation shows that in the notation of (4) and (5)

$$K(Y_1, Y_1) = 1, \quad K(Y_j, Y_j) = -1 \quad (2 \leq j \leq q+1).$$

Since K is symmetric and nondegenerate there exists a unique left invariant pseudo-Riemannian structure $\widetilde{K}$ on G^- such that $\widetilde{K}_e = K$. Moreover, since K is invariant under the conjugation $Y \to gYg^{-1}$ of $\mathfrak{o}(1, q+1)$, $\widetilde{K}$ is also right invariant. Let $\widetilde{\square}$ denote the corresponding Laplace-Beltrami operator on G^-. Then $\widetilde{\square}$ is invariant under all left and right translations on G^-. Let $u = \mathcal{D}(X)$. Since $\widetilde{\square}\tilde{u}$ is invariant under all right translations from H there is a unique function $v \in \mathcal{E}(X)$ such that $\widetilde{\square}\tilde{u} = \tilde{v}$. The mapping $u \to v$ is a differential operator which at the origin must coincide with $\square$, that is $\widetilde{\square}\tilde{u}(e) = \square u(o)$. Since, in addition, both $\square$ and the operator $u \to v$ are invariant under the action of G^- on X it follows that they coincide. This proves $\widetilde{\square}\tilde{u} = (\square u)^{\sim}$.

The case $X = G^+/H$ is handled in the same manner. For the flat case $X = G^0/H$ let

$$Y_j = (0, \ldots, 1, \ldots, 0),$$

the jth coordinate vector on $\mathbf{R}^{1+q}$. Then $\square = Y_1^2 - Y_2^2 - \cdots - Y_{q+1}^2$. Since $\mathbf{R}^{1+q}$ is naturally embedded in the Lie algebra of G^0 we can extend Y_j to a left invariant vector field $\widetilde{Y}_j$ on G^0. The operator

$$\widetilde{\square} = \widetilde{Y}_1^2 - \widetilde{Y}_2^2 - \cdots - \widetilde{Y}_{q+1}^2$$

is then a left and right invariant differential operator on G^0 and again we have $\widetilde{\square}\widetilde{u} = (\square u)^\sim$. This proves the lemma.

We can now prove Theorem 2.11. If $g \in G$ let $L(g)$ and $R(g)$, respectively, denote the left and right translations $\ell \to g\ell$, and $\ell \to \ell g$ on G. If $\ell \cdot o = x, x \in \mathbf{S}_r(o)$ $(r > 0)$ and $g \cdot o = y$ then

$$(M^r u)(y) = \int_H \widetilde{u}(gh\ell)\, dh$$

because of (11). As g and ℓ run through sufficiently small compact neighborhoods the integration takes place within a fixed compact subset of H as remarked earlier. Denoting by subscript the argument on which a differential operator is to act we shall prove the following result.

Lemma 2.13.

$$\widetilde{\square}_\ell \left(\int_H \widetilde{u}(gh\ell)\, dh \right) = \int_H (\widetilde{\square}\widetilde{u})(gh\ell)\, dh = \widetilde{\square}_g \left(\int_H \widetilde{u}(gh\ell)\, dh \right).$$

Proof. The first equality sign follows from the left invariance of $\widetilde{\square}$. In fact, the integral on the left is

$$\int_H (\widetilde{u} \circ L(gh))(\ell)\, dh$$

so

$$\widetilde{\square}_\ell \left(\int_H \widetilde{u}(gh\ell)\, dh \right) = \int_H \left[\widetilde{\square}(\widetilde{u} \circ L(gh)) \right](\ell)\, dh$$

$$= \int_H \left[(\widetilde{\square}\widetilde{u}) \circ L(gh) \right](\ell)\, dh = \int_H (\widetilde{\square}\widetilde{u})(gh\ell)\, dh.$$

The second equality in the lemma follows similarly from the right invariance of $\widetilde{\square}$. But this second equality is just the commutativity statement in Theorem 2.11.

Lemma 2.13 also implies the following analog of the Darboux equation in Lemma 3.2, Ch. I.

Corollary 2.14. *Let* $u \in \mathcal{D}(X)$ *and put*

$$U(y, z) \;=\; (M^r u)(y) \quad \text{if } z \in \mathbf{S}_r(o).$$

Then

$$\square_y(U(y, z)) \;=\; \square_z(U(y, z)).$$

Remark 2.15. In $\mathbf{R}^n$ the solutions to the Laplace equation $Lu = 0$ are characterized by the spherical mean-value theorem $M^r u = u$ (all r). This can be stated equivalently: $M^r u$ is a constant in r. In this latter form the mean value theorem holds for the solutions of the wave equation $\square u = 0$ in an isotropic Lorentzian manifold: *If u satisfies $\square u = 0$ and if u is suitably small at ∞ then $(M^r u)(o)$ is constant in r.* For a precise statement and proof see Helgason [1959], p. 289. For $\mathbf{R}^2$ such a result had also been noted by Ásgeirsson.

§3   Generalized Riesz Potentials

In this section we generalize part of the theory of Riesz potentials (Ch. VII, §6) to isotropic Lorentz spaces.

Consider first the case

$$X = Q_{-1} = G^-/H = \mathbf{O}_o(1, n)/\mathbf{O}_o(1, n-1)$$

of dimension n and let $f \in \mathcal{D}(X)$ and $y \in X$. If $z = \mathrm{Exp}_y Y$ $(Y \in D_y)$ we put $r_{yz} = g(Y, Y)^{1/2}$ and consider the integral

$$(13) \qquad (I_-^\lambda f)(y) = \frac{1}{H_n(\lambda)} \int\limits_{D_y} f(z) \sinh^{\lambda-n}(r_{yz})\, dz,$$

where dz is the volume element on X, and

$$(14) \qquad H_n(\lambda) = \pi^{(n-2)/2} 2^{\lambda-1} \Gamma(\lambda/2)\, \Gamma((\lambda + 2 - n)/2).$$

The integral converges for $\mathrm{Re}\,\lambda \geq n$. We transfer the integral in (13) over to D_y via the diffeomorphism $\mathrm{Exp}(= \mathrm{Exp}_y)$. Since

$$dz = dr\,d\mathbf{w}^r = dr \left(\frac{\sinh r}{r}\right)^{n-1} d\omega^r$$

and since $dr \, d\omega^r$ equals the volume element dZ on D_y we obtain

$$(I^\lambda f)(y) = \frac{1}{H_n(\lambda)} \int_{D_y} (f \circ \mathrm{Exp})(Z) \left(\frac{\sinh r}{r}\right)^{\lambda - 1} r^{\lambda - n} \, dZ \,,$$

where $r = g(Z, Z)^{1/2}$. This has the form

(15) $$\frac{1}{H_n(\lambda)} \int_{D_y} h(Z, \lambda) r^{\lambda - n} \, dZ \,,$$

where $h(Z, \lambda)$, as well as each of its partial derivatives with respect to the first argument, is holomorphic in λ and h has compact support in the first variable. The methods of Riesz [1949], Ch. III, can be applied to such integrals (15). In particular we find that the function $\lambda \to (I^\lambda f)(y)$ which by its definition is holomorphic for $\operatorname{Re} \lambda > n$ admits a holomorphic continuation to the entire λ-plane and that its value at $\lambda = 0$ is $h(0, 0) = f(y)$. (In Riesz' treatment $h(Z, \lambda)$ is independent of λ, but his method still applies.) Denoting the holomorphic continuation of (13) by $(I^\lambda_-)f(y)$ we have thus obtained

(16) $$I^0_- f = f \,.$$

We would now like to differentiate (13) with respect to y. For this we write the integral in the form $\int_F f(z) K(y, z) \, dz$ over a bounded region F which properly contains the intersection of the support of f with the closure of $\mathbf{D}_y$. The kernel $K(y, z)$ is defined as $\sinh^{\lambda - n} r_{yz}$ if $z \in \mathbf{D}_y$, otherwise 0. For $\operatorname{Re} \lambda$ sufficiently large, $K(y, z)$ is twice continuously differentiable in y so we can deduce for such λ that $I^\lambda_- f$ is of class C^2 and that

(17) $$(\Box I^\lambda_- f)(y) = \frac{1}{H_n(\lambda)} \int_{\mathbf{D}_y} f(z) \Box_y (\sinh^{\lambda - n} r_{yz}) \, dz \,.$$

Moreover, given $m \in \mathbf{Z}^+$ we can find k such that $I^\lambda_- f \in C^m$ for $\operatorname{Re} \lambda > k$ (and all f). Using Lemma 2.10 and the relation

$$\frac{1}{A(r)} \frac{dA}{dr} = (n - 1) \coth r$$

we find

$$\Box_y (\sinh^{\lambda - n} r_{yz}) = \Box_z (\sinh^{\lambda - n} r_{yz})$$
$$= (\lambda - n)(\lambda - 1) \sinh^{\lambda - n} r_{yz} + (\lambda - n)(\lambda - 2) \sinh^{\lambda - n - 2} r_{yz} \,.$$

We also have

$$H_n(\lambda) = (\lambda - 2)(\lambda - n) H_n(\lambda - 2),$$

so substituting into (17) we get

$$\Box I_-^\lambda f = (\lambda - n)(\lambda - 1)I_-^\lambda f + I_-^{\lambda-2} f \,.$$

Still assuming $\operatorname{Re}\lambda$ large we can use Green's formula to express the integral

$$(18) \qquad \int_{\mathbf{D}_y} \left[f(z)\Box_z(\sinh^{\lambda-n} r_{yz}) - \sinh^{\lambda-n} r_{yz}(\Box f)(z) \right] dz$$

as a surface integral over a part of $\mathbf{C}_y$ (on which $\sinh^{\lambda-n} r_{yz}$ and its first order derivatives vanish) together with an integral over a surface inside $\mathbf{D}_y$ (on which f and its derivatives vanish). Hence the expression (18) vanishes so we have proved the relations

$$(19) \qquad \Box(I_-^\lambda f) \;=\; I_-^\lambda(\Box f)$$
$$(20) \qquad I_-^\lambda(\Box f) \;=\; (\lambda - n)(\lambda - 1)I_-^\lambda f + I_-^{\lambda-2} f$$

for $\operatorname{Re}\lambda > k$, k being some number (independent of f).

Since both sides of (20) are holomorphic in λ this relation holds for all $\lambda \in \mathbf{C}$. We shall now deduce that for each $\lambda \in \mathbf{C}$, we have $I_-^\lambda f \in \mathcal{E}(X)$ and (19) holds. For this we observe by iterating (20) that for each $p \in \mathbf{Z}^+$

$$(21) \qquad I_-^\lambda f = I_-^{\lambda+2p}(Q_p(\Box)f) \,,$$

Q_p being a certain p^{th}-degree polynomial. Choosing p arbitrarily large we deduce from the remark following (17) that $I_-^\lambda f \in \mathcal{E}(X)$; secondly (19) implies for $\operatorname{Re}\lambda + 2p > k$ that

$$\Box I_-^{\lambda+2p}(Q_p(\Box)f) = I_-^{\lambda+2p}(Q_p(\Box)\Box f) \,.$$

Using (21) again this means that (19) holds for all λ.

Putting $\lambda = 0$ in (20) we get

$$(22) \qquad I_-^{-2} = \Box f - nf \,.$$

Remark 3.1. In Riesz' paper [1949], p. 190, an analog I^α of the potentials in Ch. V, §5, is defined for any analytic Lorentzian manifold. These potentials I^α are however different from our I_-^λ and satisfy the equation $I^{-2}f = \Box f$ in contrast to (22).

We consider next the case

$$X = Q_{+1} = G^+/H = \mathbf{O}_o(2, n-1)/\mathbf{O}_o(1, n-1)$$

and we define for $f \in \mathcal{D}(X)$

$$(23) \qquad (I_+^\lambda f)(y) = \frac{1}{H_n(\lambda)} \int_{\mathbf{D}_y} f(z) \sin^{\lambda-n}(r_{yz}) \, dz \,.$$

Again $H_n(\lambda)$ is given by (14) and dz is the volume element. In order to bypass the difficulties caused by the fact that the function $z \to \sin r_{yz}$ vanishes on $\mathbf{S}_\pi$ we assume that f has support disjoint from $\mathbf{S}_\pi(o)$. Then the support of f is disjoint from $\mathbf{S}_\pi(y)$ for all y in some neighborhood of o in X. We can then prove just as before that

$$(24) \qquad (I_+^0 f)(y) \;=\; f(y)$$

$$(25) \qquad (\Box I_+^\lambda f)(y) \;=\; (I_+^\lambda \Box f)(y)$$

$$(26) \qquad (I_+^\lambda \Box f)(y) \;=\; -(\lambda - n)(\lambda - 1)(I_+^\lambda f)(y) + (I_+^{\lambda-2} f)(y)$$

for all $\lambda \in \mathbf{C}$. In particular

$$(27) \qquad\qquad I_+^{-2} f = \Box f + n f\,.$$

Finally we consider the flat case

$$X = \mathbf{R}^n = G^0/H = \mathbf{R}^n \cdot \mathbf{O}_o(1, n-1)/\mathbf{O}_o(1, n-1)$$

and define

$$(I_o^\lambda f)(y) = \frac{1}{H_n(\lambda)} \int\limits_{\mathbf{D}_y} f(z) r_{yz}^{\lambda-n}\, dz\,.$$

These are the potentials defined by Riesz in [1949], p. 31, who proved

$$(28) \qquad\qquad I_o^0 f = f, \;\; \Box I_o^\lambda f = I_o^\lambda \Box f = I_o^{\lambda-2} f\,.$$

§4   Determination of a Function from Its Integral over Lorentzian Spheres

In a Riemannian manifold a function is determined in terms of its spherical mean values by the simple relation $f = \lim_{r \to 0} M^r f$. We shall now solve the analogous problem for an even-dimensional isotropic Lorentzian manifold and express a function f in terms of its orbital integrals $M^r f$. Since the spheres $\mathbf{S}_r(y)$ do not shrink to a point as $r \to 0$ the formula (cf. Theorem 4.1) below is quite different.

For the solution of the problem we use the geometric description of the wave operator $\Box$ developed in §2, particularly its commutation with the orbital integral M^r, and combine this with the results about the generalized Riesz potentials established in §3.

We consider first the negatively curved space $X = G^-/H$. Let $n = \dim X$ and assume n even. Let $f \in \mathcal{D}(X)$, $y \in X$ and put $F(r) = (M^r f)(y)$. Since the volume element dz on $\mathbf{D}_y$ is given by $dz = dr\, d\mathbf{w}^r$ we obtain from (12) and Lemma 2.9 ,

$$(29) \qquad\qquad (I_-^\lambda f)(y) = \frac{1}{H_n(\lambda)} \int\limits_0^\infty \sinh^{\lambda-1} r F(r)\, dr\,.$$

Let $Y_1, \ldots, Y_n$ be a basis of X_y such that the Lorentzian structure is given by

$$g_y(Y) = y_1^2 - y_2^2 - \cdots - y_n^2, \quad Y = \sum_1^n y_i Y_i.$$

If $\theta_1, \ldots, \theta_{n-2}$ are geodesic polar coordinates on the unit sphere in $\mathbf{R}^{n-1}$ we put

$$
\begin{aligned}
y_1 &= -r \cosh \zeta \quad (0 \le \zeta < \infty, 0 < r < \infty) \\
y_2 &= r \sinh \zeta \cos \theta_1 \\
&\vdots \\
y_n &= r \sinh \zeta \sin \theta_1 \ldots \sin \theta_{n-2}.
\end{aligned}
$$

Then $(r, \zeta, \theta_1, \ldots, \theta_{n-2})$ are coordinates on the retrograde cone D_y and the volume element on $S_r(y)$ is given by

$$d\omega^r = r^{n-1} \sinh^{n-2} \zeta \, d\zeta \, d\omega^{n-2}$$

where $d\omega^{n-2}$ is the volume element on the unit sphere in $\mathbf{R}^{n-1}$. It follows that

$$d\mathbf{w}^r = \sinh^{n-1} r \sinh^{n-2} \zeta \, d\zeta \, d\omega^{n-2}$$

and therefore

$$(30) \qquad F(r) = \iint (f \circ \mathrm{Exp})(r, \zeta, \theta_1, \ldots, \theta_{n-2}) \sinh^{n-2} \zeta d\zeta d\omega^{n-2},$$

where for simplicity

$$(r, \zeta, \theta_1, \ldots, \theta_{n-2})$$

stands for

$$(-r \cosh \zeta, r \sinh \zeta \cos \theta_1, \ldots, r \sinh \zeta \sin \theta_1 \ldots \sin \theta_{n-2}).$$

Now select A such that $f \circ \mathrm{Exp}$ vanishes outside the sphere $y_1^2 + \cdots + y_n^2 = A^2$ in X_y. Then, in the integral (30), the range of ζ is contained in the interval $(0, \zeta_o)$ where

$$r^2 \cosh^2 \zeta_o + r^2 \sinh^2 \zeta_o = A^2.$$

Then

$$r^{n-2} F(r) = \int_{\mathbf{S}^{n-2}} \int_0^{\zeta_o} (f \circ \mathrm{Exp})(r, \zeta, (\theta))(r \sinh \zeta)^{n-2} \, d\zeta \, d\omega^{n-2}.$$

Since

$$|r \sinh \zeta| \le r e^\zeta \le 2A \text{ for } \zeta \le \zeta_o$$

this implies

$$(31) \qquad |r^{n-2}(M^r f)(y)| \leq CA^{n-2} \sup |f|,$$

where C is a constant independent of r. Also substituting $t = r \sinh \zeta$ in the integral above, the ζ-integral becomes

$$\int_0^k \varphi(t) t^{n-2} (r^2 + t^2)^{-1/2} \, dt,$$

where $k = [(A^2 - r^2)/2]^{1/2}$ and φ is bounded. Thus if $n > 2$ the limit

$$(32) \qquad a = \lim_{r \to 0} \sinh^{n-2} r F(r) \quad n > 2$$

exist and is $\neq 0$. Similarly, we find for $n = 2$ that the limit

$$(33) \qquad b = \lim_{r \to 0} (\sinh r) F'(r) \qquad (n = 2)$$

exists.

Consider now the case $n > 2$. We can rewrite (29) in the form

$$(I_-^\lambda f)(y) = \frac{1}{H_n(\lambda)} \int_0^A \sinh^{n-2} r F(r) \sinh^{\lambda-n+1} r \, dr,$$

where $F(A) = 0$. We now evaluate both sides for $\lambda = n - 2$. Since $H_n(\lambda)$ has a simple pole for $\lambda = n - 2$ the integral has at most a simple pole there and the residue is

$$\lim_{\lambda \to n-2} (\lambda - n + 2) \int_0^A \sinh^{n-2} r F(r) \sinh^{\lambda-n+1} r \, dr.$$

Here we can take λ real and greater than $n - 2$. This is convenient since by (32) the integral is then absolutely convergent and we do not have to think of it as an implicitly given holomorphic extension. We split the integral in two parts

$$(\lambda - n + 2) \int_0^A (\sinh^{n-2} r F(r) - a) \sinh^{\lambda-n+1} r \, dr$$

$$+ a(\lambda - n + 2) \int_0^A \sinh^{\lambda-n+1} r \, dr.$$

For the last term we use the relation

$$\lim_{\mu \to 0+} \mu \int_0^A \sinh^{\mu-1} r \, dr = \lim_{\mu \to 0+} \mu \int_0^{\sinh A} t^{\mu-1} (1 + t^2)^{-1/2} \, dt = 1$$

by (38) in Chapter VII. For the first term we can for each $\epsilon > 0$ find a $\delta > 0$ such that

$$|\sinh^{n-2} r F(r) - a| < \epsilon \quad \text{for } 0 < r < \delta.$$

If $N = \max |\sinh^{n-2} r F(r)|$ we have for $n - 2 < \lambda < n - 1$ the estimate

$$\left|(\lambda - n + 2) \int_{\delta}^{A} (\sinh^{n-2} r F(r) - a) \sinh^{\lambda - n + 1} r \, dr\right|$$

$$\leq (\lambda - n + 2)(N + |a|)(A - \delta)(\sinh \delta)^{\lambda - n + 1};$$

$$\left|(\lambda + n - 2) \int_{0}^{\delta} (\sinh^{n-2} r F(r) - a) \sinh^{\lambda - n + 1} r \, dr\right|$$

$$\leq \epsilon(\lambda - n + 2) \int_{0}^{\delta} r^{\lambda - n + 1} \, dr = \epsilon \delta^{\lambda - n + 2}.$$

Taking $\lambda - (n - 2)$ small enough the right hand side of each of these inequalities is $< 2\epsilon$. We have therefore proved

$$\lim_{\lambda \to n-2} (\lambda - n + 2) \int_{0}^{\infty} \sinh^{\lambda - 1} r F(r) \, dr = \lim_{r \to 0} \sinh^{n-2} r \, F(r).$$

Taking into account the formula for $H_n(\lambda)$ we have proved for the integral (29):

$$(34) \qquad I_-^{n-2} f = (4\pi)^{(2-n)/2} \frac{1}{\Gamma((n-2)/2)} \lim_{r \to 0} \sinh^{n-2} r \, M^r f.$$

On the other hand, using formula (20) recursively, we obtain for $u \in \mathcal{D}(X)$,

$$I_-^{n-2}(Q(\Box)u) = u,$$

where

$$Q(\Box) = (\Box + (n - 3)2)(\Box + (n - 5)4) \cdots (\Box + 1(n - 2)).$$

We combine this with (34) and use the commutativity $\Box M^r = M^r \Box$. This gives

$$(35) \qquad u = (4\pi)^{(2-n)/2} \frac{1}{\Gamma((n-2)/2)} \lim_{r \to 0} \sinh^{n-2} r \, Q(\Box) M^r u.$$

Here we can for simplicity replace $\sinh r$ by r.

For the case $n = 2$ we have by (29)

$$(36) \qquad (I_-^2 f)(y) = \frac{1}{H_2(2)} \int_0^\infty \sinh r F(r)\, dr .$$

This integral, which in effect only goes from 0 to A, is absolutely convergent because our estimate (31) shows (for $n = 2$) that $rF(r)$ is bounded near $r = 0$. But using (20), Lemma 2.10, Theorem 2.11 and Cor. 2.14, we obtain for $u \in \mathcal{D}(X)$,

$$
\begin{aligned}
u &= I_-^2 \Box u = \tfrac{1}{2} \int_0^\infty \sinh r M^r \Box u \, dr \\[2mm]
&= \tfrac{1}{2} \int_0^\infty \sinh r \Box M^r u \, dr = \tfrac{1}{2} \int_0^\infty \sinh r \left(\frac{d^2}{dr^2} + \coth r \frac{d}{dr} \right) M^r u \, dr \\[2mm]
&= \tfrac{1}{2} \int_0^\infty \frac{d}{dr} \left(\sinh r \frac{d}{dr} M^r u \right) dr = -\tfrac{1}{2} \lim_{r \to 0} \sinh r \frac{d(M^r u)}{dr} .
\end{aligned}
$$

This is the substitute for (35) in the case $n = 2$.

The spaces G^+/H and G°/H can be treated in the same manner. We have thus proved the following principal result of this chapter.

Theorem 4.1. *Let X be one of the isotropic Lorentzian manifolds G^-/H, G°/H, G^+/H. Let κ denote the curvature of X ($\kappa = -1, 0, +1$) and assume $n = \dim X$ to be even, $n = 2m$. Put*

$$Q(\Box) = (\Box - \kappa(n-3)2)(\Box - \kappa(n-5)4) \cdots (\Box - \kappa 1(n-2)) .$$

Then if $u \in \mathcal{D}(X)$

$$
\begin{aligned}
u &= c \lim_{r \to 0} r^{n-2} Q(\Box)(M^r u), && (n \neq 2) \\[2mm]
u &= \tfrac{1}{2} \lim_{r \to 0} r \frac{d}{dr}(M^r u) && (n = 2) .
\end{aligned}
$$

Here $c^{-1} = (4\pi)^{m-1}(m-2)!$ and $\Box$ is the Laplace-Beltrami operator on X.

§5   Orbital Integrals and Huygens' Principle

We shall now write out the limit in (35) and thereby derive a statement concerning Huygens' principle for $\Box$. As $r \to 0$, $\mathbf{S}_r(o)$ has as limit the boundary $C_R = \partial \mathbf{D}_o - \{o\}$ which is still an H-orbit. The limit

$$(37) \qquad \lim_{r \to 0} r^{n-2}(M^r v)(o), \qquad v \in C_c(X - o),$$

is by (31)–(32) a positive H-invariant functional with support in the H-orbit C_R, which is closed in $X - o$. Thus the limit (37) only depends on the restriction $v|C_R$. Hence it is "the" H-invariant measure on C_R and we denote it by μ. Thus

$$\text{(38)} \qquad \lim_{r \to 0} r^{n-2} (M^r v)(o) = \int_{C_R} v(z) \, d\mu(z) \, .$$

To extend this to $u \in \mathcal{D}(X)$, let $A > 0$ be arbitrary and let φ be a "smoothed out" characteristic function of $\operatorname{Exp} B_A$. Then if

$$u_1 = u\varphi \, , \ u_2 = u(1 - \varphi)$$

we have

$$\left| r^{n-2}(M^r u)(o) - \int_{C_R} u(z) \, d\mu(z) \right|$$

$$\leq \left| r^{n-2}(M^r u_1)(o) - \int_{C_R} u_1(z) \, d\mu(z) \right| + \left| r^{n-2}(M^r u_2)(o) - \int_{C_R} u_2(z) \, d\mu(z) \right| .$$

By (31) the first term on the right is $O(A)$ uniformly in r and by (38) the second tends to 0 as $r \to 0$. Since A is arbitrary (38) holds for $u \in \mathcal{D}(X)$.

Proposition 5.1 (Huygens' Principle). *Let $n = 2m \, (m > 1)$ and δ the delta distribution at o. Then*

$$\text{(39)} \qquad \delta = c \, Q(\square)\mu \, ,$$

where $c^{-1} = (4\pi)^{m-1}(m - 2)!$.

In fact, by (35), (38) and Theorem 2.11

$$u = c \lim_{r \to 0} r^{n-2}(M^r Q(\square)u)(o) = c \int_{C_R} (Q(\square)u)(z) \, d\mu(z)$$

and this is (39).

Remark 5.2. Formula (39) shows that each factor

$$\text{(40)} \qquad \square_k = \square - \kappa(n - k)(k - 1) \quad k = 3, 5, \ldots, n - 1$$

in $Q(\square)$ has fundamental solution supported on the *retrograde conical surface* $\overline{C}_R$. This is known to be the equivalent to the validity of Huygens' principle for the Cauchy problem for the equation $\square_k u = 0$ (see Günther [1991] and [1988], Ch. IV, Cor. 1.13). For a recent survey on Huygens' principle see Berest [1998].

Bibliographical Notes

§1.   The construction of the constant curvature spaces (Theorems 1.3 and 1.5) was given by the author ([1959], [1961]). The proof of Lemma 1.4 on the connectivity is adapted from Boerner [1955]. For more information on isotropic manifolds (there is more than one definition) see Tits [1955], p. 183 and Wolf [1967].

§§2-4.   This material is based on Ch. IV in Helgason [1959]. Proposition 5.1 with a different proof and the subsequent remark were shown to me by Schlichtkrull. See Schimming and Schlichtkrull [1994] (in particular Lemma 6.2) where it is also shown that the constants $c_k = -\kappa(n-k)(k-1)$ in (40) are *the only ones* for which $\square + c_k$ satisfies Huygens' principle. Here it is of interest to recall that in the flat Lorentzian case $\mathbf{R}^{2m}$, $\square + c$ satisfies Huygens' principle only for $c = 0$. Theorem 4.1 was extended to pseudo-Riemannian manifolds of constant curvature by Orloff [1985], [1987]. For more recent representative work on orbital integrals see e.g. Bouaziz [1995], Flicker [1996], Harinck [1998], Renard [1997].

For further work on Lorentzian manifolds with contact with Theorem 4.1 see Kurusa [1997] and [2000].

CHAPTER VI

THE MEAN-VALUE OPERATOR

§1 An Injectivity Result

It is a consequence of a well-known theorem of Godement [1952] that if X is a two-point homogeneous space and u a harmonic function on X, i.e., $Lu = 0$, then u has the mean-value property

$$(1) \qquad\qquad u = M^r u \qquad \text{for each } r \,.$$

Conversely (1) implies that u is harmonic. Suitably restated the result holds for open subsets of X, thus also for compact X.

We shall now see that harmonicity of u for X compact follows from the validity of (2) for a *single* r. Let d denote the diameter of X.

Theorem 1.1. *Let $X = U/K$ be a compact two-point homogeneous space.*

(i)   If $f \in \mathcal{E}(X)$ satisfies the mean-value relation

$$(2) \qquad\qquad M^\rho f = f$$

for a single ρ, $0 < \rho < d$ then f is a constant.

Proof. We may assume f real-valued. Let x_0 be a maxiumum point for f. Then

$$(3) \qquad\qquad (M^\rho(f(x_0) - f))(x_0) = 0$$

so since the integrand in this equation is positive on $S_\rho(x_0)$ we deduce

$$(4) \qquad\qquad f = f(x_0) \qquad \text{on } S_\rho(x_0) \,.$$

Let $y \in X$ be arbitrary and let S be a sphere in Theorem 1.5, Ch. IV, totally geodesic in X containing x_0 and y. We can join x_0 to y by geodesic arcs in S each of length ρ. Let $x_0, x_1, \ldots, x_k = y$ be the endpoints of these arcs. Then we have from (4)

$$f(x_1) = f(x_0)$$

and since $x_{i+1} \in S_\rho(x_i)$ we obtain inductively,

$$(5) \qquad\qquad f(x_{i+1}) = f(x_0)$$

and finally

$$(6) \qquad\qquad f(y) = f(x_0) \,,$$

proving the result.

S. Helgason, *Integral Geometry and Radon Transforms*,
DOI 10.1007/978-1-4419-6055-9_6, © Springer Science+Business Media, LLC 2010

We consider now a noncompact two-point homogeneous space X.

Theorem 1.2. *For a fixed $r > 0$ the transform*

$$f(x) \to (M^r f)(x)$$

is injective on $L^1(X)$.

Proof. As in Ch. IV we represent X as G/K and the mean-value operator is for $g \in G$

$$(M^r f)(g \cdot o) = \int_K f(gk \cdot z) \, dk \, ,$$

where $z \in S_r(o)$. The operator M^r is symmetric, that is

(7) $$\int_X (M^r f)(x)\varphi(x) \, dx = \int_X f(x)(M^r \varphi)(x) \, dx \, .$$

Selecting $h \in G$ such that $h \cdot o = z$ the left hand side equals

$$\int_G \left(\int_K f(gkh \cdot o) \, dk \right) \varphi(g \cdot o) \, dg \, .$$

Interchanging integrations this reduces to

$$\int_G f(g \cdot o)\varphi(gh^{-1} \cdot o) \, dg \, ,$$

which can be written

$$\int_G f(g \cdot o)\left(\int_K \varphi(gkh^{-1} \cdot o) \, dk \right) dg \, .$$

Since $d(h^{-1} \cdot o, o) = r$, d denoting distance, (7) follows.

Let μ_z denote the normalized invariant measure on the orbit $K \cdot z$. By (24) in Chapter II,

$$(f \times \mu_z)(\varphi) = \int_G f(g \cdot o)\mu_z(\varphi^{\tau(g^{-1})}) \, dg$$

and

$$\mu_z(\varphi^{\tau(g^{-1})}) = \int_K \varphi(gk \cdot z) \, dk = (M^r \varphi)(g \cdot o) \, .$$

Thus (7) implies

(8) $$f \times \mu_z = M^r f \, .$$

Thus we must show that the convolution relation

$$f \times \mu_z \equiv 0$$

implies $f = 0$ almost everywhere.

Since $f \in L^1(X)$ we can consider its Fourier transform (Helgason [2005], Theorem 8.1) and conclude since μ_z is K-invariant that

$$\widetilde{f}(\lambda, b)\widetilde{\mu}_z(\lambda) = 0 \qquad \lambda \in \mathbf{R},$$

for almost all $b \in K/M$. (Here M is the centralizer of $\mathfrak{a}$ in K.)

Since $\widetilde{\mu}_z$ is holomorphic on $\mathbf{C}$ and $\not\equiv 0$ this implies $\widetilde{f}(\lambda, b) = 0$ for $\lambda \in \mathbf{R}$ so by *loc. cit.* $f = 0$.

§2 Ásgeirsson's Mean-Value Theorem Generalized

In his paper [1937], Ásgeirsson proved the following result. Let $u \in C^2(\mathbf{R}^{2n})$ satisfy the ultrahyperbolic equation

(9)
$$\frac{\partial^2 u}{\partial x_1^2} + \cdots + \frac{\partial^2 u}{\partial x_n^2} = \frac{\partial^2 u}{\partial y_1^2} + \cdots + \frac{\partial^2 u}{\partial y_n^2}.$$

Then for each $r \geq 0$ and each point

$$(x_0, y_0) = (x_1^0, \ldots, x_n^0, y_1^0, \ldots, y_n^0) \in \mathbf{R}^{2n}$$

the following identity holds for the $(n-1)$-dimensional spherical integrals:

(10)
$$\int_{S_r(x_0)} u(x, y_0) \, d\omega(x) = \int_{S_r(y_0)} u(x_0, y) \, d\omega(y).$$

The theorem was proved in the quoted paper for u of class C^2 in a suitable region of $\mathbf{R}^{2n}$. We now state and prove a generalization to a two-point homogeneous space.

Theorem 2.1. *Let X be a symmetric space of rank one (or $\mathbf{R}^n$) and let the function $u \in C^2(X \times X)$ satisfy the differential equation*

(11)
$$L_x(u(x, y)) = L_y(u(x, y))$$

on $X \times X$. Then for each $r \geq 0$ and each $(x_0, y_0) \in X \times X$ we have

(12)
$$\int_{S_r(x_0)} u(x, y_0) \, d\omega(x) = \int_{S_r(y_0)} u(x_0, y) \, d\omega(y).$$

Proof. If $n = 1$, the solutions to (11) have the form

$$u(x, y) = \varphi(x + y) + \psi(x - y)$$

and (12) reduces to the obvious relation

$$u(x_0 - r, y_0) + u(x_0 + r, y_0) = u(x_0, y_0 - r) + u(x_0, y_0 + r).$$

Hence we may assume that $n \geq 2$.

First, let $u = C^2(X \times X)$ be arbitrary. We form the function

$$(13) \qquad U(r, s) = (M_1^r M_2^s u)(x, y).$$

The subscripts 1 and 2 indicate that we consider the first and second variables, respectively; for example,

$$(M_1^r u)(x, y) = \frac{1}{A(r)} \int_{S_r(x)} u(s, y) \, d\omega(s).$$

Let $o \in X$ be the origin; write $X = G/K$, where $G = I_o(X)$, the identity component of $I(X)$, and K is the isotropy subgroup of G at o. If $g, h \in G$, $r = d(o, h \cdot o)$ (d = distance) we have as before

$$(14) \qquad (M^r f)(g \cdot o) = \int_K f(gkh \cdot o) \, dk, \quad f \in C(X),$$

where dk is the normalized Haar measure on K. Thus we have by interchanging the order of integration

$$(15) \qquad U(r, s) = (M_1^r M_2^s u)(x, y) = (M_2^s M_1^r u)(x, y).$$

Keeping y and s fixed, we have since L and M^r commute

$$\frac{\partial^2 U}{\partial r^2} + \frac{1}{A(r)} A'(r) \frac{\partial U}{\partial r} = L_x (M_1^r M_2^s u)(x, y)$$

$$= (M_1^r L_1 M_2^s u)(x, y) = (M_1^r M_2^s L_1 u)(x, y),$$

the last identity resulting from the fact that L_1 and M_2^s act on different arguments. Similarly, by using (15),

$$\frac{\partial^2 U}{\partial s^2} + \frac{1}{A(s)} A'(s) \frac{\partial U}{\partial s} = (M_2^s M_1^r L_2 u)(x, y).$$

Assuming now u is a solution to (11), we obtain from the above

$$\frac{\partial^2 U}{\partial r^2} + \frac{1}{A(r)} A'(r) \frac{\partial U}{\partial r} = \frac{\partial^2 U}{\partial s^2} + \frac{1}{A(s)} A'(r) \frac{\partial U}{\partial s}.$$

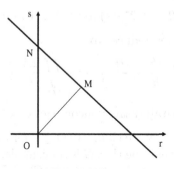

FIGURE VI.1.

Now putting $F(r, s) = U(r, s) - U(s, r)$, we obtain the relations

(16) $$\frac{\partial^2 F}{\partial r^2} + \frac{1}{A(r)}A'(r)\frac{\partial F}{\partial r} - \frac{\partial^2 F}{\partial s^2} - \frac{1}{A(s)}A'(s)\frac{\partial F}{\partial s} = 0,$$

$$F(r, s) = -F(s, r).$$

After multiplication of (16) by $2A(r)\partial F/\partial s$ and some manipulation we obtain

$$-A(r)\frac{\partial}{\partial s}\left[\left(\frac{\partial F}{\partial r}\right)^2 + \left(\frac{\partial F}{\partial s}\right)^2\right]$$

$$+ 2\frac{\partial}{\partial r}\left(A(r)\frac{\partial F}{\partial r}\frac{\partial F}{\partial s}\right) - \frac{2A(r)}{A(s)}\frac{dA}{ds}\left(\frac{\partial F}{\partial s}\right)^2 = 0.$$

Consider the line MN with equation $r + s = $ const. in the (r, s)-plane and integrate the last expression over the right-angled triangle OMN (see Fig. VI,1). Using the divergence theorem (Chapter I, §2, (26)) we then obtain, if $\mathbf{n}$ denotes the outgoing unit normal, $d\ell$ the element of arc length and dot the inner product

$$\int_{OMN} \left(2A(r)\frac{\partial F}{\partial r}\frac{\partial F}{\partial s}, -A(r)\left[\left(\frac{\partial F}{\partial r}\right)^2 + \left(\frac{\partial F}{\partial s}\right)^2\right]\right) \cdot \mathbf{n}\, d\ell$$

$$= \iint_{OMN} \frac{2A(r)}{A(s)}\frac{dA}{ds}\left(\frac{\partial F}{\partial s}\right)^2 dr\, ds.$$

On OMN we have $s \geq r$ so that for small r, s

$$\frac{A(r)}{A(s)}A'(s) \leq Cr^{n-1}s^{-1} \leq Cr^{n-2} \qquad (C = \text{const.}),$$

so that the last integral does indeed exist since we have assumed $n \geq 2$. We now use the following data:

On OM: $\quad \mathbf{n} = (2^{-1/2}, -2^{-1/2})$, $F(r,r) = 0$, so that $\dfrac{\partial F}{\partial r} + \dfrac{\partial F}{\partial s} = 0$.

On MN: $\quad \mathbf{n} = (2^{-1/2}, 2^{-1/2})$, on ON: $A(r) = 0$.

The formula above then reduces to

$$(17) \quad 2^{-1/2} \int_{MN} A(r) \left(\frac{\partial F}{\partial r} - \frac{\partial F}{\partial s} \right)^2 d\ell + \iint_{OMN} \frac{2A(r)}{A(s)} \frac{dA}{ds} \left(\frac{\partial F}{\partial s} \right)^2 dr\, ds = 0.$$

Now we consider separately the noncompact case and the compact case.

(I) X *noncompact.* Here we claim $A'(r) \geq 0$ for all $r \geq 0$. If $X = \mathbf{R}^n$, this is obvious, so we may assume that G is semisimple. But then the statement is clear from the analog of Lemma 2.1 in Ch. III for our noncompact X because sin has just to be replaced by sinh. Consequently, both terms in (17) vanish, so we can conclude that $F \equiv 0$. In particular, $U(r,0) = U(0,r)$, and this is the desired formula (12).

(II) X *compact.* In this case we know that $A'(s) \geq 0$ for s in a certain interval $0 \leq s \leq r_0$. As before, we can conclude that $U(r,0) = U(0,r)$ for r in this interval $0 \leq r \leq r_0$. In order to extend this to all r we approximate the solution u to (11) by analytic solutions. Let φ, ψ be analytic functions on the compact Lie group G with Haar measure dg and consider the convolution

$$(18) \quad u_{\varphi,\psi}(x,y) = \iint_{GG} u(g_1^{-1} \cdot x, g_2^{-1} \cdot y)\, \varphi(g_1)\, \varphi(g_2)\, dg_1\, dg_2$$

Then

$$L_x(u_{\varphi,\psi}(x,y)) = \iint_{GG} (L_1 u)(g_1^{-1} \cdot x, g_2^{-1} \cdot y)\, \varphi(g_1)\, \psi(g_2)\, dg_1\, dg_2$$

$$= \int (L_2 u)(g_1^{-1} \cdot x, g_2^{-1} \cdot y)\, \varphi(g_1)\, \psi(g_2)\, dg_1\, dg_2$$

$$= L_y(u_{\varphi,\psi}(x,y)),$$

so

$$(19) \quad \int_{S_r(x_0)} u_{\varphi,\psi}(x, y_0)\, d\omega(x) = \int_{S_r(y_0)} u_{\varphi,\psi}(x_0, y)\, d\omega(y)$$

for $0 \leq r \leq r_0$. For $f \in C(X \times Y)$ define the function $\widetilde{f}$ on $G \times G$ by $\widetilde{f}(g_1, g_2) = f(g_1 \cdot o, g_2 \cdot o)$; then (19) can be written

$$(20) \quad \int_K \widetilde{u}_{\varphi,\psi}(g_1 kh, g_2)\, dk = \int_K \widetilde{u}_{\varphi,\psi}(g_1, g_2 kh)\, dk$$

for all $h \in G$ such that $d(o, h \cdot o) \leq r_0$.

Changing variables in the integral (18), we see that $u_{\varphi,\psi}$ is analytic on $X \times X$, so $\tilde{u}_{\varphi,\psi}$ is analytic on $G \times G$. Since (20) holds for h varying in an open subset of G, it holds for all $h \in G$. Substituting the definition (18) into (20), we see that for each fixed $h \in G$, the function

$$(21) \qquad (z, w) \to \int_K \left[\tilde{u}(z^{-1}g_1kh, w^{-1}g_2) - \tilde{u}(z^{-1}g_1, w^{-1}g_2kh)\right] dk$$

on $G \times G$ is orthogonal to all functions of the form $\varphi(z)\psi(w)$, φ and ψ being analytic. By the Peter–Weyl theorem this remains true for $\varphi, \psi \in \mathcal{E}(G)$. But by the Stone–Weierstrass theorem these functions $\varphi(z)\psi(w)$ span a uniformly dense subspace of $C(G \times G)$. Thus the function (21) is identically 0 so (12) is proved for all r.

Definition. Let X be a two-point homogeneous space. The *Mean-Value transform* is the map $f \to \hat{f}$ where $f \in C^2(X)$ and

$$(22) \qquad \hat{f}(r, x) = (M^r f)(x),$$

the mean value of f on $S_r(x)$.

This transform could be considered for any Riemannian manifold. One can also specify for x to lie in a certain set S and investigate whether $(M^r f)(x)$ for $x \in S$ and $r \geq 0$ determines f.

Theorem 2.2. *The range of the map (22) is the null space of the Darboux operator*

$$(23) \qquad \frac{\partial^2}{\partial r^2} + \frac{A'(r)}{A(r)} \frac{\partial}{\partial r} - L.$$

For $\mathbf{H}^n$ we have shown (Ch. III, (33) and (36)) that the operator (23) annihilates $\hat{f}$. The proof depends only on the commutation of M^r and L and is thus valid for the present space X.

On the other hand, suppose $\varphi(r, x)$ is in the kernel of (23), φ assumed smooth and even in r. Defining $F(y, x) = \varphi(r, x)$ if $r = d(o, y)$ we have $L_y F = L_x F$ so Theorem 2.1 applies. Putting $f(x) = \varphi(o, x)$ the relation states $(M^r f)(x_0) = \varphi(r, x_o)$ so $\hat{f} = \varphi$.

§3   John's Identities

In his book, [1955] John introduced the iterated spherical mean on $\mathbf{R}^n$ by

$$M(z, r, s) = \frac{1}{\Omega_n^2} \int_{S^{n-1}} \int_{S^{n-1}} f(z + r\zeta + s\xi) \, d\omega_\zeta \, d\omega_\xi, \qquad f \in C(\mathbf{R}^n),$$

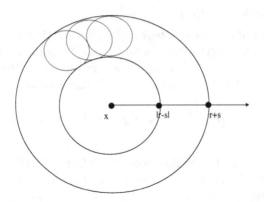

FIGURE VI.2

and proved the identity

$$(24) \qquad M(z,r,s) = \frac{2\Omega_{n-1}}{\Omega_n (2rs)^{n-2}} \cdot$$

$$\times \int_{|r-s|}^{r+s} [(t+r-s)(t+r+s)(t-r+s)(-t+r+s)]^{\frac{n-3}{2}} t(M^t f)(z)\, dt .$$

This formula reflects the fact that the torus can be swept out by spheres with center x and radii t $|r-s| < t < r+s$ and also by spheres of radius s with centers on the spheres $S_r(x)$, Fig. VI, 1.

The identity is easily proved by group theory using the formula

$$(M^x f)(z) = M^r f(z) = \int_K f(z + k \cdot x)\, dk \quad \text{for} \quad |x| = r .$$

Then

$$M(z,r,s) = (M^y M^x f)(z) = \int_K \int_K f(z + \ell \cdot x + k \cdot y)\, d\ell\, dk$$

if $|y| = s, |x| = r$. This equals

$$\int_K \int_K f(z + \ell \cdot x + \ell k \cdot y)\, dk\, d\ell = \int_K (M^{x+k \cdot y} f)(z)\, dk .$$

Taking $y = se_n$, $x = re_n$ the integral is

$$\int_{\mathbf{S}^{n-1}(0)} (M^{x+sw} f)(z)\, dw .$$

Let θ denote the angle between e_n and ω. In the integral we first let ω vary in the section of $\mathbf{S}^{n-1}(0)$ with the plane $(e_n, y) = -\cos\theta$. Since $|x + s\omega|^2 = r^2 + s^2 - 2rs\cos\theta$ the last integral equals

$$\frac{\Omega_{n-1}}{\Omega_n} \int_0^\pi \left(M^{(r^2 + s^2 - 2rs\cos\theta)^{1/2}} f\right)(z) \sin^{n-2}\theta \, d\theta.$$

Using the substitution

$$t = r^2 + s^2 - 2rs\cos\theta$$

formula (24) follows by direct computation.

John has also (*loc. cit.*, Ch. IV) proved formulas describing $f(x)$ explicitly in terms of the iterated mean value.

Exercises and Further Results

Support and Injectivity Results.

1.   Given $a \in \mathbf{R}^n$ let $D(a) = B_{|a|/2}\left(\frac{a}{2}\right)$. Let $G \subset \mathbf{R}$ be an open connected set containing O and let $\widetilde{G} = \bigcup_{a \in G} D(a)$. Assume $f \in C(\widetilde{G})$ is C^∞ at 0 and that

$$\int_{D(a)} f(x) \, dx = 0 \quad \text{for all } a \in \widetilde{G}.$$

Then $f \equiv 0$ on G. (Grinberg-Quinto [1999].)

2.   Let $\xi \subset \mathbf{R}^n$ be a hyperplane and assume $f \in C(\mathbf{R}^n)$ has the property

$$(M^r f)(x) = 0 \quad \text{for} \ \ x \in \xi, \quad r > 0.$$

Then f is odd relative to ξ. (Courant-Hilbert [1937].)

3.   A subset $S \subset \mathbf{R}^n$ is a *set of injectivity* for the mean value operator M if

$$(M^r f)(x) = 0 \quad \text{for all} \ \ x \in S, \quad r > 0$$

and some $f \in C_c(M)$ implies $f \equiv 0$.

For $n = 2$ the following results hold. (Agranovsky, Quinto [1996].)

Given $N \in \mathbf{Z}^+$ let $\sum_N$ denote the set of lines $L_0, \ldots, L_{N-1}$ in $\mathbf{R}^2$ where

$$L_k = \{te^{\pi ik/N} : t \subset \mathbf{R}\}.$$

The L_k pass through 0 and through $2N^{\text{th}}$ root of unity. Then

(i)   A set of $S \subset \mathbf{R}^2$ is a set of *injectivity* for M if and only if S is not contained in any set of the form

$$\sigma\left(\Sigma_N\right) \cup F,$$

where $\sigma \in \mathbf{M}(2)$ and F a finite set.

(ii)   By duality this property is equivalent to the density in $C(\mathbf{R}^2)$ of the span of the radial functions $g(x) = F(|x - a|)$, $a \in S$.

(iii)   Given $f \in C_c(\mathbf{R}^2)$ and $k \in \mathbf{Z}^+$, let

$$Q_k[f] = f * |x|^{2k}, \quad S[f] = \{a \in \mathbf{R}^2 : (M^r f)(a) = 0, \text{ all } r\}.$$

Then

(a) $S[f] = \bigcap_{k=0}^{\infty} N(Q_k[f])$ where $N(Q_k)$ is the zero set of the polynomial Q_k.

(b) Assume $f \neq 0$. Then $Q_k[f] \neq 0$ for some k. If k_0 is the smallest such k then $Q_{k_0}[f] = f * |x|^{2k_0}$ is a harmonic polynomial. In fact, $a \in S[f]$ if and only if $(M^r f)(a) = 0$ for all r. We have the relation

$$\int_0^{\infty} (M^r f)(a) r^{2k} \, dr = (f * |x|^{2k})(a) = Q_k[f](a),$$

proving (a). We then have by Ch. VII, Lemma 6.3,

$$L Q_{k_0}[f] = f * L|x|^{2k_0} = cf * |x|^{2k_0 - 2} = 0$$

proving (b).

(iv)   Show by (i) or (iii) that every closed curve in $\mathbf{R}^2$ is a set of injectivity for M.

4.   A very general two radius theorem is given by Zhou [2001] and Zhou-Quinto [2000]:

Let M be a real analytic manifold with injectivity radius $I_M > 0$. Let $0 < a < b < a + b < 2I_M$ and assume a/b irrational. Suppose $f \in C(M)$ satisfies

(i) $M^{\frac{a}{2}} f = M^{\frac{b}{2}} f = 0$.

(ii) $f \equiv 0$ on some sphere S_0 of radius $r < I_M$.

Then $f \equiv 0$ on M.

5.   The double fibration in (7), Ch. II, §2 has the Radon transform

$$\widehat{f}(\xi) = \frac{1}{\Omega_n} \int_{\xi} f(x) \, dm(x) \qquad \xi \text{ a unit sphere, } dm \text{ Euclidean measure}.$$

Writing $\xi = y + \xi_0$ where $\xi_0 = S_1(0)$

$$\widehat{f}(y + \xi_0) = \frac{1}{\Omega_n} \int_{\xi_0} f(y + x)\, dm(x) = (M^1 f)(y)\,.$$

(i)   Assume $n = 3$. Show using (25) that $f \to \widehat{f}$ is injective on the space of $f \in C(\mathbf{R}^3)$ satisfying

$$|x| f(x) \to 0 \qquad \text{as } |x| \to \infty\,.$$

(ii)   If $\widehat{f} \equiv 0$ and $f(x) = 0$ for $|x| < 1 + \epsilon$ then $f \equiv 0$.

(iii)   Assuming $\widehat{f}(x)$ and its derivatives $O\left(\frac{1}{|x|^3}\right)$

$$f(x) = -\sum_0^\infty \frac{1}{2\pi(2n+1)} \int_{|y|=2n+1} L_x \widehat{f}(x + y)\, dm(y)\,.$$

For (i), (ii) and (iii) see John [1955], Ch. VI.

Bibliographical Notes

While Theorem 1.2 might be new, Theorem 1.1 is contained in Günther [1966]. Theorem 2.1, generalizing Ásgeirsson's theorem is from Helgason [1959], even for arbitrary Riemannian homogeneous spaces. The transform (22), with x restricted to a subset, has been extensively investigated, see e.g. Agranovski–Quinto [1996] and Agranovski–Berenstein–Kuchment [1996] for a sample.

CHAPTER VII

FOURIER TRANSFORMS AND DISTRIBUTIONS.
A RAPID COURSE

§1 The Topology of Spaces $\mathcal{D}(\mathbf{R}^n)$, $\mathcal{E}(\mathbf{R}^n)$, and $\mathcal{S}(\mathbf{R}^n)$

Let $\mathbf{R}^n = \{x = (x_1, \ldots, x_n) : x_i \in \mathbf{R}\}$ and let ∂_i denote $\partial/\partial x_i$. If $(\alpha_1, \ldots, \alpha_n)$ is an n-tuple of integers $\alpha_i \geq 0$ we put $\alpha! = \alpha_1! \cdots \alpha_n!$,

$$D^\alpha = \partial_1^{\alpha_1} \ldots \partial_n^{\alpha_n}, \quad x^\alpha = x_1^{\alpha_1} \ldots x_n^{\alpha_n}, \quad |\alpha| = \alpha_1 + \cdots + \alpha_n.$$

For a complex number c, $\operatorname{Re} c$ and $\operatorname{Im} c$ denote respectively, the real part and the imaginary part of c. For a given compact set $K \subset \mathbf{R}^n$ let

$$\mathcal{D}_K = \mathcal{D}_K(\mathbf{R}^n) = \{f \in \mathcal{D}(\mathbf{R}^n) : \operatorname{supp}(f) \subset K\},$$

where supp stands for support. The space $\mathcal{D}_K$ is topologized by the semi-norms

$$(1) \qquad \|f\|_{K,m} = \sum_{|\alpha| \leq m} \sup_{x \in K} |(D^\alpha f)(x)|, \quad m \in \mathbf{Z}^+.$$

The topology of $\mathcal{D} = \mathcal{D}(\mathbf{R}^n)$ is defined as the largest locally convex topology for which all the embedding maps $\mathcal{D}_K \to \mathcal{D}$ are continuous. This is the so-called **inductive limit** topology. More explicitly, this topology is characterized as follows:

A *convex* set $C \subset \mathcal{D}$ is a neighborhood of 0 in $\mathcal{D}$ if and only if for each compact set $K \subset \mathbf{R}^n$, $C \cap \mathcal{D}_K$ is a neighborhood of 0 in $\mathcal{D}_K$.

A fundamental system of neighborhoods in $\mathcal{D}$ can be characterized by the following theorem. If B_R denotes the ball $|x| < R$ in $\mathbf{R}^n$ then

$$(2) \qquad \mathcal{D} = \cup_{j=0}^\infty \mathcal{D}_{\overline{B}_j}.$$

Theorem 1.1. *Given two monotone sequences*

$$\{\epsilon\} = \epsilon_0, \epsilon_1, \epsilon_2, \ldots, \qquad \epsilon_i \to 0$$
$$\{N\} = N_0, N_1, N_2, \ldots, \qquad N_i \to \infty \quad N_i \in \mathbf{Z}^+$$

let $V(\{\epsilon\}, \{N\})$ denote the set of functions $\varphi \in \mathcal{D}$ satisfying for each j the conditions

$$(3) \qquad |(D^\alpha \varphi)(x)| \leq \epsilon_j \quad for \ |\alpha| \leq N_j, \quad x \notin B_j.$$

Then the sets $V(\{\epsilon\}, \{N\})$ form a fundamental system of neighborhoods of 0 in $\mathcal{D}$.

S. Helgason, *Integral Geometry and Radon Transforms*,
DOI 10.1007/978-1-4419-6055-9_7, © Springer Science+Business Media, LLC 2010

Proof. It is obvious that each $V(\{\epsilon\}, \{N\})$ intersects each $\mathcal{D}_K$ in a neighborhood of 0 in $\mathcal{D}_K$. Conversely, let W be a *convex* subset of $\mathcal{D}$ intersecting each $\mathcal{D}_K$ in a neighborhood of 0. For each $j \in \mathbf{Z}^+$, $\exists N_j \in \mathbf{Z}^+$ and $\eta_j > 0$ such that each $\varphi \in \mathcal{D}$ satisfying

$$|D^\alpha \varphi(x)| \leq \eta_j \text{ for } |\alpha| \leq N_j, \quad \operatorname{supp}(\varphi) \subset \overline{B}_{j+2}$$

belongs to W. Fix a sequence (β_j) with

$$\beta_j \in \mathcal{D}, \beta_j \geq 0, \ \Sigma\beta_j = 1, \ \operatorname{supp}(\beta_j) \subset \overline{B}_{j+2} - B_j$$

and write for $\varphi \in \mathcal{D}$,

$$\varphi = \sum_j \frac{1}{2^{j+1}} (2^{j+1} \beta_j \varphi) .$$

Then by the convexity of W, $\varphi \in W$ if each function $2^{j+1} \beta_j \varphi$ belongs to W. However, $D^\alpha(\beta_j\varphi)$ is a finite linear combination of derivatives $D^\beta \beta_j$ and $D^\gamma \varphi$, $(|\beta|, |\gamma| \leq |\alpha|)$. Since (β_j) is fixed and only values of φ in $\overline{B}_{j+2} - B_j$ enter, $\exists$ constant k_j such that the condition

$$|(D^\alpha\varphi)(x)| \leq \epsilon_j \text{ for } |x| \geq j \text{ and } |\alpha| \leq N_j$$

implies

$$|2^{j+1} D^\alpha(\beta_j\varphi)(x)| \leq k_j\epsilon_j \quad \text{for } |\alpha| \leq N_j, \quad \text{all } x .$$

Choosing the sequence $\{\epsilon\}$ such that $k_j\epsilon_j \leq \eta_j$ for all j we deduce for each j

$$\varphi \in V(\{\epsilon\}, \{N\}) \Rightarrow 2^{j+1}\beta_j\varphi \in W ,$$

whence $\varphi \in W$. This proves Theorem 1.1.

The space $\mathcal{E} = \mathcal{E}(\mathbf{R}^n)$ is topologized by the seminorms (1) for the varying K. Thus the sets

$$V_{j,k,\ell} = \{\varphi \in \mathcal{E}(\mathbf{R}^n) : \|\varphi\|_{\overline{B}_j, k} < 1/\ell\} \qquad j, k, \ell \in \mathbf{Z}^+$$

form a fundamental system of neighborhoods of 0 in $\mathcal{E}(\mathbf{R}^n)$. This system being countable the topology of $\mathcal{E}(\mathbf{R}^n)$ is defined by sequences: A point $\varphi \in \mathcal{E}(\mathbf{R}^n)$ belongs to the closure of a subset $A \subset \mathcal{E}(\mathbf{R}^n)$ if and only if φ is the limit of a sequence in A. It is important to realize that this fails for the topology of $\mathcal{D}(\mathbf{R}^n)$ since the family of sets $V(\{\epsilon\}, \{N\})$ is uncountable.

The space $\mathcal{S} = \mathcal{S}(\mathbf{R}^n)$ of rapidly decreasing functions on $\mathbf{R}^n$ is topologized by the seminorms (6), Ch. I. We can restrict the P in (6), Ch. I to polynomials with rational coefficients.

Lemma 1.2. $\mathcal{D}(\mathbf{R}^n)$ *is dense in* $\mathcal{S}(\mathbf{R}^n)$.

Proof. Let $f \in \mathcal{S}$. Select $\psi \in \mathcal{D}$ such that $\psi \equiv 1$ on $B_1(0)$. For $\delta > 0$ put $f_\delta(x) = f(x)\psi(\delta x)$. Then $f_\delta \in \mathcal{D}$. Then

$$f_\delta(x) - f(x) = f(x)(\psi(\delta x) - 1) \equiv 0 \quad \text{for } |x| < \frac{1}{\delta}.$$

We have

$$(4) \qquad x^\beta \left(D^\alpha f_\delta(x) - D^\alpha f(x) \right) = x^\beta D^\alpha f(x)(\psi(\delta x) - 1) + F_\delta(x),$$

where $F_\delta(x)$ is a linear combination of terms containing a power of δ. Given $\epsilon \in 0$ three exist a $\delta > 0$ such that the first term on the right in (4) is 0 for $|x| \leq \frac{1}{\delta}$ and less than ϵ for $|x| > \frac{1}{\delta}$ (since $f \in \mathcal{S}$). Also $|F_\delta(x)| < \epsilon$ for all x if δ is small enough.

In contrast to the space $\mathcal{D}$ the spaces $\mathcal{D}_K$, $\mathcal{E}$ and $\mathcal{S}$ are Fréchet spaces, they are complete and their topologies are given by a countable family of seminorms.

The spaces $\mathcal{D}_K(M)$, $\mathcal{D}(M)$ and $\mathcal{E}(M)$ can be topologized similarly if M is a manifold.

§2   Distributions

A **distribution** by definition is a member of the dual space $\mathcal{D}'(\mathbf{R}^n)$ of $\mathcal{D}(\mathbf{R}^n)$. By the definition of the topology of $\mathcal{D}$, $T \in \mathcal{D}'$ if and only if the restriction $T|\mathcal{D}_K$ is continuous for each compact set $K \subset \mathbf{R}^n$. Thus a linear form on $\mathcal{D}(\mathbf{R}^n)$ is a distribution if and only if for each K, $T(\varphi_i) \to 0$ for each sequence $(\varphi_i) \subset \mathcal{D}_K(\mathbf{R}^n)$ converging to 0. Each locally integrable function F on $\mathbf{R}^n$ gives rise to a distribution $\varphi \to \int \varphi(x) F(x)\, dx$ denoted T_F. A measure on $\mathbf{R}^n$ is also a distribution. The notion of a distribution extends in an obvious way to manifolds M.

The **derivative** $\partial_i T$ of a distribution T is by definition the distribution $\varphi \to -T(\partial_i \varphi)$. If $F \in C^1(\mathbf{R}^n)$ then the distributions $T_{\partial_i F}$ and $\partial_i(T_F)$ coincide (integration by parts).

A **tempered distribution** by definition is a member of the dual space $\mathcal{S}'(\mathbf{R}^n)$. Since the imbedding $\mathcal{D} \to \mathcal{S}$ is continuous the restriction of a $T \in \mathcal{S}'$ to $\mathcal{D}$ is a distribution; since $\mathcal{D}$ is dense in $\mathcal{S}$ two tempered distributions coincide if they coincide on $\mathcal{D}$. In this sense we have $\mathcal{S}' \subset \mathcal{D}'$.

Since distributions generalize measures it is sometimes convenient to write

$$T(\varphi) = \int \varphi(x)\, dT(x)$$

for the value of a distribution T on the function φ. A distribution T is said to be 0 on an open set $U \subset \mathbf{R}^n$ if $T(\varphi) = 0$ for each $\varphi \in \mathcal{D}$ with support contained in U. Let U be the union of all open sets $U_\alpha \subset \mathbf{R}^n$ on which

T is 0. Then $T = 0$ on U. In fact, if $f \in \mathcal{D}(U)$, $\operatorname{supp}(f)$ can be covered by finitely many U_α, say $U_1, \ldots, U_r$. Then $U_1, \ldots, U_r, \mathbf{R}^n - \operatorname{supp}(f)$ is a covering of $\mathbf{R}^n$. If $1 = \sum_1^{r+1} \varphi_i$ is a corresponding partition of unity we have $f = \sum_1^r \varphi_i f$ so $T(f) = 0$. The complement $\mathbf{R}^n - U$ is called the **support of** T, denoted $\operatorname{supp}(T)$.

A distribution T of compact support extends to a unique element of $\mathcal{E}'(\mathbf{R}^n)$ by putting

$$T(\varphi) = T(\varphi \varphi_0), \quad \varphi \in \mathcal{E}(\mathbf{R}^n)$$

if φ_0 is any function in $\mathcal{D}$ which is identically 1 on a neighborhood of $\operatorname{supp}(T)$. Since $\mathcal{D}$ is dense in $\mathcal{E}$, this extension is unique. On the other hand let $\tau \in \mathcal{E}'(\mathbf{R}^n)$, T its restriction to $\mathcal{D}$. Then T is a distribution. Also $\operatorname{supp}(T)$ is compact. Otherwise we could for each j find $\varphi_j \in \mathcal{E}$ such that $\varphi_j \equiv 0$ on $\overline{B}_j$ but $T(\varphi_j) = 1$. Then $\varphi_j \to 0$ in $\mathcal{E}$, yet $\tau(\varphi_j) = 1$ which is a contradiction.

This identifies $\mathcal{E}'(\mathbf{R}^n)$ with the space of distributions of compact support and we have the following canonical inclusions:

$$\begin{array}{ccccc} \mathcal{D}(\mathbf{R}^n) & \subset & \mathcal{S}(\mathbf{R}^n) & \subset & \mathcal{E}(\mathbf{R}^n) \\ \cap & & \cap & & \cap \\ \mathcal{E}'(\mathbf{R}^n) & \subset & \mathcal{S}'(\mathbf{R}^n) & \subset & \mathcal{D}'(\mathbf{R}^n). \end{array}$$

§3 Convolutions

For f and g in $L^1(\mathbf{R}^n)$ the convolution $f * g$ is defined by

$$(f * g)(x) = \int_{\mathbf{R}^n} f(x - y)g(y) \, dy.$$

We now state some simple results (Prop. 3.1–Prop. 3.4) for the extension of this operation to distributions, referring to Hörmander [1963] Chapter I or [1983], Chapter 4 for quick and easy proofs.

Proposition 3.1. *Given* $\varphi \in \mathcal{D}(\mathbf{R}^n)$ *and* $T \in \mathcal{D}'(\mathbf{R}^n)$ *the function* $\varphi * T$ *given by*

$$(\varphi * T)(x) = \int_{\mathbf{R}^n} \varphi(x - y) \, dT(y)$$

belongs to $\mathcal{E}(\mathbf{R}^n)$. *Moreover,*

$$\operatorname{supp}(\varphi * T) \subset \operatorname{supp}(\varphi) + \operatorname{supp}(T)$$

and

$$D^\alpha(\varphi * T) = D^\alpha \varphi * T = \varphi * D^\alpha T.$$

Proposition 3.2. *Given* $\varphi, \psi \in \mathcal{D}(\mathbf{R}^n)$, $T \in \mathcal{D}'(\mathbf{R}^n)$ *we have*

$$(5) \qquad\qquad (T * \varphi) * \psi = T * (\varphi * \psi).$$

Proposition 3.3. *Given* $S, T \in \mathcal{D}'(\mathbf{R}^n)$, *one of which has compact support, there exists a unique distribution, denoted* $S * T$, *such that*

$$S * (T * \varphi) = (S * T) * \varphi, \qquad \varphi \in \mathcal{D}(\mathbf{R}^n).$$

This convolution is commutative,

$$(6) \qquad\qquad S * T = T * S,$$

and associative,

$$(7) \qquad\qquad T_1 * (T_2 * T_3) = (T_1 * T_2) * T_3,$$

if all T_i, except at most one, has compact support. Moreover, as justified below,

$$(8) \qquad\qquad \operatorname{supp}(T * S) \subset \operatorname{supp} T + \operatorname{supp}(S).$$

Let $\varphi \in \mathcal{D}(\mathbf{R}^n)$ be ≥ 0, $\int \varphi(x)\, dx = 1$ and $\operatorname{supp}(\varphi) \subset B_1(0)$. The following **regularization** holds.

Proposition 3.4. *Let* $\varphi_\epsilon(x) = \epsilon^{-n} \varphi\left(\frac{x}{\epsilon}\right)$. *Then if* $T \in \mathcal{D}'(\mathbf{R}^n)$ *and* $\epsilon \to 0$,

$$(\varphi_\epsilon * T)(\varphi) \to T(\varphi) \qquad \text{for } \varphi \in \mathcal{D}(\mathbf{R}^n).$$

Let $\check{f}(x) = f(-x)$. Then

$$(\varphi * T)(0) = T(\check{\varphi}).$$

Let $f \in \mathcal{D}$ and $T \in \mathcal{D}'$ and put $g = f * T$ so

$$g(x) = \int f(x - y)\, dT(y)$$

Note that $g(x) = 0$ unless $x - y \in \operatorname{supp}(f)$ for some $y \in \operatorname{supp}(T)$. Thus $\operatorname{supp}(g) \subset \operatorname{supp}(f) + \operatorname{supp} T$. More generally,

$$\operatorname{supp}(S * T) \subset \operatorname{supp}(S) + \operatorname{supp}(T),$$

as one sees from the special case $S = T_g$ by approximating S by functions $S * \varphi_\epsilon$ with $\operatorname{supp}(\varphi_\epsilon) \subset B_\epsilon(0)$.

The convolution can be defined for more general S and T, for example if $S \in \mathcal{S}$, $T \in \mathcal{S}'$ then $S * T \in \mathcal{S}'$ (see Prop. 4.4). Also $S \in \mathcal{E}'$, $T \in \mathcal{S}'$ implies $S * T \in \mathcal{S}'$.

§4   The Fourier Transform

For $f \in L^1(\mathbf{R}^n)$ the **Fourier transform** is defined by

$$(9) \qquad \widetilde{f}(\xi) = \int_{\mathbf{R}^n} f(x) e^{-i\langle x,\xi\rangle} \, dx, \quad \xi \in \mathbf{R}^n .$$

If f has compact support we can take $\xi \in \mathbf{C}^n$. For $f \in \mathcal{S}(\mathbf{R}^n)$ one proves quickly

$$(10) \qquad i^{|\alpha|+|\beta|} \xi^\beta (D^\alpha \widetilde{f})(\xi) = \int_{\mathbf{R}^n} D^\beta (x^\alpha f(x)) e^{-i\langle x,\xi\rangle} \, dx$$

and this together with Theorem 4.2(i) below implies easily the following result.

Theorem 4.1. *The Fourier transform is a linear homeomorphism of $\mathcal{S}$ onto $\mathcal{S}$.*

The function $\psi(x) = e^{-x^2/2}$ on $\mathbf{R}$ satisfies $\psi'(x) + x\psi = 0$. It follows from (10) that $\widetilde{\psi}$ satisfies the same differential equation and thus is a constant multiple of $e^{-\xi^2/2}$. Since $\widetilde{\psi}(0) = \int e^{-\frac{x^2}{2}} \, dx = (2\pi)^{1/2}$ we deduce $\widetilde{\psi}(\xi) = (2\pi)^{1/2} e^{-\xi^2/2}$. More generally, if $\psi(x) = e^{-|x|^2/2}$, $(x \in \mathbf{R}^n)$ then by product integration

$$(11) \qquad \widetilde{\psi}(\xi) = (2\pi)^{n/2} e^{-|\xi|^2/2} .$$

Theorem 4.2. *The Fourier transform has the following properties.*

(i) $f(x) = (2\pi)^{-n} \int \widetilde{f}(\xi) e^{i\langle x,\xi\rangle} \, d\xi$ for $f \in \mathcal{S}$.

(ii) $f \to \widetilde{f}$ extends to a bijection of $L^2(\mathbf{R}^n)$ onto itself and

$$\int_{\mathbf{R}^n} |f(x)|^2 = (2\pi)^{-n} \int_{\mathbf{R}^n} |\widetilde{f}(\xi)|^2 \, d\xi .$$

(iii) $(f_1 * f_2)^\sim = \widetilde{f_1}\widetilde{f_2}$ for $f_1, f_2 \in \mathcal{S}$.

(iv) $(f_1 f_2)^\sim = (2\pi)^{-n} \widetilde{f_1} * \widetilde{f_2}$ for $f_1, f_2 \in \mathcal{S}$.

Proof. (i) The integral on the right equals

$$\int e^{i\langle x,\xi\rangle} \left(\int f(y) e^{-i\langle y,\xi\rangle} \, dy \right) d\xi$$

but here we cannot exchange the integrations. Instead we consider for $g \in \mathcal{S}$ the integral

$$\int e^{i\langle x,\xi\rangle} g(\xi) \left(\int f(y) e^{-i\langle y,\xi\rangle} \, dy \right) d\xi ,$$

which equals the expressions

$$(12) \qquad \int \widetilde{f}(\xi)g(\xi)e^{i\langle x,\xi\rangle}\, d\xi = \int f(y)\widetilde{g}(y-x)\, dy = \int f(x+y)\widetilde{g}(y)\, dy\,.$$

Replace $g(\xi)$ by $g(\epsilon\xi)$ whose Fourier transform is $\epsilon^{-n}\widetilde{g}(y/\epsilon)$. Then we obtain

$$\int \widetilde{f}(\xi)g(\epsilon\xi)e^{i\langle x,\xi\rangle}\, d\xi = \int \widetilde{g}(y)f(x+\epsilon y)\, dy\,,$$

which upon letting $\epsilon \to 0$ gives

$$g(0)\int \widetilde{f}(\xi)e^{i\langle x,\xi\rangle}\, d\xi = f(x)\int \widetilde{g}(y)\, dy\,.$$

Taking $g(\xi)$ as $e^{-|\xi|^2/2}$ and using (11) Part (i) follows. The identity in (ii) follows from (12) (for $x = 0$) and (i). It implies that the image $L^2(\mathbf{R}^n)^\sim$ is closed in $L^2(\mathbf{R}^n)$. Since it contains the dense subspace $\mathcal{S}(\mathbf{R}^n)$ (ii) follows. Formula (iii) is an elementary computation and now (iv) follows taking (i) into account. Part (iii) shows that the convolution $*$ is a continuous map from $\mathcal{S} \times \mathcal{S}$ into $\mathcal{S}$.

If $T \in \mathcal{S}'(\mathbf{R}^n)$ its Fourier transform is the linear form $\widetilde{T}$ on $\mathcal{S}(\mathbf{R}^n)$ defined by

$$(13) \qquad\qquad\qquad \widetilde{T}(\varphi) = T(\widetilde{\varphi})\,.$$

Then by Theorem 4.1, $\widetilde{T} \in \mathcal{S}'$. Note that

$$(14) \qquad\qquad \int \varphi(\xi)\widetilde{f}(\xi)\, d\xi = \int \widetilde{\varphi}(x)f(x)\, dx$$

for all $f \in L^1(\mathbf{R}^n)$, $\varphi \in \mathcal{S}(\mathbf{R}^n)$. Consequently

$$(15) \qquad\qquad\qquad (T_f)^\sim = T_{\widetilde{f}} \quad \text{for } f \in L^1(\mathbf{R}^n)$$

so the definition (13) extends the old one (9).

If $S, T \in \mathcal{E}'(\mathbf{R}^n)$ then tensor product $S \otimes T \in \mathcal{E}'(\mathbf{R}^n \times \mathbf{R}^n)$ is defined by

$$(S \otimes T)(\varphi) = \int_{\mathbf{R}^n \times \mathbf{R}^n} \varphi(x,y)\, dS(x)\, dT(y)\,, \quad \varphi \in \mathcal{D}(\mathbf{R}^n \times \mathbf{R}^n)\,.$$

Since the space spanned by functions of the form $\varphi_1(x)\varphi_2(y)$ $(\varphi_i \in \mathcal{D}(\mathbf{R}^n))$ is dense in $\mathcal{D}(\mathbf{R}^n \times \mathbf{R}^n)$ we have

$$(16) \qquad\qquad\qquad (S \otimes T)(\varphi) = (T \otimes S)(\varphi)$$

for all $\varphi \in \mathcal{D}(\mathbf{R}^n \times \mathbf{R}^n)$ and by continuity for all $\varphi \in \mathcal{E}(\mathbf{R}^n \times \mathbf{R}^n)$. Note that

$$(S * T)(\varphi) = \int \varphi(x+y)\, d(S \otimes T)(x,y)\,.$$

In fact

$$\int \left(\int \varphi(x+y)\, dS(x) \right) dT(y) = \int (S * \check{\varphi})(-y)\, dT(y)$$
$$= T(\check{S} * \varphi) = (T * (\check{S} * \varphi)^{\vee})(0)$$
$$= (T * S * \check{\varphi})(0) = (T * S)(\varphi).$$

Proposition 4.3. *If* $T \in \mathcal{E}'(\mathbf{R}^n)$ *then the distribution* $\widetilde{T}$ *equals the function*

(17)
$$\xi \to \int_{\mathbf{R}^n} e^{-i\langle x,\xi \rangle}\, dT(x), \quad \xi \in \mathbf{R}^n.$$

Taking ξ *in* $\mathbf{C}^n$ *this function becomes an entire function on* $\mathbf{C}^n$.

Proof. Let $\varphi \in \mathcal{D}(\mathbf{R}^n)$. Then

$$\widetilde{T}(\varphi) = T(\widetilde{\varphi}) = \int \widetilde{\varphi}(\xi)\, dT(\xi) = \left(\int e^{-i\langle x,\xi \rangle} \varphi(x)\, dx \right) dT(\xi)$$

which by (16) equals

$$\int_{\mathbf{R}^n} \varphi(x) \left(\int_{\mathbf{R}^n} e^{-i\langle x,\xi \rangle}\, dT(\xi) \right) dx,$$

showing that $\widetilde{T}$ equals the function (17). Let $e_{i\xi}(x) = e^{i\langle x,\xi \rangle}$. If $\xi_j \to \xi$ the function $e_{i\xi_j} \to e_{i\xi}$ in $\mathcal{E}(\mathbf{R}^n)$ so the function (17) is continuous. Also with φ_ϵ as in Prop. 3.4,

$$(T * \varphi_\epsilon)(e_{-i\xi}) = T * (\varphi_\epsilon * e_{i\xi})(0) = T * (e_{i\xi}\widetilde{\varphi}_\epsilon(\xi))(0)$$
$$= \widetilde{\varphi}(\epsilon\xi)T(e_{-i\xi}).$$

The left hand side is holomorphic in ξ and $\widetilde{\varphi}(\epsilon\xi) \to \widetilde{\varphi}(0) = 1$ uniformly on compact subsets of $\mathbf{C}^n$. Hence $T(e_{-i\xi})$ is holomorphic.

Proposition 4.4. *Let* $T \in \mathcal{S}'(\mathbf{R}^n)$ *and* $f \in \mathcal{S}(\mathbf{R}^n)$. *Then*

$$f * T \in \mathcal{S}'(\mathbf{R}^n).$$

Proof. For $z \in \mathbf{R}^n$ let $f^z(x) = f(x-z)$. Then for some C and N

$$|(f * T)(x)| = |T_y(\check{f}(y-x))| = T((\check{f})^x)|$$

$$\leq C \sup_y \sum_{|\alpha| \leq N} |(1+|y|)^N (D^\alpha \check{f})(y-x)|$$

$$\leq C \sup_z \sum_{|\alpha| \leq N} |(1+|x+z|)^N (D^\alpha f)(z)|$$

$$\leq C(1+|x|)^N \sup_z \sum_{|\alpha| \leq N} |(1+|z|)^N (D^\alpha f)(z)|,$$

so $f * T$ is bounded by a polynomial, hence tempered. If $z_n \to z$ then, by Theorem 4.1 $f^{z_n} \to f^z$ in $\mathcal{S}$ so $f * T$ is continuous.

We now have an analog to Prop. 3.2.

Proposition 4.5. *Let* $T \in \mathcal{S}'(\mathbf{R}^n)$, $\varphi, \psi \in \mathcal{S}(\mathbf{R}^n)$. *Then*

$$(T * \varphi) * \psi = T * (\varphi * \psi).$$

Proof. First let $\psi \in \mathcal{D}$ and let $\varphi_i \in \mathcal{D}$ be a sequence converging to φ in $\mathcal{S}$. Since $(T * \varphi)(0) = T(\check{\varphi})$ we see that $(T * \varphi)(x) = \lim_i (T * \varphi_i)(x)$ and $(T * \varphi)(\psi) = \lim_i (T * \varphi_i)(\psi)$. Using Prop. 3.2 we conclude

$$(T * \varphi) * \check{\psi} = \lim_i (T * \varphi_i) * \check{\psi} = \lim_i T * (\varphi_i * \check{\psi})$$

$$= T * (\varphi * \check{\psi})$$

by continuity of $*$ in $\mathcal{S}$. Taking a sequence $(\psi_j) \subset \mathcal{D}$ converging to $\psi \in \mathcal{S}$ the result follows.

Theorem 4.6. *Let* $f \in \mathcal{S}(\mathbf{R}^n)$, $T \in \mathcal{S}'(\mathbf{R}^n)$. *Then*

$$(T * f)^\sim = \widetilde{f}\,\widetilde{T}.$$

Proof. We have for $g \in \mathcal{S}(\mathbf{R}^n)$ using Prop. 4.5,

$$(T * f)^\sim(g) = (T * f)(\widetilde{g}) = (T * f * (\widetilde{g})^\vee)(0)$$

$$= T * ((\check{f} * \widetilde{g})^\vee)(0) = T(\check{f} * \widetilde{g}) = (2\pi)^n T((\widetilde{f})^\sim * \widetilde{g})$$

$$= T((\widetilde{f}g)^\sim) = \widetilde{T}(\widetilde{f}g) = (\widetilde{f}\widetilde{T})(g).$$

The classical Paley–Wiener theorem gave an intrinsic description of $L^2(0, 2\pi)^\sim$. We now prove an extension to a characterization of $\mathcal{D}(\mathbf{R}^n)^\sim$ and $\mathcal{E}'(\mathbf{R}^n)^\sim$.

Theorem 4.7. *(i)   A holomorphic function* $F(\zeta)$ *on* $\mathbf{C}^n$ *is the Fourier transform of a distribution with support in* $\overline{B}_R$ *if and only if for some constants* C *and* $N \geq 0$ *we have*

(18) $$|F(\zeta)| \leq C(1 + |\zeta|^N) e^{R|\mathrm{Im}\,\zeta|}.$$

(ii) $F(\zeta)$ *is the Fourier transform of a function in* $\mathcal{D}_{\bar{B}_R}(\mathbf{R}^n)$ *if and only if for each* $N \in \mathbf{Z}^+$ *there exists a constant* C_N *such that*

$$(19) \qquad |F(\zeta)| \le C_N(1+|\zeta|)^{-N}e^{R|\operatorname{Im}\zeta|}.$$

Proof. First we prove that (18) is necessary. Let $T \in \mathcal{E}'$ have support in $\bar{B}_R$ and let $\chi \in \mathcal{D}$ have support in $\bar{B}_{R+1}$ and be identically 1 in a neighborhood of $\bar{B}_R$. Since $\mathcal{E}(\mathbf{R}^n)$ is topologized by the semi-norms (1) for varying K and m we have for some $C_0 \ge 0$ and $N \in \mathbf{Z}^+$

$$|T(\varphi)| = |T(\chi\varphi)| \le C_0 \sum_{|\alpha|\le N} \sup_{x \in \bar{B}_{R+1}} |(D^\alpha(\chi\varphi))(x)|.$$

Computing $D^\alpha(\chi\varphi)$ we see that for another constant C_1

$$(20) \qquad |T(\varphi)| \le C_1 \sum_{|\alpha|\le N} \sup_{x\in\mathbf{R}^n} |D^\alpha\varphi(x)|, \quad \varphi \in \mathcal{E}(\mathbf{R}^n).$$

Let $\psi \in \mathcal{E}(\mathbf{R})$ such that $\psi \equiv 1$ on $(-\infty, \frac{1}{2})$, and $\equiv 0$ on $(1,\infty)$. Then if $\zeta \ne 0$ the function

$$\varphi_\zeta(x) = e^{-i\langle x,\zeta\rangle}\psi(|\zeta|(|x| - R))$$

belongs to $\mathcal{D}$ and equals $e^{-i\langle x,\zeta\rangle}$ in a neighborhood of $\bar{B}_R$. Hence

$$|\widetilde{T}(\zeta)| = |T(\varphi_\zeta)| \le C_1 \sum_{|\alpha|\le N} \sup |D^\alpha\varphi_\zeta|.$$

Now $\operatorname{supp}(\varphi_\zeta) \subset \bar{B}_{R+|\zeta|^{-1}}$ and for x on this ball

$$|e^{-i\langle x,\zeta\rangle}| \le e^{|x|\,|\operatorname{Im}\zeta|} \le e^{(R+|\zeta|^{-1})|\operatorname{Im}\zeta|} \le e^{R|\operatorname{Im}\zeta|+1}.$$

Furthermore $D^\alpha\varphi_\zeta$ is a linear combination of terms

$$D_x^\beta(e^{-i\langle x,\zeta\rangle})D_x^\gamma(\psi(|\zeta|\,|x| - R)).$$

The first factor has estimate on this ball

$$|\zeta|^{|\beta|}|e^{-i\langle x,\zeta\rangle}| \le |\zeta|^{|\beta|}e^{R|\operatorname{Im}\zeta|+1}$$

and the second factor is bounded by a power of $|\zeta|$. Thus $\widetilde{T}(\zeta)$ satisfies (18).

The necessity of (19) is an easy consequence of (10).

Next we prove the sufficiency of (19). Let

$$(21) \qquad f(x) = (2\pi)^{-n}\int_{\mathbf{R}^n} F(\xi)e^{i\langle x,\xi\rangle}\,d\xi.$$

Because of (19) we can shift the integration in (21) to the complex domain so that for any fixed $\eta \in \mathbf{R}^n$,

$$(22) \qquad f(x) = (2\pi)^{-n} \int_{\mathbf{R}^n} F(\xi + i\eta) e^{i\langle x, \xi + i\eta \rangle} \, d\xi .$$

We use (19) for $N = n + 1$ to estimate this integral and this gives

$$|f(x)| \leq C_N e^{R|\eta| - \langle x, \eta \rangle} (2\pi)^{-n} \int_{\mathbf{R}^n} (1 + |\xi|)^{-(n+1)} \, d\xi .$$

Taking now $\eta = tx$ and letting $t \to +\infty$ we deduce $f(x) = 0$ for $|x| > R$.

For the sufficiency of (18) we note first that F as a distribution on $\mathbf{R}^n$ is tempered. Thus $F = \tilde{f}$ for some $f \in \mathcal{S}'(\mathbf{R}^n)$. Convolving f with a $\varphi_\epsilon \in \mathcal{D}_{\overline{B}_\epsilon}$ we see that $(f * \varphi_\epsilon)^{\sim}$ satisfies estimates (19) with R replaced by $R + \epsilon$. Thus $\text{supp}(f * \varphi_\epsilon) \subset \overline{B}_{R+\epsilon}$. Letting $\epsilon \to 0$ we deduce $\text{supp}(f) \subset \overline{B}_R$, concluding the proof.

Let $\mathcal{H}_R = \mathcal{H}_R(\mathbf{C})$ denote the space of holomorphic functions satisfying (19) with the topology defined by the seminorms

$$\|\|\varphi\|\|_N = \sup_{\zeta \in \mathbf{C}^n} (1 + |\zeta|)^N e^{-R|\text{Im}\,\zeta|} |\varphi(\zeta)| < \infty .$$

Corollary 4.8. *The Fourier transform $f \to \tilde{f}$ is a homeomorphism of $\mathcal{D}_{\overline{B}_R}(\mathbf{R}^n)$ onto $\mathcal{H}_R$.*

The range statement is Theorem 4.7 and the continuity statements follow from (10) extended from ξ to ζ and by differentiating (22).

We shall now prove a refinement of Theorem 4.7 in that the topology of $\mathcal{D}$ is described in terms of $\widetilde{\mathcal{D}}$. This has important applications to differential equations as we shall see in the next section.

Theorem 4.9. *A convex set $V \subset \mathcal{D}$ is a neighborhood of 0 in $\mathcal{D}$ if and only if there exist positive sequences*

$$M_0, M_1, \ldots, \quad \delta_0, \delta_1, \ldots$$

such that V contains all $u \in \mathcal{D}$ satisfying

$$(23) \qquad |\tilde{u}(\zeta)| \leq \sum_{k=0}^{\infty} \delta_k \frac{1}{(1 + |\zeta|)^{M_k}} e^{k|\text{Im}\,\zeta|}, \quad \zeta \in \mathbf{C}^n .$$

The proof is an elaboration of that of Theorem 4.7. Instead of the contour shift $\mathbf{R}^n \to \mathbf{R}^n + i\eta$ used there one now shifts $\mathbf{R}^n$ to a contour on which the two factors on the right in (19) are comparable.

Let $W(\{\delta\}, \{M\})$ denote the set of $u \in \mathcal{D}$ satisfying (23). Given k the set

$$W_k = \{u \in \mathcal{D}_{\overline{B}_k} : |\tilde{u}(\zeta)| \leq \delta_k (1 + |\zeta|)^{-M_k} e^{k|\text{Im}\,\zeta|}\}$$

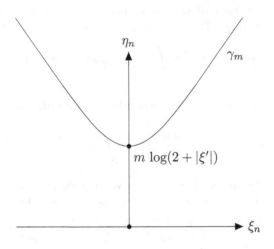

FIGURE VII.1.

is contained in $W(\{\delta\}, \{M\})$. Thus if V is a convex set containing $W(\{\delta\}, \{M\})$ then $V \cap \mathcal{D}_{\overline{B}_k}$ contains W_k which by Corollary 4.8 is a neighborhood of 0 in $\mathcal{D}_{\overline{B}_k}$. Thus V is a neighborhood of 0 in $\mathcal{D}$.

Proving the converse amounts to proving that given $V(\{\epsilon\}, \{N\})$ in Theorem 1.1 there exist $\{\delta\}, \{M\}$ such that

$$W(\{\delta\}, \{M\}) \subset V(\{\epsilon\}, \{N\}) \,.$$

For this we shift the contour in (22) to others where the two factors in (19) are comparable. Let

$$x = (x_1, \ldots, x_n), \qquad x' = (x_1, \ldots, x_{n-1})$$
$$\zeta = (\zeta_1, \ldots, \zeta_n) \qquad \zeta' = (\zeta_1, \ldots, \zeta_{n-1})$$
$$\zeta = \xi + i\eta, \qquad \xi, \eta \in \mathbf{R}^n \,.$$

Then

$$(24) \qquad \int_{\mathbf{R}^n} \widetilde{u}(\xi) e^{i\langle x, \xi \rangle} \, d\xi = \int_{\mathbf{R}^{n-1}} e^{i\langle x', \xi' \rangle} \, d\xi' \int_{\mathbf{R}} e^{ix_n \xi_n} \widetilde{u}(\xi', \xi_n) \, d\xi_n \,.$$

In the last integral we shift from $\mathbf{R}$ to the contour in $\mathbf{C}$ given by

$$(25) \qquad \gamma_m : \zeta_n = \xi_n + im \log(2 + (|\xi'|^2 + \xi_n^2)^{1/2})$$

$m \in \mathbf{Z}^+$ being fixed.

We claim that (cf. Fig. VII.1)

$$(26) \qquad \int_{\mathbf{R}} e^{ix_n \xi_n} \widetilde{u}(\xi', \xi_n) \, d\xi_n = \int_{\gamma_m} e^{ix_n \zeta_n} \widetilde{u}(\xi', \zeta_n) \, d\zeta_n \,.$$

Since (19) holds for each N, $\tilde{u}$ decays between ξ_n–axis and γ_m faster than any $|\zeta_n|^{-M}$ with $M > 0$. Also

$$\left|\frac{d\zeta_n}{d\xi_n}\right| = \left|1 + im\frac{1}{2 + |\xi|} \cdot \frac{\partial(|\xi|)}{\partial\xi_n}\right| \leq 1 + m.$$

Thus (26) follows from Cauchy's theorem in **one** variable. Putting

$$\Gamma_m = \{\zeta \in \mathbf{C}^n : \zeta' \in \mathbf{R}^{n-1}, \zeta_n \in \gamma_m\}$$

we thus have with $d\zeta = d\xi_1 \ldots d\xi_{n-1}\,d\zeta_n$,

$$(27) \qquad u(x) = (2\pi)^{-n} \int_{\Gamma_m} \tilde{u}(\zeta)e^{i\langle x, \zeta\rangle}\,d\zeta.$$

Now suppose the sequences $\{\epsilon\}$, $\{N\}$ and $V(\{\epsilon\}, \{N\})$ are given as in Theorem 1.1. We have to construct sequences $\{\delta\}$ $\{M\}$ such that (23) implies (3). By rotational invariance we may assume $x = (0, \ldots, 0, x_n)$ with $x_n > 0$. For each n-tuple α we have

$$(D^\alpha u)(x) = (2\pi)^{-n} \int_{\Gamma_m} \tilde{u}(\zeta)(i\zeta)^\alpha e^{i\langle x, \zeta\rangle}\,d\zeta.$$

It is important to note that here we can choose m at will. Starting with positive sequences $\{\delta\}$, $\{M\}$ we shall modify them successively such that $(23) \Rightarrow (3)$. Note that for $\zeta \in \Gamma_m$

$$e^{k|\mathrm{Im}\,\zeta|} \leq (2 + |\xi|)^{km}, \qquad |\zeta^\alpha| \leq |\zeta|^{|\alpha|} \leq ([|\xi|^2 + m^2(\log(2 + |\xi|))^2]^{1/2})^{|\alpha|}.$$

For (3) with $j = 0$ we take $x_n = |x| \geq 0$, $|\alpha| \leq N_0$ so

$$|e^{i\langle x, \zeta\rangle}| = e^{-\langle x, \mathrm{Im}\,\zeta\rangle} \leq 1 \qquad \text{for } \zeta \in \Gamma_m.$$

Thus if u satisfies (23) we have by the above estimates

$$(28) \quad |(D^\alpha u)(x)|$$

$$\leq \sum_0^\infty \delta_k \int_{\mathbf{R}^n} (1 + [|\xi|^2 + m^2(\log(2 + |\xi|))^2]^{1/2})^{N_0 - M_k}(2 + |\xi|)^{km}(1 + m)\,d\xi.$$

We can choose sequences $\{\delta\}$, $\{M\}$ (all δ_k, $M_k > 0$) such that this expression is $\leq \epsilon_0$. This then verifies (3) for $j = 0$. We now fix δ_0 and M_0. Next we want to prove (3) for $j = 1$ by shrinking the terms in $\delta_1, \delta_2, \ldots$ and increasing the terms in $M_1, M_2, \ldots$ (δ_0, M_0 having been fixed).

Now we have $x_n = |x| \geq 1$ so

$$(29) \qquad |e^{i\langle x, \zeta\rangle}| = e^{-\langle x, \mathrm{Im}\,\zeta\rangle} \leq (2 + |\xi|)^{-m} \text{ for } \zeta \in \Gamma_m$$

so in the integrals in (28) the factor $(2+|\xi|)^{km}$ is replaced by $(2+|\xi|)^{(k-1)m}$.

In the sum we separate out the term with $k = 0$. Here M_0 has been fixed but now we have the factor $(2 + |\xi|)^{-m}$ which assures that this $k = 0$ term is $< \frac{\epsilon_1}{2}$ for a sufficiently large m which we now fix. In the remaining terms in (28) (for $k > 0$) we can now increase $1/\delta_k$ and M_k such that the sum is $< \epsilon_1/2$. Thus (3) holds for $j = 1$ and it will remain valid for $j = 0$. We now fix this choice of δ_1 and M_1.

Now the inductive process is clear. We assume $\delta_0, \delta_1, \ldots, \delta_{j-1}$ and M_0, M_1, $\ldots, M_{j-1}$ having been fixed by this shrinking of the δ_i and enlarging of the M_i.

We wish to prove (3) for this j by increasing $1/\delta_k$, M_k for $k \geq j$. Now we have $x_n = |x| \geq j$ and (29) is replaced by

$$|e^{i\langle x, \zeta \rangle}| = e^{-\langle x, \mathrm{Im}\, \zeta \rangle} \leq (2 + |\xi|)^{-jm}$$

and since $|\alpha| \leq N_j$, (28) is replaced by

$$|(D^\alpha f)(x)|$$

$$\leq \sum_{k=0}^{j-1} \delta_k \int_{\mathbf{R}^n} (1 + [|\xi|^2 + m^2(\log(2 + |\xi|))^2]^{1/2})^{N_j - M_k} (2 + |\xi|)^{(k-j)m}(1 + m)\,d\xi$$

$$+ \sum_{k \geq j} \delta_k \int_{\mathbf{R}^n} (1 + [|\xi|^2 + m^2(\log(2 + |\xi|))^2]^{1/2})^{N_j - M_k} (2 + |\xi|)^{(k-j)m}(1 + m)\,d\xi.$$

In the first sum the M_k have been fixed but the factor $(2+|\xi|)^{(k-j)m}$ decays exponentially. Thus we can fix m such that the first sum is $< \frac{\epsilon_j}{2}$.

In the latter sum the $1/\delta_k$ and the M_k can be increased so that the total sum is $< \frac{\epsilon_j}{2}$. This implies the validity of (3) for this particular j and it remains valid for $0, 1, \ldots j - 1$. Now we fix δ_j and M_j.

This completes the induction. With this construction of $\{\delta\}, \{M\}$ we have proved that $W(\{\delta\}, \{M\}) \subset V(\{\epsilon\}, \{N\})$. This proves Theorem 4.9.

§5   Differential Operators with Constant Coefficients

The description of the topology of $\mathcal{D}$ in terms of the range $\widetilde{\mathcal{D}}$ given in Theorem 4.9 has important consequences for solvability of differential equations on $\mathbf{R}^n$ with constant coefficients.

Theorem 5.1. Let $D \neq 0$ be a differential operator on $\mathbf{R}^n$ with constant coefficients. Then the mapping $f \to Df$ is a homeomorphism of $\mathcal{D}$ onto $D\mathcal{D}$.

Proof. It is clear from Theorem 4.7 that the mapping $f \to Df$ is injective on $\mathcal{D}$. The continuity is also obvious.

For the continuity of the inverse we need the following simple lemma.

Lemma 5.2. *Let $P \neq 0$ be a polynomial of degree m, F an entire function on $\mathbf{C}^n$ and $G = PF$. Then*

$$|F(\zeta)| \leq C \sup_{|z| \leq 1} |G(z + \zeta)|, \quad \zeta \in \mathbf{C}^n,$$

where C is a constant.

Proof. Suppose first $n = 1$ and that $P(z) = \sum_0^m a_k z^k (a_m \neq 0)$. Let $Q(z) = z^m \sum_0^m \bar{a}_k z^{-k}$. Then, by the maximum principle,

$$(30) \qquad |a_m F(0)| = |Q(0)F(0)| \leq \max_{|z|=1} |Q(z)F(z)| = \max_{|z|=1} |P(z)F(z)|.$$

For general n let A be an $n \times n$ complex matrix, mapping the ball $|\zeta| < 1$ in $\mathbf{C}^n$ into itself and such that

$$P(A\zeta) = a\zeta_1^m + \sum_0^{m-1} p_k(\zeta_2, \ldots, \zeta_n)\zeta_1^k, \quad a \neq 0.$$

Let

$$F_1(\zeta) = F(A\zeta), \quad G_1(\zeta) = G(A\zeta), \quad P_1(\zeta) = P(A\zeta).$$

Then

$$G_1(\zeta_1 + z, \zeta_2, \ldots, \zeta_n) = F_1(\zeta_1 + z, \zeta_2, \ldots, \zeta_n)P_1(\zeta + z, \zeta_2, \ldots, \zeta_n)$$

and the polynomial

$$z \to P_1(\zeta_1 + z, \ldots, \zeta_n)$$

has leading coefficient a. Thus by (30)

$$|aF_1(\zeta)| \leq \max_{|z|=1} |G_1(\zeta_1 + z, \zeta_2, \ldots, \zeta_n)| \leq \max_{\substack{z \in \mathbf{C}^n \\ |z| \leq 1}} |G_1(\zeta + z)|.$$

Hence by the choice of A

$$|aF(\zeta)| \leq \sup_{\substack{z \in \mathbf{C}^n \\ |z| \leq 1}} |G(\zeta + z)|,$$

proving the lemma.

For Theorem 5.1 it remains to prove that if V is a convex neighborhood of 0 in $\mathcal{D}$ then there exists a convex neighborhood W of 0 in $\mathcal{D}$ such that

$$(31) \qquad f \in \mathcal{D}, \, Df \in W \Rightarrow f \in V.$$

We take V as the neighborhood $W(\{\delta\}, \{M\})$. We shall show that if $W = W(\{\epsilon\}, \{M\})$ (same $\{M\}$) then (31) holds provided the ϵ_j in $\{\epsilon\}$ are small enough. We write $u = Df$ so $\widetilde{u}(\zeta) = P(\zeta)\widetilde{f}(\zeta)$ where P is a polynomial. By Lemma 5.2

$$(32) \qquad\qquad |\widetilde{f}(\zeta)| \le C \sup_{|z| \le 1} |\widetilde{u}(\zeta + z)|.$$

But $|z| \le 1$ implies

$$(1 + |z + \zeta|)^{-M_j} \le 2^{M_j}(1 + |\zeta|)^{-M_j}, \quad |\text{Im}(z + \zeta)| \le |\text{Im}\,\zeta| + 1,$$

so if $C\,2^{M_j}e^j\epsilon_j \le \delta_j$ then (31) holds.

$$\text{Q.E.D.}$$

Corollary 5.3. *Let $D \ne 0$ be a differential operator on $\mathbf{R}^n$ with constant (complex) coefficients. Then*

$$(33) \qquad\qquad DD' = D'.$$

In particular, there exists a distribution T on $\mathbf{R}^n$ such that

$$(34) \qquad\qquad DT = \delta.$$

Definition. A distribution T satisfying (34) is called a *fundamental solution* for D.

To verify (33) let $L \in \mathcal{D}'$ and consider the functional $D^*u \to L(u)$ on $D^*\mathcal{D}$ ($*$ denoting adjoint). Because of Theorem 4.7 this functional is well defined and by Theorem 5.1 it is continuous. By the Hahn-Banach theorem it extends to a distribution $S \in \mathcal{D}'$. Thus $S(D^*u) = Lu$ so $DS = L$, as claimed.

Corollary 5.4. *Given $f \in \mathcal{D}$ there exists a smooth function u on $\mathbf{R}^n$ such that*

$$(35) \qquad\qquad Du = f.$$

In fact, if T is a fundamental solution one can put $u = f * T$.

§6   Riesz Potentials

We shall now study some examples of distributions in detail. If $\alpha \in \mathbf{C}$ satisfies $\text{Re}\,\alpha > -1$ the functional

$$(36) \qquad\qquad x_+^\alpha : \varphi \to \int_0^\infty x^\alpha \varphi(x)\,dx, \quad \varphi \in \mathcal{S}(\mathbf{R}),$$

is a well-defined tempered distribution. The mapping $\alpha \to x_+^\alpha$ from the half-plane $\operatorname{Re}\alpha > -1$ to the space $\mathcal{S}'(\mathbf{R})$ of tempered distributions is holomorphic (that is $\alpha \to x_+^\alpha(\varphi)$ is holomorphic for each $\varphi \in \mathcal{S}(\mathbf{R})$). Writing

$$x_+^\alpha(\varphi) = \int_0^1 x^\alpha(\varphi(x) - \varphi(0))\, dx + \frac{\varphi(0)}{\alpha + 1} + \int_1^\infty x^\alpha \varphi(x)\, dx,$$

the function $\alpha \to x_+^\alpha$ is continued to a holomorphic function in the region $\operatorname{Re}\alpha > -2, \alpha \neq -1$. In fact

$$\varphi(x) - \varphi(0) = x \int_0^1 \varphi'(tx)\, dt,$$

so the first integral above converges for $\operatorname{Re}\alpha > -2$. More generally, $\alpha \to x_+^\alpha$ can be extended to a holomorphic $\mathcal{S}'(\mathbf{R})$-valued mapping in the region

$$\operatorname{Re}\alpha > -n - 1, \quad \alpha \neq -1, -2, \ldots, -n,$$

by means of the formula

$$(37)\ \ x_+^\alpha(\varphi) = \int_0^1 x^\alpha \left[\varphi(x) - \varphi(0) - x\varphi'(0) - \cdots - \frac{x^{n-1}}{(n-1)!} \varphi^{(n-1)}(0) \right] dx$$

$$+ \int_1^\infty x^\alpha \varphi(x)\, dx + \sum_{k=1}^n \frac{\varphi^{(k-1)}(0)}{(k-1)!(\alpha + k)}.$$

In this manner $\alpha \to x_+^\alpha$ is a meromorphic distribution-valued function on $\mathbf{C}$, with simple poles at $\alpha = -1, -2, \ldots$. We note that the residue at $\alpha = -k$ is given by

$$(38)\qquad\qquad \operatorname*{Res}_{\alpha=-k} x_+^\alpha = \lim_{\alpha \to -k} (\alpha + k)x_+^\alpha = \frac{(-1)^{k-1}}{(k-1)!} \delta^{(k-1)}.$$

Here $\delta^{(h)}$ is the hth derivative of the delta distribution δ. We note that x_+^α is always a tempered distribution.

Next we consider for $\operatorname{Re}\alpha > -n$ the distribution r^α on $\mathbf{R}^n$ given by

$$r^\alpha : \varphi \to \int_{\mathbf{R}^n} \varphi(x)|x|^\alpha\, dx, \quad \varphi \in \mathcal{D}(\mathbf{R}^n).$$

Lemma 6.1. *The mapping $\alpha \to r^\alpha$ extends uniquely to a meromorphic mapping from $\mathbf{C}$ to the space $\mathcal{S}'(\mathbf{R}^n)$ of tempered distributions. The poles are the points*

$$\alpha = -n - 2h \quad (h \in \mathbf{Z}^+)$$

and they are all simple.

Proof. We have for $\operatorname{Re} \alpha > -n$

$$(39) \qquad r^\alpha(\varphi) = \Omega_n \int_0^\infty (M^t\varphi)(0)t^{\alpha+n-1}\, dt \,.$$

Next we note that the mean value function $t \to (M^t\varphi)(0)$ extends to an even $\mathcal{C}^\infty$ function on $\mathbf{R}$, and its odd order derivatives at the origin vanish. Each even order derivative is nonzero if φ is suitably chosen. Since by (39)

$$(40) \qquad r^\alpha(\varphi) = \Omega_n t_+^{\alpha+n-1}(M^t\varphi)(0)$$

the first statement of the lemma follows. The possible (simple) poles of r^α are by the remarks about x_+^α given by $\alpha + n - 1 = -1, -2, \ldots$. However if $\alpha + n - 1 = -2, -4, \ldots$, formula (38) shows, since $(M^t\varphi(0))^{(h)} = 0$, $(h \text{ odd})$ that $r^\alpha(\varphi)$ is holomorphic at the points $\alpha = -n - 1, -n - 3, \ldots$.

The remark about the even derivatives of $M^t\varphi$ shows on the other hand, that the points $\alpha = -n - 2h$ $(h \in \mathbf{Z}^+)$ are genuine poles. We note also from (38) and (40) that

$$(41) \qquad \operatorname{Res}_{\alpha=-n} r^\alpha = \lim_{\alpha \to -n} (\alpha + n)r^\alpha = \Omega_n \delta \,.$$

We recall now that the Fourier transform $T \to \widetilde{T}$ of a tempered distribution T on $\mathbf{R}^n$ is defined by

$$\widetilde{T}(\varphi) = T(\widetilde{\varphi}) \qquad \varphi = \mathcal{S}(\mathbf{R}^n) \,.$$

We shall now calculate the Fourier transforms of these tempered distributions r^α.

Lemma 6.2. *We have the following identity*

$$(42) \qquad (r^\alpha)^\sim = 2^{n+\alpha}\pi^{\frac{n}{2}}\frac{\Gamma((n+\alpha)/2)}{\Gamma(-\alpha/2)}r^{-\alpha-n}, \qquad -\alpha - n \notin 2\mathbf{Z}^+ \,.$$

For $\alpha = 2h$ $(h \in \mathbf{Z}^+)$ the singularity on the right is removable and (42) takes the form

$$(43) \qquad (r^{2h})^\sim = (2\pi)^n(-L)^h\delta, \qquad h \in \mathbf{Z}^+ \,.$$

Proof. We use a method due to Deny based on the fact that if $\psi(x) = e^{-|x|^2/2}$ then $\widetilde{\psi}(u) = (2\pi)^{\frac{n}{2}}e^{-|u|^2/2}$ so by the formula $\int f\widetilde{g} = \int \widetilde{f}g$ we obtain for $\varphi \in \mathcal{S}(\mathbf{R}^n)$, $t > 0$,

$$\int \widetilde{\varphi}(x)e^{-t|x|^2/2}\, dx = (2\pi)^{n/2}t^{-n/2}\int \varphi(u)e^{-|u|^2/2t}\, du \,.$$

We multiply this equation by $t^{-1-\alpha/2}$ and integrate with respect to t. On the left we obtain the expression

$$\Gamma(-\alpha/2)2^{-\frac{\alpha}{2}}\int \widetilde{\varphi}(x)|x|^\alpha\, dx \,,$$

using the formula

$$\int_0^\infty e^{-t|x|^2/2} t^{-1-\alpha/2}\, dt = \Gamma(-\tfrac{\alpha}{2}) 2^{-\frac{\alpha}{2}} |x|^\alpha\,,$$

which follows from the definition

$$\Gamma(x) = \int_0^\infty e^{-t} t^{x-1}\, dt\,.$$

On the right we similarly obtain

$$(2\pi)^{\frac{n}{2}} \Gamma((n+\alpha)/2)\, 2^{\frac{n+\alpha}{2}} \int \varphi(u)|u|^{-\alpha-n}\, du\,.$$

The interchange of the integrations is valid for α in the strip $-n < \operatorname{Re}\alpha < 0$ so (42) is proved for these α. For the remaining α it follows by analytic continuation. Finally, (43) is immediate from the definitions and (6).

By the analytic continuation, the right hand sides of (42) and (43) agree for $\alpha = 2h$. Since

$$\operatorname*{Res}_{\alpha=2h} \Gamma(-\alpha/2) = -2(-1)^h/h!$$

and since by (40) and (38),

$$\operatorname*{Res}_{\alpha=2h} r^{-\alpha-n}(\varphi) = -\Omega_n \frac{1}{(2h)!} \left[\left(\frac{d}{dt}\right)^{2h} (M^t\varphi) \right]_{t=0}$$

we deduce the relation

$$\left[\left(\frac{d}{dt}\right)^{2h} (M^t\varphi) \right]_{t=0} = \frac{\Gamma(n/2)}{\Gamma(h+n/2)} \frac{(2h)!}{2^{2h} h!} (L^h\varphi)(0)\,.$$

This gives the expansion

(44) $$M^t = \sum_{h=0}^\infty \frac{\Gamma(n/2)}{\Gamma(h+n/2)} \frac{(t/2)^{2h}}{h!} L^h$$

on the space of analytic functions so M^t is a modified Bessel function of $tL^{1/2}$. This formula can also be proved by integration of Taylor's formula.

Lemma 6.3. *The action of the Laplacian is given by*

(45) $\quad Lr^\alpha \;=\; \alpha(\alpha+n-2)r^{\alpha-2}, \quad (-\alpha-n+2 \notin 2\mathbf{Z}^+),$

(46) $\quad Lr^{2-n} \;=\; (2-n)\Omega_n\delta \quad (n \neq 2).$

For $n = 2$ this 'Poisson equation' is replaced by

(47) $$L(\log r) = 2\pi\delta\,.$$

Proof. For $\operatorname{Re}\alpha$ sufficiently large (45) is obvious by computation. For the remaining α it follows by analytic continuation. For (46) we use the Fourier transform and the fact that for a tempered distribution S,

$$(-LS)^\sim = r^2 \widetilde{S}.$$

Hence, by (42),

$$(-Lr^{2-n})^\sim = 4\frac{\pi^{\frac{n}{2}}}{\Gamma(\frac{n}{2}-1)} = \frac{2\pi^{\frac{n}{2}}}{\Gamma(\frac{n}{2})}(n-2)\widetilde{\delta}.$$

Finally, we prove (47). If $\varphi \in \mathcal{D}(\mathbf{R}^2)$ we have, putting $F(r) = (M^r\varphi)(0)$,

$$(L(\log r))(\varphi) = \int_{\mathbf{R}^2} \log r (L\varphi)(x)\, dx = \int_0^\infty (\log r) 2\pi r (M^r L\varphi)(0)\, dr.$$

Using Lemma 3.2 in Chapter I this becomes

$$\int_0^\infty \log r\; 2\pi r (F''(r) + r^{-1}F'(r))\, dr,$$

which by integration by parts reduces to

$$\left[\log r(2\pi r)F'(r)\right]_0^\infty - 2\pi\int_0^\infty F'(r)\, dr = 2\pi F(0).$$

This proves (47).

Another method is to write (45) in the form $L(\alpha^{-1}(r^\alpha - 1)) = \alpha r^{\alpha-2}$. Then (47) follows from (41) by letting $\alpha \to 0$.

We shall now define fractional powers of L, motivated by the formula

$$(-Lf)^\sim(u) = |u|^2\widetilde{f}(u),$$

so that formally we should like to have a relation

(48) $$((-L)^p f)^\sim(u) = |u|^{2p}\widetilde{f}(u).$$

Since the Fourier transform of a convolution is the product of the Fourier transforms, formula (42) (for $2p = -\alpha - n$) suggests defining

(49) $$(-L)^p f = I^{-2p}(f),$$

where I^γ is the **Riesz potential**

(50) $$(I^\gamma f)(x) = \frac{1}{H_n(\gamma)}\int_{\mathbf{R}^n} f(y)|x-y|^{\gamma-n}\, dy$$

with

(51) $$H_n(\gamma) = 2^\gamma \pi^{\frac{n}{2}} \frac{\Gamma(\frac{\gamma}{2})}{\Gamma(\frac{n-\gamma}{2})}.$$

Note that if $-\gamma \in 2\mathbf{Z}^+$ the poles of $\Gamma(\gamma/2)$ cancel against the poles of $r^{\gamma-n}$ because of Lemma 6.1. Thus if $\gamma - n \notin 2\mathbf{Z}^+$ we can write

(52) $$(I^\gamma f)(x) = (f * (H_n(\gamma)^{-1} r^{\gamma-n}))(x), \quad f \in \mathcal{S}(\mathbf{R}^n).$$

By Theorem 4.6 and Lemma 6.2 we then have

(53) $$(I^\gamma f)^\sim(u) = |u|^{-\gamma} \tilde{f}(u), \quad \gamma - n \notin 2\mathbf{Z}^+$$

as tempered distributions. Thus we have the following result.

Lemma 6.4. *If $f \in \mathcal{S}(\mathbf{R}^n)$ then $\gamma \to (I^\gamma f)(x)$ extends to a holomorphic function in the set $\mathbf{C}_n = \{\gamma \in \mathbf{C} : \gamma - n \notin 2\mathbf{Z}^+\}$. Also*

(54) $$I^0 f = \lim_{\gamma \to 0} I^\gamma f = f,$$

(55) $$I^\gamma L f = L I^\gamma f = -I^{\gamma-2} f.$$

We now prove an important property of the Riesz' potentials. Here it should be observed that $I^\gamma f$ is defined for all f for which (50) is absolutely convergent and $\gamma \in \mathbf{C}_n$.

Proposition 6.5. *The following identity holds:*

$$I^\alpha(I^\beta f) = I^{\alpha+\beta} f \text{ for } f \in \mathcal{S}(\mathbf{R}^n), \quad \operatorname{Re}\alpha, \operatorname{Re}\beta > 0, \quad \operatorname{Re}(\alpha+\beta) < n,$$

$I^\alpha(I^\beta f)$ being well defined. The relation is also valid if

$$f(x) = 0(|x|^{-p}) \text{ for some } p > \operatorname{Re}\alpha + \operatorname{Re}\beta.$$

Proof. We have

$$
\begin{aligned}
I^\alpha(I^\beta f)(x) &= \frac{1}{H_n(\alpha)} \int |x - z|^{\alpha-n} \left(\frac{1}{H_n(\beta)} \int f(y)|z - y|^{\beta-n} \, dy \right) dz \\
&= \frac{1}{H_n(\alpha) H_n(\beta)} \int f(y) \left(\int |x - z|^{\alpha-n}|z - y|^{\beta-n} \, dz \right) dy.
\end{aligned}
$$

The substitution $v = (x - z)/|x - y|$ reduces the inner integral to the form

(56) $$|x - y|^{\alpha+\beta-n} \int_{\mathbf{R}^n} |v|^{\alpha-n}|w - v|^{\beta-n} \, dv,$$

where w is the unit vector $(x - y)/|x - y|$. Using a rotation around the origin we see that the integral in (56) equals the number

(57) $$c_n(\alpha, \beta) = \int_{\mathbf{R}^n} |v|^{\alpha-n}|e_1 - v|^{\beta-n} \, dv,$$

where $e_1 = (1, 0, \ldots, 0)$. The assumptions made on α and β insure that this integral converges. By the Fubini theorem the exchange order of integrations above is permissible and

$$(58) \qquad I^\alpha(I^\beta f) = \frac{H_n(\alpha + \beta)}{H_n(\alpha)H_n(\beta)} c_n(\alpha, \beta) I^{\alpha+\beta} f \,.$$

It remains to calculate $c_n(\alpha, \beta)$. For this we use the following lemma, which was already used in Chapter I, §2. As there, let $\mathcal{S}^*(\mathbf{R}^n)$ denote the set of functions in $\mathcal{S}(\mathbf{R}^n)$ which are orthogonal to all polynomials.

Lemma 6.6. *Each $I^\alpha(\alpha \in \mathbf{C}_n)$ leaves the space $\mathcal{S}^*(\mathbf{R}^n)$ invariant.*

Proof. We recall that (53) holds in the sense of tempered distributions. Suppose now $f \in \mathcal{S}^*(\mathbf{R}^n)$. We consider the sum in the Taylor formula for $\widetilde{f}$ in $|u| \leq 1$ up to order m with $m > |\alpha|$. Since each derivative of $\widetilde{f}$ vanishes at $u = 0$ this sum consists of terms

$$(\beta!)^{-1} u^\beta (D^\beta \widetilde{f})(u^*) \,, \qquad |\beta| = m$$

where $|u^*| \leq 1$. This shows that

$$(59) \qquad \lim_{u \to 0} |u|^{-\alpha} \widetilde{f}(u) = 0 \,.$$

Iterating this argument with $\partial_i(|u|^{-\alpha}\widetilde{f}(u))$ etc. we conclude that the limit relation (59) holds for each derivative $D^\beta(|u|^{-\alpha}\widetilde{f}(u))$. Because of (59), relation (53) can be written

$$(60) \qquad \int_{\mathbf{R}^n} (I^\alpha f)^\sim(u) g(u) \, du = \int_{\mathbf{R}^n} |u|^{-\alpha} \widetilde{f}(u) g(u) \, du \,, \quad g \in \mathcal{S} \,,$$

so (53) holds as an identity for functions $f \in \mathcal{S}^*(\mathbf{R}^n)$. The remark about $D^\beta(|u|^{-\alpha}\widetilde{f}(u))$ thus implies $(I^\alpha f)^\sim \in \mathcal{S}_0$ so $I^\alpha f \in \mathcal{S}^*$ as claimed.

We can now finish the proof of Prop. 6.5. Taking $f_o \in \mathcal{S}^*$ we can put $f = I^\beta f_o$ in (53) and then

$$\begin{aligned} (I^\alpha(I^\beta f_0))^\sim(u) &= (I^\beta f_0)^\sim(u)|u|^{-\alpha} = \widetilde{f}_0(u)|u|^{-\alpha-\beta} \\ &= (I^{\alpha+\beta} f_0)^\sim(u) \,. \end{aligned}$$

This shows that the scalar factor in (58) equals 1 so Prop. 6.5 is proved.

In the process we have obtained the evaluation

$$\int_{\mathbf{R}^n} |v|^{\alpha-n} |e_1 - v|^{\beta-n} \, dv = \frac{H_n(\alpha)H_n(\beta)}{H_n(\alpha+\beta)} \,.$$

In recent literature this is often left as an exercise. An entirely different proof is in Riesz [1949], using the nonabsolutely convergent integral $\int_o^\infty e^{ir\cos\theta} r^{\alpha-1} \, dr$.

We shall now prove a refinement of Lemma 6.4. Let C_N denote the space of continuous functions on $\mathbf{R}^n$ satisfying

$$f(x) = 0(|x|^{-N}), \quad \text{for } x \in \mathbf{R}^n,$$

and put $\mathcal{E}_N(\mathbf{R}^n) = C_N \cap \mathcal{E}(\mathbf{R}^n)$.

Theorem 6.7. *If $f \in \mathcal{E}_N(\mathbf{R}^n)$ the Riesz potential*

$$(I^\gamma f)(x) = \frac{1}{H_n(\gamma)} \int\limits_{\mathbf{R}^n} f(y)|x - y|^{\gamma-n}\,dy$$

is holomorphic for

$$\{\gamma \in \mathbf{C} - \{n + 2\mathbf{Z}^+\} : \operatorname{Re}\gamma < N + n - 1\}$$

with simple poles at most in $\{n + 2\mathbf{Z}^+\}$.

Proof. If $f \in \mathcal{E}_N(\mathbf{R}^n)$ the function $t \to (M^t f)(0)$ belongs to $\mathcal{E}_N(R)$. For $\varphi \in \mathcal{E}_N(\mathbf{R})$ we can define $t_+^\alpha(\varphi)$ by (37) for $\operatorname{Re}\alpha < N - 1$ and then define $r^\alpha(f)$ by (40) for $f \in \mathcal{E}_N(\mathbf{R}^n)$. Then by (37) the function

$$\alpha \to r^\alpha(f) = \Omega_n t_+^{\alpha+n-1}(M^t f)(0)$$

is holomorphic for $\operatorname{Re}\alpha < N - 1$ with simple poles at most at

$$\alpha + n - 1 = -1, -2, \ldots .$$

Again since the derivative $(M^t f(0))^{(h)}$ vanishes at $t = 0$ for h odd we have holomorphy in α at the points

$$\alpha + n - 1 = -2, -4, \ldots .$$

Thus if $f \in \mathcal{E}_N(\mathbf{R}^n)$ the function $\alpha \to r^\alpha(f)$ is holomorphic in $\operatorname{Re}\alpha < N-1$ with (simple) poles at most at

$$\alpha = -n, -n - 2, -n - 4, \ldots .$$

The integral

$$\int\limits_{\mathbf{R}^n} f(y)|x - y|^{\gamma-n}\,dy = r^{\gamma-n}(f_x), \quad f_x(y) = f(x + y)$$

is thus holomorphic in $\operatorname{Re}\gamma < N + n - 1$ with simple poles at most at $\gamma - n = -n, -n-2, \ldots$ i.e., $\gamma = 0, -2, -4, \ldots$. Since these poles are canceled by $\Gamma(\frac{n}{2})$ the theorem follows.

We now prove a mild extension of Proposition 6.5.

Proposition 6.8. *Let $0 < k < n$. Then*

$$I^{-k}(I^k f) = f \qquad f \in \mathcal{E}(\mathbf{R}^n),$$

if $f(x) = 0(|x|^{-N})$ for some $N > n$.

Proof. By Prop. 6.5 we have if $f(y) = 0(|y|^{-N})$,

(61) $$\qquad\qquad I^\alpha(I^k f) = I^{\alpha+k} f \quad \text{for } 0 < \operatorname{Re}\alpha < n - k.$$

We shall prove that the function $\varphi = I^k f$ satisfies

(62) $$\qquad\qquad \sup_x |\varphi(x)|\, |x|^{n-k} < \infty.$$

For an $N > n$ we have an estimate $|f(y)| \le C_N (1 + |y|)^{-N}$ where C_N is a constant. We then have

$$\left(\int_{\mathbf{R}^n} f(y)|x - y|^{k-n}\, dy \right) \le C_N \int_{|x-y|\le \frac{1}{2}|x|} (1 + |y|)^{-N}|x - y|^{k-n}\, dy$$

$$+ C_N \int_{|x-y|\ge \frac{1}{2}|x|} (1 + |y|)^{-N}|x - y|^{k-n}\, dy.$$

In the second integral, $|x - y|^{k-n} \le (\frac{|x|}{2})^{k-n}$ so since $N > n$ this second integral satisfies (62). In the first integral we have $|y| \ge \frac{|x|}{2}$ so the integral is bounded by

$$\left(1 + \frac{|x|}{2}\right)^{-N} \int_{|x-y|\le \frac{|x|}{2}} |x - y|^{k-n}\, dy = \left(1 + \frac{|x|}{2}\right)^{-N} \int_{|z|\le \frac{|x|}{2}} |z|^{k-n}\, dz$$

which is $0(|x|^{-N}|x|^k)$. Thus (62) holds also for this first integral. This proves (62) provided

$$f(x) = 0(|x|^{-N}) \text{ for some } N > n.$$

Next we observe that $I^\alpha(\varphi) = I^{\alpha+k}(f)$ is holomorphic for $0 < \operatorname{Re}\alpha < n-k$. For this note that by (39)

$$(I^{\alpha+k} f)(0) = \frac{1}{H_n(\alpha + k)} \int_{\mathbf{R}^n} f(y)|y|^{\alpha+k-n}\, dy$$

$$= \frac{1}{H_n(\alpha + k)} \Omega_n \int_0^\infty (M^t f)(0) t^{\alpha+k-1}\, dt.$$

Since the integrand is bounded by a constant multiple of $t^{-N} t^{\alpha+k-1}$, and since the factor in front of the integral is harmless for $0 < k + \operatorname{Re}\alpha < n$, the holomorphy statement follows.

We claim now that $I^\alpha(\varphi)(x)$, which as we saw is holomorphic for $0 < \operatorname{Re}\alpha < n - k$, extends to a holomorphic function in the half-plane $\operatorname{Re}\alpha < n-k$. It suffices to prove this for $x = 0$. We decompose $\varphi = \varphi_1 + \varphi_2$, where φ_1 is a smooth function identically 0 in a neighborhood $|x| < \epsilon$ of 0, and $\varphi_2 \in \mathcal{S}(\mathbf{R}^n)$. Since φ_1 satisfies (62) we have for $\operatorname{Re}\alpha < n - k$,

$$\left| \int \varphi_1(x) |x|^{\operatorname{Re}\alpha - n} \, dx \right| \leq C \int_\epsilon^\infty |x|^{k-n} |x|^{\operatorname{Re}\alpha - n} |x|^{n-1} d|x|$$

$$= C \int_\epsilon^\infty |x|^{\operatorname{Re}\alpha + k - n - 1} d|x| < \infty$$

so $I^\alpha \varphi_1$ is holomorphic in this half-plane. On the other hand $I^\alpha \varphi_2$ is holomorphic for $\alpha \in \mathbf{C}_n$ which contains this half-plane. Now we can put $\alpha = -k$ in (61). Thus

$$I^{-k}(I^k f) = I^o f .$$

Since f is not in $\mathcal{S}(\mathbf{R}^n)$ we must still prove $I^o f = f$. It suffices to prove this at $x = 0$. By Theorem 6.7 $(I^o f)(0) = \lim_{\gamma \to 0+}(I^\gamma f)(0)$ and

$$(I^\gamma f)(0) = \frac{1}{H_n(\gamma)} \Omega_n t_+^{\gamma-1}(M^t f)(0), \quad 0 < \gamma < \epsilon .$$

Putting $\varphi(t) = (M^t f)(0)$ we have from (37)

$$\frac{\gamma}{2} t_+^{\gamma-1}(\varphi(t)) = \frac{\gamma}{2} \int_0^1 t^{\gamma-1} [\varphi(t) - \varphi(0)] + \frac{\gamma}{2} \int_1^\infty t^{\gamma-1} \varphi(t) \, dt + \frac{\varphi(0)}{2} .$$

Since $\Gamma(\gamma)\gamma \to 1$ as $\gamma \to 0$ we see that

$$\lim_{\gamma \to 0+} (I^\gamma f)(0) = \frac{\Gamma\left(\frac{n}{2}\right)}{\pi^{n/2}} \Omega_n \frac{\varphi(0)}{2} = \varphi(0) = f(0)$$

as claimed.

Denoting again by C_N the class of continuous functions f on $\mathbf{R}^n$ satisfying $f(x) = 0(|x|^{-N})$ we proved in (62) that if $N > n$, $0 < k < n$, then

(63) $$I^k C_N \subset C_{n-k} .$$

More generally, we have the following result.

Proposition 6.9. *If $N > 0$ and $0 < \operatorname{Re}\gamma < N$, then*

$$I^\gamma C_N \subset C_s ,$$

where $s = \min(n, N) - \operatorname{Re}\gamma \quad (n \neq N)$.

Proof. Modifying the proof of Prop. 6.8 we divide the integral

$$I = \int (1 + |y|)^{-N} |x - y|^{\operatorname{Re} \gamma - n} \, dy$$

into integrals I_1, I_2 and I_3 over the disjoint sets

$$A_1 = \{ y : |y - x| \leq \tfrac{1}{2} |x| \}, \qquad A_2 = \{ y : |y| < \tfrac{1}{2} |x| \,,$$

and the complement $A_3 = \mathbf{R}^n - A_1 - A_2$. On A_1 we have $|y| \geq \tfrac{1}{2} |x|$ so

$$I_1 \leq \left(1 + \frac{|x|}{2} \right)^{-N} \int_{A_1} |x - y|^{\operatorname{Re} \gamma - n} \, dy = \left(1 + \frac{|x|}{2} \right)^{-N} \int_{|z| \leq |x|/2} |z|^{\operatorname{Re} \gamma - n} \, dz$$

so

(64) $$I_1 = 0(|x|^{-N + \operatorname{Re} \gamma}).$$

On A_2 we have $|x| + \tfrac{1}{2} |x| \geq |x - y| \geq \tfrac{1}{2} |x|$ so

$$|x - y|^{\operatorname{Re} \gamma - n} \leq C |x|^{\operatorname{Re} \gamma - n}, \quad C = \text{const.} .$$

Thus

$$I_2 \leq C |x|^{\operatorname{Re} \gamma - n} \int_{A_2} (1 + |y|)^{-N} .$$

If $N > n$ then

$$\int_{A_2} (1 + |y|)^{-N} \, dy \leq \int_{\mathbf{R}^n} (1 + |y|)^{-N} \, dy < \infty .$$

If $N < n$ then

$$\int_{A_2} (1 + |y|)^{-N} \, dy \leq C |x|^{n - N} .$$

In either case

(65) $$I_2 = 0(|x|^{\operatorname{Re} \gamma - \min(n, N)}) .$$

On A_3 we have $(1 + |y|)^{-N} \leq |y|^{-N}$. The substitution $y = |x| u$ gives (with $e = x/|x|$)

(66) $\quad I_3 \leq |x|^{\operatorname{Re} \gamma - N} \displaystyle\int_{|u| \geq \frac{1}{2}, |e - u| \geq \frac{1}{2}} |u|^{-N} |e - u|^{\operatorname{Re} \gamma - n} \, du = 0(|x|^{\operatorname{Re} \gamma - N}) .$

Combining (64)–(66) we get the result.

Lemma 6.10. *Let $0 < k < N \le n$. Let $f \in \mathcal{E}_N(\mathbf{R}^n)$. Then $I^k f$ is smooth and $I^k f \in \mathcal{E}_{N-k}(\mathbf{R}^n)$.*

Proof. From Prop. 6.9 we see that $I^k f \in C_{N-k}$. For the smoothness let $B = B_R(0)$ To prove $I^k f$ smooth on B we split $f = f_1 + f_2$ where $f_1 \in \mathcal{D}(\mathbf{R}^n)$ and $f_2 \equiv 0$ on B. Then $I^k f$ splits into two pieces $I_1(x)$ and $I_2(x)$. Then $I_1(x)$ is the convolution of f_1 with a tempered distribution so is smooth by Proposition 3.1. For I_2 we have $f_2(x) = 0(|x|^{-N})$ and the integral

$$\int\limits_{|y| \ge R} |y|^{-N} |x - y|^{k-n} \, dy$$

is smooth for $|x| < R$.

The following result extending Prop. 6.8 has a significant application to Radon transform theory.

Theorem 6.11. *Let $0 < k < N \le n$ where k is an integer. Suppose $f \in \mathcal{E}_N(\mathbf{R}^n)$. Then*

$$I^{-k}(I^k f) = f \, .$$

Proof. Since $I^k f \in \mathcal{E}_{N-k}(\mathbf{R}^n)$, the mapping $\alpha \to I^\alpha(I^k f)(x)$ is by Theorem 6.7 holomorphic in

(67) $$\{\alpha \in \mathbf{C}_n : \operatorname{Re}\alpha < N - k + n - 1\} \, .$$

Also $\alpha \to (I^{\alpha+k} f)(x)$ is composed of the maps $\alpha \to \alpha + k$ and $\gamma \to (I^\gamma f)(x)$ and by Theorem 6.7 holomorphic for

(68) $$\begin{aligned} \{\alpha | \alpha + k \in \mathbf{C}_n : \operatorname{Re}(\alpha + k) < N + n - 1\} = \\ \{\alpha \in \mathbf{C}_n - k : \operatorname{Re}\alpha < N - k + n - 1\} \, . \end{aligned}$$

The regions (67) and (68) have a half plane in common where $I^\alpha(I^k f)(x)$ and $(I^{\alpha+k} f)(x)$ are both analytic in α. According to Proposition 6.5 they coincide if $N > k + \operatorname{Re}\alpha$ that is, $\operatorname{Re}\alpha < N - k$. Hence they coincide when both are holomorphic. This includes the point $\alpha = -k$ which belongs to both (67) and (68). Thus $I^{-k}(I^k f) = I^o f$. Finally $I^o f = f$ by the final argument in the proof of Proposition 6.8 because this relied only on the property $(I^o f)(o) = \lim_{\gamma \to o}(I^\gamma f)(o)$, which we know from Theorem 6.7.

Exercises and Further Results

Results on Distributions (1 – 5)

1.   Extend formula (10) §4 to $\mathcal{S}'(\mathbf{R}^n)$.

2.   Let H be the Heaviside function $H(x) = +1$ for $x > 0$, 0 for $x \leq 0$. Let S be Cauchy Principal value (Ch. I, §3). Then

$$\widetilde{H} = -iS + \pi\delta\,.$$

3.   Here and in **4** we list various examples from Gelfand-Shilov [1960], Hörmander [1983] and Schwartz [1966].

(i)   Let x_+^α be defined as in §6 ($\alpha \neq -1, -2, \ldots$) and define x_-^α by

$$x_-^\alpha(\varphi) = \int_0^\infty x^\alpha \varphi(-x)\,dx\,, \qquad \varphi \in \mathcal{D}(\mathbf{R})\,.$$

Prove that

$$\frac{d}{dx}x_+^\alpha = \alpha x_+^{\alpha-1}\,, \qquad \frac{d}{dx}x_-^\alpha = -\alpha x_-^{\alpha-1}\,,$$

$$\frac{d}{d\alpha}x_+^\alpha = x_+^\alpha \log_+\,, \qquad \frac{d}{d\alpha}x_-^\alpha = x_-^\alpha \log x_-\,,$$

where by definition,

$$(x_+^\alpha \log x_+)(\varphi) = \int_0^\infty x^\alpha \log x\,\varphi(x)\,dx$$

$$(x_-^\alpha \log x_-)(\varphi) = \int_0^\infty x^\alpha \log x\,\varphi(-x)\,dx\,.$$

(ii)   The distributions

$$\chi_+^\alpha = \frac{x_+^\alpha}{\Gamma(\alpha+1)}\,, \qquad \chi_-^\alpha = \frac{x_-^\alpha}{\Gamma(\alpha+1)}$$

exist for all $\alpha \in \mathbf{C}$ (as residues).

(iii)   As usual let Log z denote the principal branch of the logarithm, holomorphic in the slit plane with the negative real axis $x \leq 0$, removed. With H and S as in Exercise 2, the distribution

$$\log(x \pm i0) = \lim_{y \to +O} \text{Log}\,(x \pm iy)\,, \qquad x \neq 0$$

equals

$$\log(x \pm i0) = \log|x| \pm i\pi H(-x)$$

and

$$\frac{d}{dx}\log(x \pm i0) = S \pm i\delta.$$

(iv) With z^α defined as $e^{\alpha \operatorname{Log} z}$ put

$$(x \pm i0)^\alpha = \lim_{y \to +0}(x \pm iy)^\alpha \qquad \alpha \in \mathbf{C}, x \neq 0.$$

Then

$$(x \pm i0)^\alpha = x_+^\alpha + e^{\pm i\alpha\pi}x_-^\alpha \qquad (\alpha \neq -1, -2, \ldots).$$

4.   With $\chi_\pm^\alpha$ as in 3(ii) its Fourier transform is

$$e^{\mp i\pi(\alpha+1)/2}(\xi \mp i0)^{-\alpha-1}$$

and

$$\chi_+^{\alpha-1} * \chi_+^{\beta-1} = \chi_+^{\alpha+\beta-1} \quad \text{for } \alpha > 0, \beta > 0.$$

5.   In $\mathcal{S}'(\mathbf{R}^2)$ we have

$$\left(\frac{1}{x_1 + ix_2}\right)^{\sim} = \frac{2\pi i}{\xi_1 + i\xi_2}.$$

6.   Fourier Series. Let G be a compact abelian group with character group $\widehat{G}$. Let A denote the Banach algebra $L^p(G)(1 \leq p \leq 2)$. In addition to the usual norm $\| \ \|_p$, each $f \in L^p(G)$ has a spectral norm

$$\|f\|_{sp} = \sup_{\chi \in \widehat{G}} |\widehat{f}(\chi)| = \|\widehat{f}\|_\infty,$$

the supremum of the absolute value of the Fourier coefficients $\widehat{f}(\chi)$. Let A_0 denote the *derived algebra*, that is the algebra of $f \in A$ such that the convolution $\varphi \to f * \varphi$ on A is continuous from the $\| \ \|_{sp}$ topology to the $\| \ \|_p$ topology. Then (cf. Helgason [1956], [1957] for (i)–(iv)).

(i) A_0 is a Banach algebra under the norm

$$\|f\|_0 = \sup_\varphi\{\|f * \varphi\|_p : \|\varphi\|_{sp} \leq 1\}.$$

(ii)   If $A = L^\infty(G)$ with the uniform norm and A_0 defined as in (i) then A_0 is the algebra of functions with absolutely convergent Fourier series. Also

$$\|f\|_0 = \sum_{\chi \in \widehat{G}} |\widehat{f}(\chi)|.$$

(iii) If $A = L^P(G)$ $(1 \leq p \leq 2)$ then $A_0 = L^2(G)$. Also

$$\|f\|_2 \leq \sqrt{2} \sup_{\varphi} \{\|f * \varphi\|_1 : \|\varphi\|_{sp} \leq 1\} \leq \sqrt{2}\,\|f\|_2\,.$$

(iv) Let G be nonabelian and compact and consider for $\varphi \in L^1(G)$ the Peter–Weyl expansion

$$\varphi(g) \sim \sum_{\lambda \in \widehat{G}} d_\lambda \operatorname{Tr}(A_\lambda U_\lambda(g))\,,$$

where $\widehat{G}$ is the unitary dual of G, d_λ the degree of λ, U_λ a member of the class λ and A_λ the Fourier coefficient defined by

$$A_\lambda = \int\limits_G \varphi(g) U_\lambda(g^{-1})\, dg\,.$$

Again we put

$$\|\varphi\|_{sp} = \sup_{\lambda \in \widehat{G}} \|A_\lambda\| \qquad \|\,,\| = \text{ operator norm.}$$

The inequality in (iii) is still valid.

(v)   For G abelian the constant $\sqrt{2}$ in (iii) can be replaced by $h = 2/\sqrt{\pi}$ and this is the best constant (Edwards and Ross [1973] and Sawa [1985]). For G nonabelian the best constant is apparently not known, not even whether it depends on G.

Bibliographical Notes

§1-2 contain an exposition of the basics of distribution theory following Schwartz [1966]. The range theorems (4.1, 4.7) are also from there but we have used the proofs from Hörmander [1963]. Theorem 4.9 describing the topology of $\mathcal{D}$ in terms of $\widetilde{\mathcal{D}}$ is from Hörmander [1983], Vol. II, Ch. XV. The idea of a proof of this nature involving a contour like Γ_m appears already in Ehrenpreis [1956] although not correctly carried out in details there. In the proof we specialize Hörmander's convex set K to a ball; it simplifies the proof a bit and requires Cauchy's theorem only in a single variable. The consequence, Theorem 5.1, and its proof were shown to me by Hörmander in 1972. The theorem appears in Ehrenpreis [1956].

§6 contains an elementary treatment of the results about Riesz potentials used in the book. The examples x_+^λ are discussed in detail in Gelfand-Shilov [1959]. The potentials I^λ appear there and in Riesz [1949] and Schwartz [1966]. In the proof of Proposition 6.8 we have used a suggestion by R. Seeley and the refinement in Proposition 6.9 was shown to me by Schlichtkrull.

A thorough study of the composition formula (Prop. 6.5) was carried out by Ortner [1980] and a treatment of Riesz potentials on L^p-spaces (Hardy-Littlewood-Sobolev inequality) is given in Hörmander [1983], Vol. I, §4.

As shown in Jensen [2004], Theorem 6.11 holds for $f \in C_N$ just of class C^1. An L^p-version is proved by Rubin [2004].

CHAPTER VIII

LIE TRANSFORMATION GROUPS AND DIFFERENTIAL OPERATORS

Since the theory of the Radon transform and its variations becomes much richer in the context of manifolds and Lie groups we give here a short self-contained account of the relevant background in the theory of Lie transformation groups. More details can be found in [DS], [GGA], but this should not be needed.

§1 Manifolds and Lie Groups

Let X be a manifold, $\mathcal{E}(X)$ and $\mathcal{D}(X)$, respectively, the spaces of complex-valued C^∞ functions (respectively C^∞ functions of compact support) on X. For $p \in X$ let X_p denote the tangent space at p. If Ξ is another manifold and $\Phi : X \to \Xi$ a differentiable mapping its differential $d\Phi_p$ at p is the linear map of X_p into $\Xi_{\Phi(p)}$ given by

$$(d\Phi_p(A)f) = A(f \circ \Phi) \text{ for } A \in X_p,$$

f being any C^∞ function on Ξ. Here $\circ$ denotes composition. Geometrically, if $t \to \gamma(t)$ is a curve in M with tangent vector $\dot{\gamma}(t_0)$ at $p = \gamma(t_0)$ then $d\Phi_p(\dot{\gamma}(t_0))$ equals the tangent vector to the image curve $\Phi(\gamma(t))$ at $\Phi(p)$. A differentiable mapping $\Phi : X \to \Xi$ is a **diffeomorphism** if it is injective, surjective and has a differentiable inverse. A linear map $D : \mathcal{D}(X) \to \mathcal{D}(X)$ is a **differential operator** if for each local chart (U, φ) on M there exists for each open relatively compact set W with $\overline{W} \subset U$, a finite family of functions $a_\alpha \in \mathcal{E}(W)$ such that

$$Df = \sum_\alpha a_\alpha(D^\alpha(f \circ \varphi^{-1})) \circ \varphi, \quad f \in \mathcal{D}(W),$$

with D^α as in VII, §1. Let $\mathbf{E}(X)$ denote the algebra of all differential operators on X.

A **vector field** Y on X is a derivation of the algebra $\mathcal{E}(X)$. Equivalently, it is a smooth family of tangent vectors to X. In a local chart (U, φ), $\varphi(q) = (x_1(q), \ldots, x_m(q))$, Y can be written $\sum_i Y^i \frac{\partial}{\partial x_i}$ ($Y^i \in \mathcal{E}(U)$) and Y is thus a differential operator on X. Here $\frac{\partial}{\partial x_i}$ denotes the vector field $f(x) \to \left(\frac{\partial}{\partial x_i}(f \circ \varphi^{-1})\right)(\varphi(x))$, $f \in \mathcal{E}(U)$, on U.

Given a point $p \in U$ an *integral curve* to Y through p is a curve $t \to \gamma(t)$, $t \in I$ satisfying

S. Helgason, *Integral Geometry and Radon Transforms*,
DOI 10.1007/978-1-4419-6055-9_8, © Springer Science+Business Media, LLC 2010

(A) $Y_{\gamma(t)} = \dot{\gamma}(t)$ $\gamma(0) = p$.

In the local chart above let $a_i = x_i(p)$, $1 \leq i \leq m$, and put $x_i(t) = x_i(\gamma(t))$. Then (A) takes the form

(B) $\frac{dx_i}{dt} = (Y^i \circ \varphi^{-1})(x_1(t), \ldots, x_m(t))$, $x_i(0) = a_i$, $(1 \leq i \leq m)$.

By standard theory, (B) has a solution smooth in t and $(a_1, \ldots, a_m)$.

Let τ be a diffeomorphism of X. If $f \in \mathcal{E}(X)$, $T \in \mathcal{D}'(X)$, $E \in \mathbf{E}(X)$, we put for $\varphi \in \mathcal{D}(X)$

(1) $f^\tau = f \circ \tau^{-1}$, $T^\tau(\varphi) = T(\varphi^{\tau^{-1}})$, $E^\tau f = (E(f^{\tau^{-1}}))^\tau$.

Then E^τ is again in $\mathbf{E}(X)$. We say E is *invariant* under τ if $E^\tau = E$, that is

(2) $E(f \circ \tau) = (Ef) \circ \tau$, $f \in \mathcal{E}(X)$.

A Lie group G is an analytic manifold which is also a group such that the group operations are analytic. Let $e \in G$ denote the identity element. Let L_g (or $L(g)$) and R_g (or $R(g)$), respectively, denote the translations $h \to gh$ and $h \to hg$ and $(dL_g)_h$, $(dR_g)_h$ their differentials at h. For each $Y \in G_e$ consider the vector field $\widetilde{Y}$ on G given by $\widetilde{Y}_g = (dL_g)_e(Y)$. Let $\Gamma(t)$ $(|t| \leq \epsilon)$ be the integral curve of $\widetilde{Y}$ through e with $\Gamma(0) = e$. Thus

$$\dot{\Gamma}(t) = \widetilde{Y}_{\Gamma(t)} \text{ for } |t| \leq \epsilon.$$

Defining successively for $n \in \mathbf{Z}$,

$$\Gamma(t) = \Gamma(n\epsilon)\Gamma(t - n\epsilon), \quad n\epsilon \leq t \leq (n+1)\epsilon,$$

Γ extends to $\mathbf{R}$. On the interval $n\epsilon \leq t \leq (n+1)\epsilon$ we have

$$\Gamma \;=\; L_{\Gamma_{(n\epsilon)}} \circ \Gamma \circ L_{-n\epsilon}$$

so

$$\dot{\Gamma}(t) \;=\; d\Gamma\left(\frac{d}{dt}\right)_t = dL_{\Gamma(n\epsilon)} \circ d\Gamma \circ dL_{-n\epsilon}\left(\frac{d}{dt}\right)_t$$

$$=\; dL_{\Gamma(n\epsilon)}\widetilde{Y}_{\Gamma(t-n\epsilon)} = \widetilde{Y}_{\Gamma(t)}.$$

Thus $\dot{\Gamma}(t) = \widetilde{Y}_{\Gamma(t)}$ for all $t \in \mathbf{R}$. Fix $s \in \mathbf{R}$ and consider the curve

$$\Gamma_s : t \to \Gamma(s)\Gamma(t - s) = L_{\Gamma(s)} \circ \Gamma \circ L_{-s}(t).$$

Again we have

$$d\Gamma_s\left(\frac{d}{dt}\right)_t \;=\; dL_{\Gamma(s)}d\Gamma\left(\frac{d}{dt}\right)_{t-s} = dL_{\Gamma(s)}\dot{\Gamma}(t-s) = dL_{\Gamma(s)}\widetilde{Y}_{\Gamma(t-s)}$$

$$=\; \widetilde{Y}_{\Gamma(s)\Gamma(t-s)} = \widetilde{Y}_{\Gamma_s(t)}$$

so

$$\dot{\Gamma}_s(t) = \widetilde{Y}_{\Gamma_s}(t).$$

Thus Γ_s is another integral curve of $\widetilde{Y}$ and since it agrees with Γ at the point $t = s$ we have $\Gamma_s \equiv \Gamma$ so Γ is a one-parameter subgroup. We put

$$\exp Y = \Gamma(1).$$

Then for each $s \in \mathbf{R}$, $t \to \Gamma(st)$ is a one-parameter group with tangent vector sY at $t = 0$. Thus by the last formula

$$\exp sY = \Gamma(s).$$

By the left invariance of $\widetilde{Y}$ we have

(3) $$(\widetilde{Y}f)(g) = \left\{ \frac{d}{dt} f(g \exp(tY)) \right\}_{t=0}, \quad f \in \mathcal{E}(G).$$

The bracket $[\widetilde{Y}_1, \widetilde{Y}_2] = \widetilde{Y}_1 \circ \widetilde{Y}_2 - \widetilde{Y}_2 \circ \widetilde{Y}_1$ is another left invariant vector field so we define $[Y_1, Y_2] \in G_e$ by

(4) $$[Y_1, Y_2] = [\widetilde{Y}_1, \widetilde{Y}_2]_e \qquad Y_1, Y_2 \in G_e.$$

The vector space G_e with this rule of composition $(Y_1, Y_2) \to [Y_1, Y_2]$ is by definition the **Lie algebra** $\mathfrak{g}$ of G. If we had chosen the right invariant vector field $\overline{Y}$ given by $\overline{Y}_g = (dR_g)_e(Y)$ we would have

(5) $$(\overline{Y}f)(g) = \left\{ \frac{d}{dt} f(\exp tXg) \right\}_{t=0}.$$

Lemma 1.1.

(6) $$[\overline{Y}_1, \overline{Y}_2] = -[Y_1, Y_2]^-.$$

Proof. Let $J(g) = g^{-1}(g \in G)$. Then $dJ_g(\overline{Y}_g) = -\overline{Y}_{g^{-1}}$ so (6) follows since a diffeomorphism is a homomorphism of the Lie algebra of vector fields.

We shall now prove that the map $\exp$ of $\mathfrak{g}$ into G is analytic. For $Z \in \mathfrak{g}$ let $\mathfrak{g}_Z$ denote the tangent space to $\mathfrak{g}$ at Z. Consider the product group $G \times \mathfrak{g}$ and the vector field Y on $G \times \mathfrak{g}$ given by

$$Y(\sigma, Z) = (\widetilde{Z}_\sigma, Z) \in G_\sigma \oplus \mathfrak{g}.$$

The integral curve of Y through σ, Z is given by

$$\eta(t) = (\sigma \exp tZ, Z) = Y_t(\sigma, Z).$$

The differential equation (B) for $\gamma(t)$ has analytic coefficients so the solution is analytic in the initial data $(a_1, \ldots, a_m)$. Using this for $\eta(t)$ in $G \times \mathfrak{g}$, we see that Y_t is analytic on $G \times \mathfrak{g}$. in particular, the map $X \to \exp X$ is analytic. It's Jacobian at 0 is clearly nonsingular so exp is a diffeomorphism between neighborhoods N_0 of 0 in $\mathfrak{g}$ and N_e of e in G.

Iterating (3) we obtain

$$(7) \qquad (\widetilde{Y}^n f)(g \exp tY) = \frac{d^n}{dt^n} f(g \exp tY).$$

Lemma 1.2 (The Taylor formula). *If f is analytic near g, then*

$$(8) \qquad f(g \exp X) = \sum_0^\infty \frac{1}{n!} (\widetilde{X}^n f)(g)$$

for X near 0 in $\mathfrak{g}$.

Proof. Let (X_i) be a basis of $\mathfrak{g}$ and $X = \sum_1^n x_i X_i$. Then

$$f(g \exp X) = P(x_1, \ldots, x_n),$$

where P is a power series. Also for $t \leq 1$

$$f(g \exp tX) = P(tx_1, \ldots, tx_n) = \sum_0^\infty \frac{1}{m!} a_m t^m$$

and by (7)

$$a_m = (\widetilde{X}^m f)(g)$$

proving the lemma.

Let $S(\mathfrak{g})$ be the symmetric algebra over $\mathfrak{g}$ and $\mathbf{D}(G)$ the algebra of differential operators on G invariant under all left translations.

Theorem 1.3 (The Symmetrization). *There exists a unique linear bijection*

$$\lambda : S(\mathfrak{g}) \to \mathbf{D}(G)$$

such that $\lambda(X^m) = \widetilde{X}^m$ ($X \in \mathfrak{g}; m \in \mathbf{Z}^+$). For any basis X_i of $\mathfrak{g}$, $P \in S(\mathfrak{g})$,

$$(9) \quad (\lambda(P)f)(g) = \{P(\partial_1, \ldots, \partial_n) f(g \exp(t_1 X_1 + \cdots + t_n X_n))\}_{t=0},$$

for $f \in \mathcal{E}(G)$; $\partial_i = \partial/\partial t_i$.

Furthermore, $\mathbf{D}(G)$ is generated by $\widetilde{Y}$, $Y \in \mathfrak{g}$.

Proof. The formula (9) defines an operator $\lambda(P)$ on $\mathcal{E}(G)$ which is left invariant. Clearly $\lambda(X_i) = \widetilde{X}_i$ and by linearity $\lambda(X) = \widetilde{X}$. Putting

$$F(x_1, \ldots, x_n) = f(g \exp X) \quad \text{for } X = \sum x_i X_i \,,$$

we have

$$\frac{d^m}{dt^m} F(tx_1, \ldots, tx_n) = \sum x_{i_1} \ldots x_{i_m} (\partial_{i_1} \cdots \partial_{i_m} F)(tx_1, \ldots, tx_n) \,.$$

Putting $t = 0$ and using (7) we deduce

$$(\widetilde{X}^m f)(g) = \sum x_{i_1} \ldots x_{i_m} \lambda(X_{i_1} \ldots X_{i_m}) f(g) = \lambda(X^m) f(g)$$

so

$$(10) \qquad\qquad\qquad \lambda(X^m) = \widetilde{X}^m \,.$$

Thus $\lambda(P) \in \mathbf{D}(G)$. From algebra we know that the powers X^m $(X \in \mathfrak{g})$ span the space of mth degree elements in $S(\mathfrak{g})$. Thus λ is unique and is independent of the choice of basis.

Next λ is injective. Suppose $P \neq 0$ but $\lambda(P) = 0$. Let $aX_1^{m_1} \ldots X_n^{m_n}$ be a nonzero term in P. Let f be a C^∞ function near e in G such that

$$f(\exp(t_1 X_1 + \cdots + t_n X_n)) = t_1^{m_1} \ldots t_n^{m_n}$$

for small t. Here we used the above property of exp. Then $(\lambda(P)f)(e) \neq 0$ contradicting $\lambda(P) = 0$.

Finally, λ is surjective. Given $u \in \mathbf{D}(G)$ there exists a polynomial P such that

$$(uf)(e) = \{ P(\partial_1, \ldots, \partial_n) f(\exp(t_1 X_1 + \cdots + t_n X_n)) \}_{t=0} \,.$$

By the left invariance of u, we have $u = \lambda(P)$ concluding the proof. The proof showed that $\mathbf{D}(G)$ is spanned by the powers $\widetilde{Y}^m$ so the last statement of the theorem is obvious.

While (9) defines the map λ by analysis it can be described algebraically as follows:

$$\lambda(Y_1 \ldots Y_p) = \frac{1}{p!} \sum_{\sigma \in \mathfrak{S}_p} \widetilde{Y}_{\sigma(1)} \ldots \widetilde{Y}_{\sigma(p)} \,, \qquad Y_i \in \mathfrak{g} \,,$$

where $\mathfrak{S}_p$ is the symmetric group of p letters. Thus λ goes by the name **symmetrization**. The formula follows from (10) used on $(t_1 Y_1 + \cdots + t_p Y_p)^p$ by equating the coefficients to $t_1 \ldots t_p$.

Let $T(\mathfrak{g})$ denote the tensor algebra of the vector space $\mathfrak{g}$ and let J denote the two-sided ideal in $T(\mathfrak{g})$ generated by all elements of the form $X \otimes Y - Y \otimes X - [X, Y]$. The factor algebra $U(\mathfrak{g}) = T(\mathfrak{g})/J$ is called the **universal enveloping algebra** of $\mathfrak{g}$. Let $x \to x^*$ denote the natural mapping of $T(\mathfrak{g})$ onto $U(\mathfrak{g})$.

Lemma 1.4. *Given a representation ρ of $\mathfrak{g}$ on a vector space V there exists a representation ρ^* of $U(\mathfrak{g})$ on V such that*

$$\rho(Y) = \rho^*(Y^*) \qquad Y \in \mathfrak{g}.$$

Each representation of $U(\mathfrak{g})$ on V arises in this way.

The proof is straightforward.

In our basis X_i above put $X^*(t) = \sum t_i X_i^*$. Let $M = (m_1, \ldots, m_n)$ be a positive integral n-tuple, let $t^M = t_1^{m_1} \ldots t_n^{m_n}$, $|M| = m_1 + \cdots + m_n$ and let

$$X^*(M) = \text{ coefficient to } t^M \text{ in } \frac{X^*(t)^{|M|}}{|M|!}.$$

Proposition 1.5. *The elements $X^*(M)$ span $U(\mathfrak{g})$.*

Proof. We must prove that each element $X_{i_1}^* \ldots X_{i_p}^*$ $(1 \leq i_1, \ldots, i_p \leq n)$ can be expressed in the form $\sum_{|M| \leq p} a_M X^*(M)$, $a_M \in \mathbf{R}$. Consider the element

$$u_p = \frac{1}{p!} \sum_\sigma X_{i_{\sigma(1)}}^* \ldots X_{i_{\sigma(p)}}^*,$$

where σ runs over all permutations of the set $\{1, \ldots, p\}$. It is not hard to show that $u_p = c X^*(M)$ where $c \in \mathbf{R}$ and M a suitable integral n-tuple. (Actually $c = m_1! \ldots m_n!$ where $m_k = $ the number of entries in $(i_1, \ldots i_p)$ which equal k and $M = (m_1, \ldots, m_n)$.) Using the relation $X_j^* X_k^* - X_k^* X_j^* = [X_j, X_k]^*$ repeatedly we see that

$$X_{i_1}^* \ldots X_{i_p}^* - X_{i_{\sigma(1)}}^* \ldots X_{i_{\sigma(p)}}^*$$

is a linear combination of elements of the form $X_{j_1}^* \ldots X_{j_{p-1}}^*$. The desired formula follows by induction on p.

With the basis X_i we now define $\widetilde{X}(t)$ and $\widetilde{X}(M)$ just as $X^*(t)$ and $X^*(M)$ above.

Proposition 1.6. (i)   *The elements $\widetilde{X}(M)$ form a basis of $\mathbf{D}(G)$.*

(ii)   *The universal enveloping algebra $U(\mathfrak{g})$ is isomorphic to $\mathbf{D}(G)$.*

Proof. If f is analytic near g, (8) implies

$$f(g \exp X(t)) = \sum t^M (\widetilde{X}(M)f)(g).$$

Comparing with the usual Taylor formula for $F(t_1, \ldots, t_n) = f(g \exp X(t))$ we get

$$(\widetilde{X}(M)f)(g) = \frac{1}{m_1! \ldots m_n!} \left\{ \frac{\partial^{|M|}}{\partial t_1^{m_1} \ldots \partial t_n^{m_n}} f(g \exp X(t)) \right\}_{t=0},$$

and this shows that the $\widetilde{X}(M)$ are linearly independent. The mapping $\rho : X \to \widetilde{X}$ (where $\widetilde{X}$ is viewed as an operator on $\mathcal{E}(G)$) is a representation of $\mathfrak{g}$ on $\mathcal{E}(G)$. The corresponding ρ^* from Lemma 1.4 gives a homomorphism of $U(\mathfrak{g})$ into $\mathbf{D}(G)$ sending $X_i^* \ldots X_{i_p}^*$ into $\widetilde{X}_{i_1} \ldots \widetilde{X}_{i_p}$ and $X^*(M)$ into $\widetilde{X}(M)$. Since the $X^*(M)$ are linearly independent this proves the result.

Consider now the automorphism $h \to ghg^{-1}$ of G and the corresponding automorphism $\mathrm{Ad}(g)$ of $\mathfrak{g}$. It is a simple exercise to prove

$$(\mathrm{Ad}(g)X)^{\sim} = \widetilde{X}^{R(g^{-1})} \qquad X \in \mathfrak{g}$$

in the sense of (1). Thus we *define* for $D \in \mathbf{D}(G)$

(11) $$\mathrm{Ad}(g)D = D^{R(g^{-1})}.$$

Then $\mathrm{Ad}(g)$ is an automorphism of $\mathbf{D}(G)$.
    We also have for $\mathrm{ad}X : X \to [X, Z]$,

$$(\mathrm{ad}(X)(Y))^{\sim} = \widetilde{X}\widetilde{Y} - \widetilde{Y}\widetilde{X}$$

and define for $D \in \mathbf{D}(G)$

$$\mathrm{ad}X(D) = XD - DX.$$

The $\mathrm{Ad}(\exp X)$ and $e^{\mathrm{ad}X}$ are automorphisms of $\mathbf{D}(G)$ which coincide on $\widetilde{\mathfrak{g}}$, hence on all of $\mathbf{D}(G)$,

$$\mathrm{Ad}(\exp X)D = e^{\mathrm{ad}X}(D).$$

From this we deduce

$$\lim_{t \to 0} \frac{1}{t}(D^{R(\exp(-tX))} - D) = \widetilde{X}D - D\widetilde{X}$$

and if $\widetilde{X}D = D\widetilde{X}$ then $D^{R(\exp tX)} = D$, $t \in \mathbf{R}$. If $\mathbf{Z}(G)$ denotes the center of $\mathbf{D}(G)$ we thus have

Proposition 1.7. *Assume G connected, Then*

$$\mathbf{Z}(G) = \{D \in \mathbf{D}(G) : D^{R(g)} = D \text{ for all } g \in G\}.$$

The mapping $\exp$ sets up a correspondence between Lie subgroups H of G and Lie subalgebras $\mathfrak{h}$ of $\mathfrak{g}$. Given such H in G the identity map $I : H \to G$ has an injective differential $dI_e : \mathfrak{h} \to \mathfrak{g}$ making $\mathfrak{h}$ a subalgebra of $\mathfrak{g}$.
    On the other hand if $\mathfrak{h} \subset \mathfrak{g}$ is a subalgebra the (abstract) subgroup H of G generated by $\exp \mathfrak{h}$ can be made into a Lie subgroup of G with Lie

algebra $\mathfrak{h}$. By a theorem of von Neumann and Cartan each closed subgroup H of G can be given an analytic structure in which it is a topological Lie subgroup of G. Its Lie algebra is given by

$$\mathfrak{h} = \{X \in \mathfrak{g} : \exp tX \in H \text{ for } t \in \mathbf{R}\}.$$

The mapping $g \to \mathrm{Ad}(g)$ defined above is a representation of G on $\mathfrak{g}$, the *adjoint representation*. It satisfies

$$\mathrm{Ad}(\exp X) = e^{\mathrm{ad}X}, \quad \exp \mathrm{Ad}(g)X = g \exp X g^{-1}.$$

The *adjoint group* $\mathrm{Ad}(G)$ is a Lie subgroup of $\mathbf{GL}(\mathfrak{g})$. For G connected, $\mathrm{Ad}(G)$ is analytically isomorphic to G/Z where Z is the center of G. In Lie algebra terms, the adjoint group coincides with the connected Lie subgroup of $GL(\mathfrak{g})$ with Lie algebra $\mathrm{ad}\mathfrak{g} \subset \mathfrak{gl}(\mathfrak{g})$ and is then denoted $\mathrm{Int}(\mathfrak{g})$.

The Killing form of a Lie algebra $\mathfrak{g}$ is defined by

$$B(X, Y) = \mathrm{Tr}(\mathrm{ad}X\,\mathrm{ad}Y).$$

This is clearly invariant under each element of $\mathrm{Aut}(\mathfrak{g})$, the group of automorphisms of $\mathfrak{g}$, i.e., $\mathrm{Aut}(\mathfrak{g}) \subset \mathbf{O}(B)$. By definition $\mathfrak{g}$ is *semisimple* if B is nondegenerate.

Definition. A Lie algebra $\mathfrak{g}$ over $\mathbf{R}$ is *compact* if the adjoint group $\mathrm{Int}(\mathfrak{g})$ is compact.

Proposition 1.8. *(i)   Let $\mathfrak{g}$ be a semisimple Lie algebra over $\mathbf{R}$. Then $\mathfrak{g}$ is compact if and only if the Killing form of $\mathfrak{g}$ is negative definite.*

(ii)   Every compact Lie algebra is the direct sum $\mathfrak{g} = \mathfrak{z} + [\mathfrak{g}, \mathfrak{g}]$ where $\mathfrak{z}$ is the center of $\mathfrak{g}$ and the ideal $[\mathfrak{g}, \mathfrak{g}]$ is semisimple and compact.

Proof. (i) If the Killing form is negative definite then $\mathbf{O}(B)$ is compact. Thus $\mathrm{Aut}(\mathfrak{g})$ is compact. Since $\mathfrak{g}$ is semisimple, $\mathrm{Int}(\mathfrak{g})$ is known to have the same Lie algebra as $\mathrm{Aut}(\mathfrak{g})$ so equals its identity component. Thus $\mathrm{Int}(\mathfrak{g})$ is compact.

On the other hand, if $\mathrm{Int}(\mathfrak{g})$ is compact it leaves invariant a positive definite quadratic form Q on $\mathfrak{g}$. Let $(X_i)_{1 \leq i \leq n}$ be a basis such that

$$Q(X) = \sum_1^n x_i X_i \text{ if } X = \sum_1^n x_i X_i.$$

In this basis each $\sigma \in \mathrm{Int}(\mathfrak{g})$ is an orthogonal matrix so if $X \in \mathfrak{g}$, $\mathrm{ad}X$ is skewsymmetric, i.e., $^t\mathrm{ad}X = -\mathrm{ad}X$. But then

$$B(X, X) = \mathrm{Tr}(\mathrm{ad}X\,\mathrm{ad}X) = -\mathrm{Tr}(\mathrm{ad}X\,^t\mathrm{ad}X) = -\sum_{i,j} x_{ij}^2.$$

This proves (i). Part (ii) is proved similarly.

§2 Lie Transformation Groups and Radon Transforms

Let X be a manifold and G a Lie transformation group of X. To each $g \in G$ is thus associated a diffeomorphism $x \to g \cdot x$ of X such that

(i) $g_1 \cdot (g_2 \cdot x) = g_1 g_2 \cdot x$ $g_i \in G$, $x \in X$.

(ii)   The mapping $(g, x) \to g \cdot x$ from $G \times X$ to X is differentiable.

We sometimes write $\tau(g)$ for the map $x \to g \cdot x$.

If $Y \in \mathfrak{g}$, the Lie algebra of G, and $\widetilde{Y}$ the corresponding left invariant vector field on G we define the vector field $\lambda(\widetilde{Y})$ on X by

$$(12) \qquad (\lambda(\widetilde{Y}))(x) = \left\{ \frac{d}{dt} f(\exp(-tY) \cdot x) \right\}_{t=0}, \qquad x \in X.$$

Theorem 2.1 (Lie). *The mapping $\widetilde{Y} \to \lambda(\widetilde{Y})$ is a homomorphism of $\widetilde{\mathfrak{g}}$ into the Lie algebra of all vector fields on X.*

Proof. Fix $x \in X$ and consider the map $\Phi : g \in G \to g \cdot x \in X$. With $\overline{Y}$ as in (5)

$$d\Phi_g(\overline{Y}_g) f = \overline{Y}_g(f \circ \Phi) = \left\{ \frac{d}{dt}(f \circ \Phi)(\exp tY g) \right\}_{t=0}$$

$$= \left\{ \frac{d}{dt} f(\exp tY \cdot x) \right\}_{t=0} = -\lambda(\widetilde{Y})_{\Phi(g)} f,$$

which is expressed by saying that the vector fields $\overline{Y}$ and $-\lambda(\widetilde{Y})$ are Φ-**related**. The same is easily seen to hold for the bracket of $\overline{Y}_1$ and $\overline{Y}_2$, that is

$$d\Phi_g \left([\overline{Y}_1, \overline{Y}_2]_g \right) = [\lambda(\widetilde{Y}_1), \lambda(\widetilde{Y}_2)]_{\Phi(g)}.$$

By (6) the left hand side is

$$-d\Phi_g \left([Y_1, Y_2]^- \right) = \lambda \left([Y_1, Y_2]^\sim \right)_{\Phi(g)}$$

so $[\lambda(\widetilde{Y}_1), \lambda(\widetilde{Y}_2)]_{\Phi(g)} = \lambda \left([\widetilde{Y}_1, \widetilde{Y}_2] \right)_{\Phi(g)}$. Taking $g = e$ the result follows.

Theorem 2.2. *The homomorphism λ extends uniquely to a homomorphism λ of $\mathbf{D}(G)$ into $\mathbf{E}(X)$.*

Proof. The map $Y \to \lambda(\widetilde{Y})$ is a representation of $\mathfrak{g}$ on the vector space $\mathcal{E}(X)$. By Lemma 1.4 and Prop. 1.6 it extends to a homomorphism of $\mathbf{D}(G)$ into $\mathbf{E}(X)$ as stated. Explicitly, we have for $Y_1, \ldots, Y_p \in \mathfrak{g}$

$$(13)$$

$$(\lambda(\widetilde{Y}_1 \cdots \widetilde{Y}_p) f)(x) = \left\{ \frac{\partial^p}{\partial t_1 \ldots \partial t_p} f(\exp(-t_p Y_p) \ldots \exp(-t_1 Y_1) \cdot x) \right\}_{t=0}.$$

For the action of G on X we have the following result.

Proposition 2.3. *Let* $D \in \mathbf{D}(G)$, $g \in G$. *Then*

$$\lambda(D)(f^{\tau(g)}) = (\lambda(D^{R(g)})f)^{\tau(g)}, \quad f \in \mathcal{E}(G).$$

Proof. Since the maps λ and $D \to D^{R(g)}$ are homomorphisms it suffices to prove this for $D = \widetilde{Y} \in \widetilde{\mathfrak{g}}$. We have

$$\lambda(\widetilde{Y})(f^{\tau(g)})(x) = \left\{ \frac{d}{dt} f(g^{-1} \exp(-tY) \cdot x) \right\}_{t=0}$$

$$= \left\{ \frac{d}{dt} f(g^{-1} \exp(-tY)gg^{-1} \cdot x) \right\}_{t=0}$$

$$= \left\{ \frac{d}{dt} f(\exp(\mathrm{Ad}(g^{-1})(-tY))g^{-1} \cdot x) \right\}_{t=0}$$

$$= (\lambda((\mathrm{Ad}(g^{-1})Y)^{\sim})f)(g^{-1} \cdot x) = (\lambda(\widetilde{Y}^{R(g)})f)^{\tau(g)}(x)$$

as claimed.

We consider now the case when G is unimodular, separable and acts transitively on X. Thus $X = G/K$ for some K and we have the natural map $\pi : G \to G/K$ given by $\pi(g) = gK$ which intertwines the maps $L(g)$ on G and $\tau(g)$ on G/K.

Proposition 2.4. *Let* $D \in \mathbf{D}(G)$, $f \in \mathcal{E}(X)$. *Then, superscript* $*$ *denoting adjoint,*

$$(\lambda(D)f)(\pi(g)) = ((D^*)^{R(g)}(f \circ \pi))(g).$$

Proof. If $Y \in \mathfrak{g}$, $g \in G$ we have

$$(\lambda(\widetilde{Y})f)(\pi(g)) = \left\{ \frac{d}{dt} f(\exp(-tY) \cdot \pi(g)) \right\}_{t=0}$$

$$= \left\{ \frac{d}{dt} ((f \circ \pi)(g \exp \mathrm{Ad}(g^{-1})(-tY))) \right\}_{t=0}$$

so

$$(\lambda(\widetilde{Y})f)(\pi(g)) = [(\mathrm{Ad}(g^{-1})(-Y))^{\sim}(f \circ \pi)](g).$$

Let $Y_1, \dots, Y_r \in \mathfrak{g}$. We shall prove by induction on r that

(14) $\quad (\lambda(\widetilde{Y}_1 \dots \widetilde{Y}_r)f)(\pi(g)) = (-1)^r [(\mathrm{Ad}(g^{-1})(\widetilde{Y}_r \dots \widetilde{Y}_1))(f \circ \pi)](g).$

For $r = 1$ this is proved; assuming it for fixed r we have for $Y_0 \in \mathfrak{g}$,

$$(\lambda(\widetilde{Y}_0 \ldots \widetilde{Y}_r)f)(\pi(g))$$
$$= (\lambda(\widetilde{Y}_0)\lambda(\widetilde{Y}_1 \ldots \widetilde{Y}_r)f)(\pi(g))$$

$$= \left\{ \frac{d}{dt}\lambda(\widetilde{Y}_1 \ldots \widetilde{Y}_r)f(\pi(\exp(-tY_0)g)) \right\}_{t=0}$$

$$= (-1)^r \left\{ \frac{d}{dt}[\mathrm{Ad}(g^{-1})\mathrm{Ad}(\exp(tY_0))(\widetilde{Y}_r \ldots \widetilde{Y}_1)(f \circ \pi)](\exp(-tY_0)g) \right\}_{t=0}$$

$$= (-1)^r \left\{ \mathrm{Ad}(g^{-1}) \left([\widetilde{Y}_0, \widetilde{Y}_r \ldots \widetilde{Y}_1] \right) (f \circ \pi) \right\}(g)$$

$$+ (-1)^r \left\{ \mathrm{Ad}(g^{-1})(-\widetilde{Y}_0)\mathrm{Ad}(g^{-1})(\widetilde{Y}_r \ldots \widetilde{Y}_1)(f \circ \pi) \right\}(g),$$

where the unique derivation of $\mathbf{D}(G)$ extending the endomorphism $\widetilde{Y} \to [\widetilde{Y}_0, \widetilde{Y}]$ of $\widetilde{\mathfrak{g}}$ has been denoted $[\widetilde{Y}_0, D]$. However, it is clear that $[\widetilde{Y}_0, D] = \widetilde{Y}_0 D - D\widetilde{Y}_0$, so the two last expressions add up to

$$(-1)^{r+1} \left\{ \mathrm{Ad}(g^{-1})(\widetilde{Y}_r \ldots \widetilde{Y}_1\widetilde{Y}_0)(f \circ \pi) \right\}(g).$$

This proves (14). Since G is unimodular $\int_G (\widetilde{Y}F)(g)\,dg = 0$. Using this on $F_1 F_2$ we see that $\widetilde{Y}^* = -\widetilde{Y}$. This proves the proposition.

If $X = G/K$ is a reductive homogeneous space (that is $\mathrm{Ad}(K)$ acting on $\mathfrak{g}$ has an invariant subspace complementary to $\mathfrak{k}$, the Lie algebra of K) then there is a surjective homomorphism μ of $\mathbf{D}(G)$ onto $\mathbf{D}(X)$, the algebra of G-invariant differential operators on X. This μ is given by $(\mu(D)f) \circ \pi = D(f \circ \pi)$. For $D \in \mathbf{Z}(G)$, $\mu(D) = \lambda(D^*)$ by Prop. 2.3.

Consider now another homogeneous space of G, say $\Xi = G/H$ and the transform $f \to \widehat{f}$, $\varphi \to \check{\varphi}$ in Ch. II, §2 (9). They commute with the G action

$$(f^{\tau(g)})\widehat{} = (\widehat{f})^{\tau(g)}, \quad (\varphi^{\tau(g)})^\vee = (\check{\varphi})^{\tau(g)}$$

so by (13) if $D \in \mathbf{D}(G)$

(15) $\qquad (\lambda(D)f)\widehat{} = \Lambda(D)\widehat{f}, \quad (\Lambda(D)\varphi)^\vee = \lambda(D)\check{\varphi}^\vee$

for $D \in \mathbf{D}(G)$, Λ being the analog of λ for Ξ. If $\lambda(\mathbf{D}(G)) = \mathbf{D}(X)$ this would be close to an answer to Problem C (Ch. I, §2). In the case when X is a symmetric space of the noncompact type and Ξ the space of horocycles one has an isomorphism $D \to \widehat{D}$ of $\mathbf{D}(X)$ into $\mathbf{D}(\Xi)$ such that

$$Df = \widehat{D}\widehat{f}, \quad (\widehat{D}\varphi)^\vee = D\check{\varphi}, \qquad D \in \mathbf{D}(X)$$

which is a step toward the quoted Problem C.

CHAPTER IX

SYMMETRIC SPACES

Since Cartan's symmetric spaces have entered in some chapters of this book
we give here a short description of the basics of their theory but with some
proofs omitted. Detailed proofs can be found in my book [1978].

§1   Definition and Examples

A complete Riemannian manifold M is *symmetric* (in the sense of É. Cartan) if for each $p \in M$ the geodesic symmetry of s_p in p extends to an
isometry of the whole space onto itself. The Euclidean space $\mathbf{R}^n$ and the
sphere $\mathbf{S}^n$ are obvious examples. The hyperbolic space $\mathbf{H}^n$ (Ch. III, §1 (1))
is also symmetric. In fact s_0 is an isometry so by the homogeneity $\mathbf{H}^n$ is
symmetric. The two-point homogeneous spaces (Ch. III, §1) are also known
to be symmetric (Ch. III, §2, §3).

A symmetric space M is homogeneous in that the group $G = I(M)$ of
isometries acts transitively on M. In fact, given any $p, q \in M$ the symmetry
in the midpoint of a geodesic joining p to q maps p into q. Fix $o \in M$. Then
the mapping

$$\sigma : g \to s_o g s_o$$

is an automorphism of $I(M)$. The group $I(M)$ is known to be a Lie group
in the compact open topology and the subgroup K fixing o is compact.
Thus $M = G/K$. If $k \in K$ then k and $s_o k s_o$ fix o and induce the same
mapping of the tangent space M_o. Considering the geodesics from o we see
that $k = s_o k s_o$, in other words, $K \subset K_\sigma$, the fixed point group of σ. On
the other hand, if X is in the Lie algebra of K_σ, then $s_o \exp tX s_o = \exp tX$
($t \in \mathbf{R}$). Thus $\exp tX \cdot o$ is a fixed point of s_o. Since the fixed points of s_o
are isolated, $\exp tX \cdot o = o$ for t small, hence for $t \in \mathbf{R}$. Thus $\exp tX \in K$
and

$$(K_\sigma)_o \subset K \subset K_\sigma \,,$$

where the subscript o denotes identity component.

Theorem 1.1. *(i)   A symmetric space M can be written*

$$M = G/K \,, \qquad G = I(M) \,,$$

where G has an involutive automorphism σ and $(K_\sigma)_o \subset K \subset K_\sigma$.

*(ii)   Conversely, if G is a Lie group with an involutive automorphism σ
whose fixed point group K is compact then G/K is a symmetric space in
any G-invariant Riemannian metric.*

S. Helgason, *Integral Geometry and Radon Transforms*,
DOI 10.1007/978-1-4419-6055-9_9, © Springer Science+Business Media, LLC 2010

Proof. Part (i) was already done. For (ii) note first that any positive definite K-invariant quadratic form on the tangent space $(G/K)_o$ $(o = eK)$ induces a G-invariant Riemannian structure Q on G/K. If $\pi : G \to G/K$ is the natural projection we consider the map s_o of G/K onto itself given by $s_o \circ \pi = \pi \circ \sigma$. Then $(ds_o)_o = -I$.

To see that s_o is an isometry let as usual $\tau(g)$ denote the map $xK \to gxK$ of G/K. Let $p = \tau(g) \cdot o$ and $X, Y \in (G/K)_p$. Then $X_o = d\tau(g^{-1})X$, $Y_o \in d\tau(g^{-1})Y$ belong to $(G/K)_o$. The formula $s_o \circ \pi = \pi \circ \sigma$ implies for each $x \in G$,

$$s_o \circ \tau(g)(xK) = \sigma(gx)K = \sigma(g)\sigma(x)K$$
$$= (\tau(\sigma(g)) \circ s_o)(xK)$$

so

$$s_o \circ \tau(g) = \tau(\sigma(g)) \circ s_o \,.$$

Hence

$$Q(ds_o(X),\, ds_o(Y)) = Q(ds_o d\tau(g)X_o \,, ds_o \, d\tau(g)Y_o)$$
$$= Q(d\tau(\sigma(g))\, ds_o(X_o),\, d\tau(\sigma(g))\, ds_o(Y_o)) = Q(X_o, Y_o) = Q(X, Y)\,.$$

Thus s_o is an isometry and by $(ds_o)_o = -I$ it is the geodesic symmetry. Since $s_p = \tau(g) \circ s_o \circ \tau(g^{-1})$, s_p is an isometry and G/K symmetric.

With M as above, let γ be a geodesic starting at o. Let p and q be two points on γ such that $s_o(p) = q$. Let τ_o and τ denote the parallel translation (along γ) from p to o and from p to q, respectively. Let $L \in M_p$. The vectors L and $\tau_o L$ are parallel with respect to the geodesic po. Since s_o is an isometry the vectors $ds_o(L)$ and $ds_o(\tau_o L)$ are parallel with respect to the geodesic $s_o(po) = oq$. Since $ds_o(\tau_o L) = -\tau_o L$ it follows that

$$(*) \qquad\qquad\qquad ds_o(L) = -\tau L \,.$$

Again, let $\mathfrak{g}$ denote the Lie algebra of $G = I(M)$ and $\mathfrak{g} = \mathfrak{k} + \mathfrak{p}$ the decomposition of $\mathfrak{g}$ into the eigenspaces of $d\sigma$ for eigenvalues $+1$ and -1, respectively. The map $d\pi$ is a linear map of $\mathfrak{g}$ onto M_o with kernel $\mathfrak{k}$. Let $X \in \mathfrak{p}$ and $t \to p_t$ the geodesic in M from o with tangent vector $d\pi(X)$. Put $s_t = s_{p_t}$ and $T_t = s_{t/2}s_o$. By the relation $(*)$ $(dT_t)_o$ is the parallel transport along the geodesic from o to p_t. Consider $s_\tau s_o s_t$ and $s_{t+\tau}$. Both fix the point $p_{t+\tau}$:

$$s_\tau s_o s_t(p_{t+\tau}) = s_\tau s_o(p_{t-\tau}) = p_{t+\tau} \,,$$

and using $(*)$ repeatedly, we see that their differentials induce the same map of $M_{p_{t+\tau}}$. Thus $s_\tau s_o s_t = s_{t+\tau}$ so $t \to T_t$ is a one parameter subgroup of G, say $T_t = \exp tZ$, $Z \in \mathfrak{g}$, $t \in \mathbf{R}$. Now $\sigma T_t = s_o s_{t/2} = s_{-t/2} s_o = T_{-t}$ so do $Z = -Z$ whence $Z \in \mathfrak{p}$. But $\pi T_t = p_t$ so $d\pi(Z) = d\pi(X)$ and $X = Z$.

Thus the geodesics in M from o are the curves

$$\gamma_X : t \to \exp tX \cdot o, \quad X \in \mathfrak{p}.$$

The isometries T_t are called *transvections*. They slide the manifold along the geodesic γ_X. This is of course familiar for $\mathbf{R}^2$ and $\mathbf{S}^2$.

Remarkably enough, if M is a simply connected symmetric space, the isometry group $I(M)$ is the product $G_o \times G_s$ where G_o is an isometry group of a Euclidean space and G_s is semisimple. Decomposing G_s into simple components and combining the compact ones and the noncompact ones, the space M is a product of three:

$$(1) \qquad M = \mathbf{R}^n \times M_c \times M_n$$

where M_c and M_n have semisimple isometry groups, and are compact and noncompact, respectively.

§2   Symmetric Spaces of the Noncompact Type

We shall now explain how the symmetric spaces M_n arise from noncompact semisimple Lie groups G. Consider first a semisimple Lie algebra $\mathfrak{g}$ over $\mathbf{C}$. Under the isomorphism $X \to \operatorname{ad}X$ the study of $\mathfrak{g}$ amounts to a study of the family $\operatorname{ad}X$ ($X \in \mathfrak{g}$) and the first tool is of course diagonalization.

Definition. A *Cartan subalgebra* of $\mathfrak{g}$ is a subalgebra $\mathfrak{h}$ such that

(i) $\mathfrak{h} \subset \mathfrak{g}$ is a maximal abelian subalgebra; and

(ii) for each $H \in \mathfrak{g}$, $\operatorname{ad}H$ is a semisimple endomorphism of $\mathfrak{g}$.

For $\mathfrak{g} = \mathfrak{sl}(n, \mathbf{C})$ the diagonal matrices serve as a Cartan subalgebra and a modified choice can be made for the other classical simple algebras. The general proof of existence is however somewhat complicated.

Given a linear form $\alpha \neq 0$ on $\mathfrak{h}$ let

$$(2) \qquad \mathfrak{g}^\alpha = \{X \in \mathfrak{g} : \operatorname{ad}H(X) = \alpha(H)X \ \text{ for all } H \in \mathfrak{h}\}.$$

The space $\mathfrak{g}^\alpha$ is one-dimensional. The linear form α is called a *root* if $\mathfrak{g}^\alpha \neq 0$. Let Δ denote the set of all roots. By (2)

$$(3) \qquad \mathfrak{g} = \mathfrak{h} + \sum_{\alpha \in \Delta} \mathfrak{g}^\alpha \quad \text{(direct sum)}.$$

The pair $(\mathfrak{h}, \Delta)$ determines $\mathfrak{g}$ in the sense that if $\mathfrak{g}', \mathfrak{h}', \Delta'$ is another triple as above and $\varphi : \mathfrak{h} \to \mathfrak{h}'$ a linear bijection such that ${}^t\varphi$ maps Δ' onto Δ

then φ extends to an isomorphism $\tilde{\varphi}$ of $\mathfrak{g}$ onto $\mathfrak{g}'$. In particular since $\alpha \in \Delta$ implies $-\alpha \in \Delta$ the map $H \to -H$ extends to an automorphism of $\mathfrak{g}$. The Killing form B is nondegenerate on $\mathfrak{h}$ so to each α there is an assigned a vector $H_\alpha \in \mathfrak{h}$ such that $B(H, H_\alpha)$ for $H \in \mathfrak{h}$.

Using these facts the decomposition (3) can be made more explicit:

Theorem 2.1. *For each $\alpha \in \Delta$ a vector $X_\alpha \in \mathfrak{g}^\alpha$ can be chosen such that for all $\alpha, \beta \in \Delta$*

$$\begin{aligned}
[X_\alpha, X_{-\alpha}] &= H_\alpha, \quad [H, X_\alpha] = \alpha(H)X_\alpha \qquad H \in \mathfrak{h} \\
[X_\alpha, X_\beta] &= 0 \quad \text{if } \alpha + \beta \neq 0 \text{ and } \alpha + \beta \notin \Delta \\
[X_\alpha, X_\beta] &= N_{\alpha,\beta}X_{\alpha,\beta} \quad \text{if } \alpha + \beta \in \Delta,
\end{aligned}$$

where the constants $N_{\alpha,\beta}$ satisfy

$$N_{\alpha,\beta} = -N_{-\alpha-\beta}.$$

A *real form* of $\mathfrak{g}$ is a real linear subspace $\mathfrak{b}$ of $\mathfrak{g}$ which is closed under the bracket operation and satisfies $\mathfrak{g} = \mathfrak{b} + i\mathfrak{b}$ (direct sum). The mapping $X + iY \to X - iY$ is called the *conjugation* of $\mathfrak{g}$ with respect to $\mathfrak{b}$. The algebras $\mathfrak{sl}(n, \mathbf{R})$ and $\mathfrak{su}(n)$ are both real forms of $\mathfrak{sl}(n, \mathbf{C})$, the latter one a compact real form.

Using Theorem 2.1 the following basic result can be proved.

Theorem 2.2. *Each semisimple Lie algebra $\mathfrak{g}$ over $\mathbf{C}$ has a real form $\mathfrak{g}_k$ which is compact.*

In fact $\mathfrak{g}_k$ can be constructed as

$$\mathfrak{g}_k = \sum_{\alpha \in \Delta} \mathbf{R}(iH_\alpha) + \sum_{\alpha \in \Delta} \mathbf{R}(X_\alpha - X_{-\alpha}) + \sum_{\alpha \in \Delta} \mathbf{R}(i(X_\alpha + X_{-\alpha})).$$

Consider now a semisimple Lie algebra $\mathfrak{g}$ over $\mathbf{R}$ with Killing form B. There are of course many possible ways of decomposing $\mathfrak{g} = \mathfrak{g}_+ \oplus \mathfrak{g}_-$ such that B is positive definite on $\mathfrak{g}_+$, negative definite on $\mathfrak{g}_-$. We would like such a decomposition directly related to the Lie algebra structure of $\mathfrak{g}$.

Definition. A *Cartan decomposition* of $\mathfrak{g}$ is a direct decomposition $\mathfrak{g} = \mathfrak{k} + \mathfrak{p}$ such that

(i) $B < 0$ on $\mathfrak{k}$; $B > 0$ on $\mathfrak{p}$.

(ii) The mapping $\theta : T + X \to T - X$ $(T \in \mathfrak{k}, X \in \mathfrak{p})$ is an automorphism of $\mathfrak{g}$.

In this case θ is called a *Cartan involution* of $\mathfrak{g}$ and the positive definite bilinear form $(X, Y) \to -B(X, \theta Y)$ is denoted by B_θ.

We shall now establish the existence and essential uniqueness of a Cartan decomposition of $\mathfrak{g}$.

Theorem 2.3. *Suppose θ is a Cartan involution of a semisimple Lie algebra $\mathfrak{g}$ over $\mathbf{R}$ and σ an arbitrary automorphism of $\mathfrak{g}$. There then exists an automorphism φ of $\mathfrak{g}$ such that the Cartan involution $\varphi\theta\varphi^{-1}$ commutes with σ.*

Proof. The product $N = \sigma\theta$ is an automorphism of $\mathfrak{g}$ and if $X, Y \in \mathfrak{g}$,

$$-B_\theta(NX, Y) = B(NX, \theta Y) = B(X, N^{-1}\theta Y) = B(X, \theta NY)$$

so

$$B_\theta(NX, Y) = B_\theta(X, NY)$$

that is, N is symmetric with respect to the positive definite bilinear form B_θ. Let $X_1, \ldots, X_n$ be an orthonormal basis of $\mathfrak{g}$ diagonalizing N. Then $P = N^2$ has a positive diagonal, say, with elements $\lambda_1, \ldots, \lambda_n$. Take P^t $t \in \mathbf{R}$ with diagonal elements $\lambda_1^t, \ldots, \lambda_n^t$ and define the structural constants c_{ijk} by

$$[X_i, X_j] = \sum_{k=1}^{n} c_{ijk} X_k .$$

Since P is an automorphism, we conclude

$$\lambda_i \lambda_j c_{ijk} = \lambda_k c_{ijk}$$

which implies

$$\lambda_i^t \lambda_j^t c_{ijk} = \lambda_k^t c_{ijk} \qquad (t \in \mathbf{R})$$

so P^t is an automorphism. Put $\theta_t = P^t \theta P^{-t}$. Since $\theta N\theta^{-1} = N^{-1}$, we have $\theta P\theta^{-1} = P^{-1}$, that is $\theta P = P^{-1}\theta$. In matrix terms (using still the basis $X_1, \ldots, X_n$) this means (since θ is symmetric with respect to B_θ)

$$\theta_{ij}\lambda_j = \lambda_i^{-1}\theta_{ij}$$

so

$$\theta_{ij}\lambda_j^t = \lambda_i^{-t}\theta_{ij}$$

that is, $\theta P^t \theta^{-1} = P^{-t}$. Hence

$$\sigma\theta_t = \sigma P^t \theta P^{-t} = \sigma\theta P^{-2t} = N P^{-2t}$$
$$\theta_t \sigma = (\sigma\theta_t)^{-1} = P^{2t} N^{-1} = N^{-1} P^{2t}$$

so it suffices to take $\varphi = P^{1/4}$ (which could be viewed as $\sqrt{\sigma\theta}$).

Corollary 2.4. *Let $\mathfrak{g}$ be a semisimple Lie algebra over $\mathbf{R}$, $\mathfrak{g}_c = \mathfrak{g} + i\mathfrak{g}$ its complexification, $\mathfrak{u}$ any compact real form of $\mathfrak{g}_c$, σ and τ the conjugations for $\mathfrak{g}_c$ with respect to $\mathfrak{g}$ and $\mathfrak{u}$, respectively. Then there exists an automorphism of $\mathfrak{g}_c$ such that $\varphi \cdot \mathfrak{u}$ is invariant under σ.*

Proof. Let $\mathfrak{g}_c^{\mathbf{R}}$ denote the Lie algebra $\mathfrak{g}_c$ considered as a Lie algebra over $\mathbf{R}$, $B^{\mathbf{R}}$ the Killing form. It is not hard to show that $B^{\mathbf{R}}(X, Y) = 2\operatorname{Re}(B_c(X, Y))$ if B_c is the Killing form of $\mathfrak{g}_c$. Thus σ is an involution of $\mathfrak{g}_c^{\mathbf{R}}$ and τ is a Cartan involution of $\mathfrak{g}_c^{\mathbf{R}}$ and the corollary follows (note that since $\sigma\tau$ is a (complex) automorphism of $\mathfrak{g}_c$, φ is one as well).

Here we used the remark that since σ commutes with $\varphi\tau\varphi^{-1}$ the space $\varphi\mathfrak{u}$, which is the fixed point set of $\varphi\tau\varphi^{-1}$, is invariant under σ.

Corollary 2.5. *Each semisimple Lie algebra $\mathfrak{g}$ over $\mathbf{R}$ has Cartan decompositions and any two such are conjugate under an automorphism of $\mathfrak{g}$.*

Proof. Let $\mathfrak{g}_c$ denote the complexification of $\mathfrak{g}$, σ the corresponding conjugation, and $\mathfrak{u}$ a compact real form of $\mathfrak{g}_c$ invariant under σ (Theorems 2.2–2.3). Then put $\mathfrak{k} = \mathfrak{g} \cap \mathfrak{u}$, $\mathfrak{p} = \mathfrak{g} \cap i\mathfrak{u}$. Then $B < 0$ on $\mathfrak{k}$, $B > 0$ on $\mathfrak{p}$, and since $\theta : T + X \to T - X$ ($T \in \mathfrak{k}, X \in \mathfrak{p}$) is an automorphism, $B(\mathfrak{k}, \mathfrak{p}) = 0$. It follows that $\mathfrak{g} = \mathfrak{k} + \mathfrak{p}$ is a Cartan decomposition.

Finally, let σ and τ be two Cartan involutions. To prove their conjugacy we may by Theorem 2.3 assume that they commute. The corresponding Cartan decompositions

$$\mathfrak{g} = \mathfrak{k}_\sigma + \mathfrak{p}_\sigma, \qquad \mathfrak{g} = \mathfrak{k}_\tau + \mathfrak{p}_\tau$$

are thus compatible, $\sigma\mathfrak{k}_\tau \subset \mathfrak{k}_\tau$, etc. so

$$\mathfrak{g} = \mathfrak{k}_\tau \cap \mathfrak{k}_\sigma + \mathfrak{k}_\tau \cap \mathfrak{p}_\sigma + \mathfrak{k}_\sigma \cap \mathfrak{p}_\tau + \mathfrak{p}_\sigma \cap \mathfrak{p}_\tau .$$

Considering the form B we see that the two middle terms vanish so $\sigma = \tau$.

Example.

Let $\mathfrak{g} = \mathfrak{sl}(n, \mathbf{R})$, the Lie algebra of the group $\mathbf{SL}(n, \mathbf{R})$. The group $\mathbf{SO}(n)$ of orthogonal matrices is a closed subgroup, hence a Lie subgroup, and by Ch. VIII, §1, its Lie algebra, denoted $\mathfrak{su}(n)$, consists of those matrices $X \in \mathfrak{sl}(n, \mathbf{R})$ for which $\exp tX \in \mathbf{SO}(n)$ for all $t \in \mathbf{R}$. But

$$\exp tX \in \mathbf{SO}(n) \Leftrightarrow \exp tX \exp t({}^t X) = 1 \quad \det(\exp tX) = 1$$

so

$$\mathfrak{so}(n) = \{X \in \mathfrak{sl}(n, \mathbf{R}) | X + {}^t X = 0\}$$

the set of skew-symmetric $n \times n$ matrices (which are automatically of trace 0).

The mapping $\theta : X \to -{}^t X$ is an automorphism of $\mathfrak{sl}(n, \mathbf{R})$ and $\theta^2 = 1$. It is easy to prove that $B(X, X) = 2n \operatorname{Tr}(XX)$, $B(X, \theta X) < 0$. Thus θ is a Cartan involution and

$$\mathfrak{sl}(n, \mathbf{R}) = \mathfrak{so}(n) + \mathfrak{p} ,$$

where $\mathfrak{p}$ is a set of $n \times n$ symmetric matrices of trace 0, is the corresponding Cartan decomposition.

Consider now a Cartan decomposition $\mathfrak{g} = \mathfrak{k} + \mathfrak{p}$, G a connected Lie group with Lie algebra $\mathfrak{g}$, and $K \subset G$ a Lie subgroup with Lie algebra $\mathfrak{k}$. Then θ extends to an involutive automorphism of G and K is closed, connected and is the fixed point set of θ. Thus by Theorem 2.1, $X = G/K$ is symmetric and is in fact simply connected. Since θ is an automorphism

(4) $$[\mathfrak{k}, \mathfrak{k}] \subset \mathfrak{k}, \; [\mathfrak{k}, \mathfrak{p}] \subset \mathfrak{p}, \; [\mathfrak{p}, \mathfrak{p}] \subset \mathfrak{k}.$$

The Killing form B is negative definite on $\mathfrak{k}$, positive definite on $\mathfrak{p}$. Being invariant under the adjoint action of K on $\mathfrak{p}$, B induces a G-invariant Riemannian structure on $X = G/K$. Let d denote the distance, and $|Y| = B(Y, Y)^{1/2}$. As proved earlier the *geodesics* through $o = eK$ are the orbits $\exp tY \cdot o$ with $Y \in \mathfrak{p}$ and $d(o, \exp Y \cdot o) = |Y|$. More generally, the *totally geodesic submanifolds* of X through o are given by $\exp \mathfrak{s} \cdot o$ where $\mathfrak{s} \subset \mathfrak{p}$ is a *Lie triple system*. In fact, $\mathfrak{s} + [\mathfrak{s}, \mathfrak{s}]$ is a subalgebra of $\mathfrak{g}$ and if G_1 is the corresponding connected Lie subgroup the orbit $G_1 \cdot o$ equals the manifold $\exp \mathfrak{s} \cdot o$.

The curvature tensor R can be shown to be

(5) $$R(X, Y) \cdot Z = -[[X, Y], Z].$$

If e, f is an orthonormal basis of a plane section in X_o the corresponding *sectional curvature* equals

$$-B(R(e, f) \cdot e, f) = B([[e, f], e], f) = B([e, f], [e, f]) \le 0$$

by the invariance of B.

Examples discussed in the text are

$$\mathbf{H}^p = \mathbf{SO}_0(p, 1)/\mathbf{SO}(p), \qquad \mathbf{H}^2 = \mathbf{SU}(1, 1)/\mathbf{SO}(2),$$

$\mathbf{SO}_0(p, 1)$ being the identity component of the Lorentz group. Here the geodesic symmetry is an isometry. Other examples are

$$\mathbf{SO}_0(p, q)/\mathbf{SO}(p) \times \mathbf{SO}(q), \qquad \mathbf{SL}(n, \mathbf{R})/\mathbf{SO}(n)$$

with the Cartan involutions

$$g \to I_{p,q} g I_{p,q} \qquad g \to {}^t g^{-1},$$

respectively. (Here $I_{p,q}$ is the diagonal matrix with diagonal $(-1, \ldots, -1, 1, \ldots, 1)$.)

Each $\mathrm{ad} X (X \in \mathfrak{p})$ is symmetric relative to the positive definite symmetric bilinear form $B_\theta(Y, Z) = -B(Y, \theta Z)$. Thus $\mathrm{ad} X$ can be diagonalized by means of a basis of $\mathfrak{g}$. Thus if $\mathfrak{a} \subset \mathfrak{p}$ is a *maximal abelian subspace*, the commutative family $\mathrm{ad}H(H \subset \mathfrak{a})$ can be simultaneously diagonalized. For each real linear form λ on $\mathfrak{a}$ let

$$\mathfrak{g}_\lambda = \{X \in \mathfrak{g} : [H, X] = \lambda(H)X \;\text{ for }\; H \in \mathfrak{a}\}.$$

Then

(6) $$\theta\mathfrak{g}_\lambda = \mathfrak{g}_{-\lambda} \;\text{ and }\; [\mathfrak{g}_\lambda, \mathfrak{g}_\mu] \subset \mathfrak{g}_{\lambda+\mu}.$$

If $\lambda \neq 0$ and $\mathfrak{g}_\lambda \neq 0$ then λ is called a *root* of $\mathfrak{g}$ with respect to $\mathfrak{a}$. If Σ denotes the set of all roots,

$$\mathfrak{g} = \mathfrak{g}_0 + \bigoplus_{\lambda \in \Sigma} \mathfrak{g}_\lambda.$$

The maximal abelian subspaces of $\mathfrak{p}$ are all conjugate under $\mathrm{Ad}_G(K)$. The components of the *regular set*

$$\mathfrak{a}' = \{H \in \mathfrak{a} : \lambda(H) \neq 0 \;\text{ for all }\; \lambda \in \Sigma\}$$

are pyramids, called the *Weyl chambers* and they are all conjugate under the set of $\mathrm{Ad}(k)$, which leave $\mathfrak{a}$ invariant.

Let $\mathfrak{a}^+$ be a fixed component of $\mathfrak{a}'$ and Σ^+ the set of $\alpha \in \Sigma$ which are > 0 on $\mathfrak{a}^+$. Then

(7) $$\mathfrak{n} = \bigoplus_{\lambda \in \Sigma^+} \mathfrak{g}_\lambda$$

is a subalgebra and we have the Chevalley-Iwasawa decomposition

$$\mathfrak{g} = \mathfrak{k} + \mathfrak{a} + \mathfrak{n}$$

and the corresponding decomposition of G,

$$G = KAN.$$

The dimension of $\mathfrak{g}_\lambda$ is called the *multiplicity* m_λ of the root λ. For simple $\mathfrak{g}$ the triple $(\mathfrak{a}, \Sigma, m)$ determines $\mathfrak{g}$ up to isomorphism. The dimension of $\mathfrak{a}$ is called the *rank* of G/K and the *real rank* of G.

A *horocycle* in X is an orbit of a group conjugate to N. The group G permutes the horocycles transitively (so they all have the same dimension). The subgroup of G which maps the horocycle $N \cdot o$ into itself equals MN where M is the centralizer of A in K. Thus G/MN can be identified with the space of all horocycles in X.

The Radon transform and its dual for the dual homogeneous spaces

$$X = G/K, \qquad \Xi = G/MN$$

is indicated in Ch. II, §4, J. A detailed treatment is given in our book [1994b].

The manifold $\exp \mathfrak{a} \cdot o = A \cdot o$ is totally geodesic in X and it is flat. In fact, if $Y, Z \in \mathfrak{a}$ then $d(\exp Y \cdot o, \exp Z \cdot o) = d(\exp(Y - Z) \cdot o, o) = |Y - Z|$. The flat manifold $A \cdot o$ and the horocycle $N \cdot o$ are perpendicular at o.

§3   Symmetric Spaces of the Compact Type

These spaces are the coset spaces $X = U/K$ where U is a compact semisimple Lie group and K is the subgroup of fixed points of an involutive automorphism σ of U. Again $\mathfrak{u}$ decomposes into the eigenspaces of $d\sigma$,

$$(8) \qquad\qquad \mathfrak{u} = \mathfrak{k} + \mathfrak{p}_* .$$

The negative of the Killing form B is positive definite on $\mathfrak{p}_*$ (and on $\mathfrak{u}$) and thus induces a U-invariant Riemannian structure on X. The geodesics through $o = eK$ are again of the type $t \to \exp tY \cdot o$ for $Y \in \mathfrak{p}_*$. We put $\mathrm{Exp}Y = \exp Y \cdot o$ for $Y \in \mathfrak{p}_*$.

There is a simple duality between all the noncompact types G/K and all the compact types X. This is based on the correspondence

$$(9) \qquad\qquad \mathfrak{g} = \mathfrak{k} + \mathfrak{p} \leftrightarrow \mathfrak{u} = \mathfrak{k} + \mathfrak{p}_* , \quad \mathfrak{p}_* = i\mathfrak{p} ,$$

within the complexification $\mathfrak{g}_{\mathbf{c}} = \mathfrak{u}_{\mathbf{c}}$. A special case is the analogy between hyperbolic spaces and spheres. Under the correspondence (9), Lie triple systems $\mathfrak{s} \subset \mathfrak{p}$ correspond to Lie triple systems $i\mathfrak{s} = \mathfrak{s}_* \subset \mathfrak{p}_*$. Again $\exp \mathfrak{s}_* \cdot o$ a totally geodesic submanifold of X. If $\mathfrak{a}_* \subset \mathfrak{p}_*$ is maximal abelian, then just as $\mathrm{Exp}\,\mathfrak{a}$ the manifold $\mathrm{Exp}(\mathfrak{a}_*)$ is a flat totally geodesic submanifold of maximal dimension. All such are conjugate under U. Their dimension is the *rank* ℓ of X.

Assume now X simply connected and let $\delta > 0$ denote the maximal sectional curvature. Then X contains totally geodesic spheres of curvature δ. All such spheres of maximal curvature are U-conjugate. Similar conjugacy holds for the closed geodesics of minimal length (cf. Ch. IV, Theorem 1.2.) The *midpoint locus* A_o for o is the set of midpoints of these closed geodesics through o of minimal length. It is a K-orbit and a totally geodesic submanifold of X.

The root theory for $(\mathfrak{g}, \mathfrak{a})$ in §2 applies also for $(\mathfrak{u}, \mathfrak{a}_*)$, because of the duality (9) in which $\mathfrak{a}_* = i\mathfrak{a}$. Then each $\alpha \in \Sigma$ is purely imaginary on $\mathfrak{a}_*$. For each $H \in \mathfrak{a}_*$, $(\mathrm{ad}H)^2$ maps $\mathfrak{p}_*$ into itself and we denote by T_H its restriction to $\mathfrak{p}_*$ (see Ch. III, §2). We then have

$$(10) \qquad\qquad \mathfrak{p}_* = \mathfrak{a}_* + \sum_{\lambda \in \Sigma^+} \mathfrak{p}_\lambda ,$$

where

$$\mathfrak{p}_\lambda = \{X \in \mathfrak{p}_* : (\mathrm{ad}H)^2 X = \lambda(H)^2 X \ \text{ for } \ H \in \mathfrak{a}^*\} .$$

Consider now the case when the rank of X is one. As mentioned in Ch. III, §2 these are the compact two-point homogeneous spaces. Except for the space $\mathbf{P}^n(\mathbf{R})$, which is doubly covered by $\mathbf{S}^n$, they are simply connected. Now the midpoint locus A_o becomes the *antipodal set*, the set of points of maximal distance from o.

For the case of rank one, the decomposition (10) becomes

(11) $\mathfrak{p}_* = \mathfrak{a}_* + \mathfrak{p}_\alpha + \mathfrak{p}_{\alpha/2}$,

where the last term is 0 for $\mathbf{P}^n(\mathbf{R})$ and $\mathbf{S}^n$. The subspaces $\mathfrak{p}_\alpha$ have the following geometric significance:

(i) $\mathrm{Exp}(\mathfrak{a}_* + \mathfrak{p}_\alpha)$ is a maximal dimensional totally geodesic sphere in M.

(ii)   If o' is the point on this sphere antipodal to o then for the corresponding antipodal manifold
$$A_{o'} = \mathrm{Exp}(\mathfrak{p}_{\alpha/2}) \,.$$

(iii) If $\mathfrak{p}_{\alpha/2} = 0$ and if $H \in \mathfrak{a}_*$ has length the diameter of U/K then for $o' = \mathrm{Exp}H$
$$A_{o'} = \mathrm{Exp}(\mathfrak{p}_\alpha) \,.$$

The geometric picture of X ($\neq \mathbf{S}^n$) used in Chapter IV is thus the following:

Through the origin $o = eK$ we have totally geodesic spheres of maximum dimension $1 + m_\alpha$ and maximum curvature $-B(\alpha, \alpha)$. All these spaces are conjugate under K and so are the geodesics issuing from o. The midpoint locus A_o is again a K-orbit of dimension $\dim \mathfrak{p}_{\alpha/2}$. It is totally geodesic in X and is another symmetric space. For $X = \mathbf{P}^n(\mathbf{C})$ and $\mathbf{P}^n(\mathbf{H})$ the midpoint locus is a space of the same type, for $X = \mathbf{P}$ (**Cay**) it is the sphere $\mathbf{S}^8$.

Bibliography

ABOUELAZ, A.

2001   Integral geometry in the sphere $\mathbf{S}^d$ in *Harmonic Analysis and Integral Geometry*. Chapman & Hall *Res. Notes Math.* **422** Boca Raton, FL, 2001.

ABOUELAZ, A. AND DAHER, R.

1993   Sur la transformation de Radon de la sphere $\mathbf{S}^d$, *Bull. Soc. Math. France* **121** (1993), 353–382.

ABOUELAZ, A. AND FOURCHI, O.E.

2001   Horocyclic Radon Transform on Damek-Ricci spaces, *Bull. Polish Acad. Sci.* **49** (2001), 107-140.

ABOUELAZ, A. AND IHSANE, A.

2008   Diophantine Integral Geometry, *Mediterr. J. Math.* **5** (2008), 77–99.

AGRANOVSKI, M.L., BERENSTEIN, C.A. AND KUCHMENT, P.

1996   Approximation by spherical waves in L^p spaces, *J. Geom. Anal.* **6** (1996), 365–383.

AGRANOVSKI, M.L., KUCHMENT, P. AND QUINTO, E.T.

2007   Range descriptions for the spherical mean Radon transform, *J. Funct. Anal.* **248** (2007), 344–386.

AGRANOVSKI, M.L. AND QUINTO, E.T.

1996   Injectivity sets for the Radon transform over circles and complete systems of radial functions, *J. Funct. Anal.* **139** (1996), 383–414.

AGRANOVSKI, M.L., VOLCHKOV, V.V. AND ZALCMAN, L.A.

1999   Conical uniqueness sets for the spherical Radon transform, *Bull. London Math. Soc.* **31** (1999), 231–236.

AGUILAR, V., EHRENPREIS, L., AND KUCHMENT, P.

1996   Range conditions for the exponential Radon transform, *J. d'Analyse Math.* **68** (1996), 1–13.

S. Helgason, *Integral Geometry and Radon Transforms*,
DOI 10.1007/978-1-4419-6055-9, © Springer Science+Business Media, LLC 2010

AMBARTSOUMIAN, G. AND KUCHMENT, P.
2006   A range description for the planar circular Radon transform,
              SIAM J. Math. Anal. **38** (2006), 681–692.

AMEMIYA, I. AND ANDO, T.
1965   Convergence of random products of contractions in Hilbert space,
              Acta Sci. Math. (Szeged) **26** (1965), 239–244.

ANDERSSON, L.-E.
1988   On the determination of a function from spherical averages, *SIAM
              J. Math. Anal.*, **19** (1988), 214–232.

ARAKELYAN, N.U.
1964   Uniform approximation on closed sets by entire functions, *Izv.
              Akad. Nauk, SSSR Sci. Mat.* **28** (1964), 1187–1206.

ARAKI, S.I.
1962   On root systems and an infinitesimal classification of irreducible
              symmetric spaces, *J. Math. Osaka City Univ.* **13** (1962), 1–34.

ARMITAGE, D.H.
1994   A non-constant function on the plane whose integral on every line
              is 0, *Amer. Math. Monthly* **101** (1994), 892–894.

ARMITAGE, D.H. AND GOLDSTEIN, M.
1993   Nonuniqueness for the Radon transform, *Proc. Amer. Math. Soc.*
              117 (1993), 175–178.

ÁSGEIRSSON, L.
1937   Über eine Mittelwertseigenschaft von Lösungen homogener lin-
              earer partieller Differentialgleichungen 2. Ordnung mit konstan-
              ten Koeffizienten, *Math. Ann.* **113** (1937), 321–346.

BEARDON, A. F.
1983   *The Geometry of Discrete Groups*, Springer Verlag, New York,
              1983.

BERENSTEIN, C.A., KURUSA, A., AND CASADIO TARABUSI, E.
1997   Radon transform on spaces of constant curvature, *Proc. Amer.
              Math. Soc.* **125** (1997), 455–461.

BERENSTEIN, C.A. AND SHAHSHAHANI, M.
1983   Harmonic analysis and the Pompeiu problem, *Amer. J. Math.*
              105 (1983), 1217–1229.

BERENSTEIN, C.A. AND CASADIO TARABUSI, E.
1991   Inversion formulas for the k-dimensional Radon transform in real
              hyperbolic spaces, *Duke Math. J.* **62** (1991), 613–631.
1992   Radon- and Riesz transform in real hyperbolic spaces, *Contemp.
              Math.* **140** (1992), 1–18.
1993   Range of the k-dimensional Radon transform in real hyperbolic
              spaces, *Forum Math.* **5** (1993), 603–616.
1994   An inversion formula for the horocyclic Radon transform on the
              real hyperbolic space, *Lectures in Appl. Math.* **30** (1994), 1–6.

BERENSTEIN, C.A. AND WALNUT, D.E.
1994   "Local inversion of the Radon transform in even dimensions using
              wavelets," in: *Proc. Conf. 75 Years of Radon Transform*, Vienna,
              1992, International Press, Hong Kong, 1994.

BERENSTEIN, C.A. AND ZALCMAN, L.
1976   Pompeiu's problem on spaces of constant curvature, *J. Anal.
            Math.* **30** (1976), 113–130.
1980   Pompeiu's problem on symmetric spaces, *Comment. Math. Helv.*
            55 (1980), 593–621.

BEREST, Y.
1998   Hierarchies of Huygens' operators and Hadamard's conjecture,
            Acta Appl. Math. **53** (1998), 125–185.

BESSE, A.
1978   *Manifolds all of whose geodesics are closed,* Ergeb. Math. Grenz-
            geb. **93**, Springer, New York, 1978.

BOCKWINKEL, H.B.A.
1906   On the propagation of light in a biaxial crystal about a midpoint
            of oscillation, *Verh. Konink. Acad. V. Wet. Wissen. Natur.* **14**
            (1906), 636.

BOERNER, H.
1955   *Darstellungen der Gruppen,* Springer-Verlag, Heidelberg, 1955.

BOGUSLAVSKY, M. I.
2001   Radon transforms and packings, *Discrete Appl. Math.* **111** 2001,
            1–22.

BOMAN, J.
1990   On generalized Radon transforms with unknown measures, *Con-
            temp. Math.* **113** (1990), 5–15.
1991   "Helgason's support theorem for Radon transforms: A new proof
            and a generalization," in: *Mathematical Methods in Tomography,*
            Lecture Notes in Math. No. 1497, Springer-Verlag, Berlin and
            New York, 1991, 1–5.
1992   Holmgren's uniqueness theorem and support theorems for real
            analytic Radon transforms, *Contemp. Math.* **140** (1992), 23–30.
1993   An example of non-uniqueness for a generalized Radon transform,
            J. Analyse Math. **61** (1993), 395–401.

BOMAN, J. AND LINDSKOG, F.
2009   Support theorems for the Radon transform and Cramér-Wold the-
            orems, *J. of Theoretical Probability* **22** (2009), 683–710.

BOMAN, J. AND QUINTO, E.T.
1987   Support theorems for real-analytic Radon transforms, *Duke
            Math. J.* **55** (1987), 943–948.
1993   Support theorems for Radon transforms on real-analytic line
            complexes in three space, *Trans. Amer. Math. Soc.* **335** (1993),
            877–890.

BONNESEN, T. AND FENCHEL, W.
1934   *Theorie der Konvexen Körper,* Springer, 1934.

BOROVIKOV, V.A.
1959   Fundamental solutions of linear partial differential equations with
            constant coefficients, *Trudy Moscov. Mat. Obshch.* **8** (1959),
            199–257.

BOUAZIZ, A.
1995   Formule d'inversion des intégrales orbitales sur les groupes de Lie
              réductifs, *J. Funct. Anal.* **134** (1995), 100–182.

BRACEWELL, R.N. AND RIDDLE, A.C.
1967   Inversion of fan beam scan in radio astronomy, *Astrophys. J.* **150**
              (1967), 427–434.

BRANSON, T.P., ÓLAFSSON, G., AND SCHLICHTKRULL, H.
1994   A bundle-valued Radon transform with applications to invariant
              wave equations, *Quart. J. Math. Oxford 45* (1994), 429–461.

BRYLINSKI, J.-L.
1996   Radon transform and functionals on the space of curves, Gelfand
              Math. Sem., 45–73, Birkhäuser, Boston, 1996.

CARTAN, É.
1896   Le principe de dualité et certaines intégrales multiple de l'espace
              tangential et de espace reglé, *Bull. Soc. Math. France* **24** (1896),
              140–177.

1927   Sur certaines formes riemannianes remarquables des géometries
              á groupe fondamental simple. *Ann. Sci. École Norm. Sup.* **44**
              (1927), 345–467.

CARTON-LEBRUN, C.
1984   Smoothness properties of certain integrals and the range of the
              Radon transform. *Bull. Soc. Roy. Sci.* Liège, **53** (1984), 257–262.

CHEN, B.-Y.
2001   Helgason spheres of compact symmetric spaces of finite type, *Bull.
              Austr. Math. Soc.***63** (2001), 243–255.

CHERN, S.S.
1942   On integral geometry in Klein spaces, *Ann. of Math.* **43** (1942),
              178–189.

CHEVALLEY, C.
1946   *The Theory of Lie Groups, Vol.* I, Princeton University Press,
              Princeton, NJ, 1946.

CORMACK, A.M.
1963–64   Representation of a function by its line integrals, with some radi-
              ological applications I, II, *J. Appl. Phys.* **34** (1963), 2722–2727;
              35 (1964), 2908–2912.

CORMACK, A.M. AND QUINTO, E.T.
1980   A Radon transform on spheres through the origin in $\mathbf{R}^n$ and
              applications to the Darboux equation, *Trans. Amer. Math. Soc.*
              260 (1980), 575–581.

COURANT, R. AND HILBERT, D.
1937   *Methoden der Mathematischen Physik*, Berlin, Springer, 1937.

COURANT, R. AND LAX, A.
1955   Remarks on Cauchy's problem for hyperbolic partial differential
              equations with constant coefficints in several independent vari-
              ables, *Comm. Pure Appl. Math.* **8** (1955), 497–502.

COXETER, H.S.M.

1957   *Non-Euclidean Geometry,* University of Toronto Press, Toronto, 1957.

DEANS, S.R.

1983   *The Radon Transform and Some of Its Applications,* Wiley, New York, 1983.

DEBIARD, A. AND GAVEAU, B.

1983   Formule d'inversion en géométrie intégrale Lagrangienne, *C. R. Acad. Sci. Paris Sér.* I *Math.* **296** (1983), 423–425.

DROSTE, B.

1983   A new proof of the support theorem and the range characterization of the Radon transform, *Manuscripta Math.* **42** (1983), 289–296.

EDWARDS, R.E. AND ROSS, K.

1973   Helgason's number and lacunarity constants, *Bull. Austr. Math. Soc.* **9** (1973), 187–218.

EHRENPREIS, L.

1956   Solutions of some problems of division, part III, *Amer. J. Math.* **78** (1956), 685–715.

2003   *The Universality of the Radon Transform,* Oxford Univ. Press, 2003.

EHRENPREIS, L., KUCHMENT, P., AND PANCHENKO, A.

1997   The exponential X-ray transform, F. John equation, and all that I: Range description, preprint, 1997.

ERDELYI, A., MAGNUS, W., OBERHETTINGER, F. AND TRICOMI, F.G.

1953   *Higher Transcendental Functions,* Vol. I, II, III, McGraw-Hill, New York, (1953), (1953), (1955).

ESTEZET, P.

1988   Tenseurs symétriques à énergie nulle sur les variétés à courbure constante, thesis, Université de Grenoble, Grenoble, France, 1988.

FARAH, S.B. AND KAMOUN, L.

1990   Distributions coniques sur le cône des matrices de rang un et de trace nulle, *Bull. Soc. Math. France* **118** (1990), 251–272.

FARAUT, J.

1982   Un théorème de Paley-Wiener pour la transformation de Fourier sur un espace Riemannian symétrique de rang un, *J. Funct. Anal.* **49** (1982), 230–268.

FARAUT, J. AND HARZALLAH, K.

1984   Distributions coniques associées au groupe orthogonal $O(p, q)$, *J. Math. Pures Appl.* **63** (1984), 81–119.

FELIX, R.

1992   Radon Transformation auf nilpotenten Lie Gruppen, *Invent. Math.* **112** (1992), 413–443.

FINCH, D.V., HALTMEIER, M., AND RAKESH

2007   Inversion and spherical means and the wave equation in even dimension, *SIAM J. Appl. Math.* **68** (2007), 392–412.

FINCH, D.V. AND HERTLE, A.
1987   The exponential Radon transform, *Contemp. Math.* **63** (1987), 67–73.

FLENSTED-JENSEN, M.
1977   Spherical functions on a simply connected semisimple Lie group II, *Math. Ann.* **228** (1977), 65–92.

FLICKER, Y.Z.
1996   Orbital integrals on symmetric spaces and spherical characters, *J. Algebra* **184** (1996), 705–754.

FRIEDLANDER, F.C.
1975   *The Wave Equation in Curved Space,* Cambridge University Press, London and New York, 1975.

FUGLEDE, B.
1958   An integral formula, *Math. Scand.* **6** (1958), 207–212.

FUNK, P.
1913   Über Flächen mit lauter geschlossenen geodätischen Linien, *Math. Ann.* **74** (1913), 278–300.

1916   Über eine geometrische Anwendung der Abelschen Integralgleichung, *Math. Ann.* **77** (1916), 129–135.

GÅRDING, L.
1961   Transformation de Fourier des distributions homogènes, *Bull. Soc. Math. France* **89** (1961), 381–428.

GARDNER, R.J.
1995   *Geometric Tomography,* Cambridge Univ. Press, New York, 1995.

GASQUI, J. AND GOLDSCHMIDT, H.
2004   *Radon Transforms and the Rigidity of the Grassmannians,* Ann. Math. Studies, Princeton Univ. Press, 2004.

GELFAND, I.M. AND GRAEV, M.I.
1964   The geometry of homogeneous spaces, group representations in homogeneous spaces and questions in integral geometry related to them, *Amer. Math. Soc. Transl.* **37** (1964).

GELFAND, I.M., GINDIKIN, S.G., AND GRAEV, M.I.
1982   Integral geometry in affine and projective spaces, *J. Soviet Math.* **18** (1982), 39–164.

2003   *Selected Topics in Integral Geometry,* Amer. Math. Soc. Transl. Vol. 220, Providence, RI, 2003.

GELFAND, I.M., GINDIKIN, S.G., AND SHAPIRO, S.J.
1979   A local problem of integral geometry in a space of curves, *Funct. Anal. Appl.* **13** (1979), 11–31.

GELFAND, I.M. AND GRAEV, M.I.
1955   Analogue of the Plancherel formula for the classical groups, *Trudy Moscov. Mat. Obshch.* **4** (1955), 375–404.

1968a   Complexes of straight lines in the space $\mathbf{C}^n$, *Funct. Anal. Appl.* **2** (1968), 39–52.

1968b   Admissible complexes of lines in $\mathbf{CP}^n$, *Funct. Anal. Appl.* **3** (1968), 39–52.

GELFAND, I.M., GRAEV, M.I., AND SHAPIRO, S.J.
1969   Differential forms and integral geometry, *Funct. Anal. Appl.* **3**
            (1969), 24–40.
GELFAND, I.M., GRAEV, M.I., AND VILENKIN, N.
1966   *Generalized Functions, Vol. 5 : Integral Geometry and Represen-*
            tation Theory, Academic Press, New York, 1966.
GELFAND, I.M. AND SHAPIRO, S.J.
1955   Homogeneous functions and their applications, *Uspehi Mat. Nauk*
            10 (1955), 3–70.
GELFAND, I.M. AND SHILOV, G.F.
1960   *Verallgemeinerte Funktionen, Vol.* I, German Transl. VEB, Ber-
            lin, 1960.
GINDIKIN, S.G.
1975   Invariant generalized functions in homogeneous domains, *Funct.*
            Anal. Appl. **9** (1975), 50–52.
1995   Integral geometry on quadrics, *Amer. Math. Soc. Transl. Ser.* 2
            169 (1995), 23–31.
GLOBEVNIK, J.
1992   A support theorem for the X-ray transform, *J. Math. Anal. Appl.*
            165 (1992), 284–287.
1994   Local support theorems for k-plane transform in R^n, *J. Math.*
            Anal. Appl. **181** (1994), 455–461.
GODEMENT, R.
1952   Une généralisation du théorème de la moyenne pour les fonctions
            harmoniques *C.R. Acad. Sci. Paris* **234** (1952), 2137–2139.
1957   Introduction aux travaux de A. Selberg, *Séminaire Bourbaki* **144**,
            Paris, 1957.
1966   The decomposition of $L^2(G/\Gamma)$ for $\Gamma = SL(2, \mathbf{Z})$, *Proc. Symp.*
            Pure Math. **9** (1966), 211–224.
GOLDSCHMIDT, H.
1990   The Radon transform for symmetric forms on real symmetric
            spaces, *Contemp. Math.* **113** (1990), 81–96.
GONCHAROV, A.B.
1989   Integral geometry on families of k-dimensional submanifolds,
            Funct. Anal. Appl. **23** 1989, 11–23.
GONZALEZ, F.
1984   Radon transforms on Grassmann manifolds, thesis, MIT, Cam-
            bridge, MA, 1984.
1987   Radon transforms on Grassmann manifolds, *J. Funct. Anal.* **71**
            (1987), 339–362.
1988   Bi-invariant differential operators on the Euclidean motion group
            and applications to generalized Radon transforms, *Ark. Mat.* **26**
            (1988), 191–204.
1990a   Bi-invariant differential operators on the complex motion group
            and the range of the d-plane transform on C^n, *Contemp. Math.*
            113 (1990), 97–110.
1990b   Invariant differential operators and the range of the Radon d-
            plane transform, *Math. Ann.* **287** (1990), 627–635.

282   Bibliography

1991   On the range of the Radon transform and its dual, *Trans. Amer. Math. Soc.* **327** (1991), 601–619.

1994   "Range of Radon transform on Grassmann manifolds," in: *Proc. Conf. 75 Years of Radon Transform,* Vienna, 1992, International Press, Hong Kong, 1994, 81–91.

2001   "John's equation and the plane to line transform on $\mathbf{R}^3$", in *Harmonic Analysis and Integral Geometry Safi* (1998), 1–7. *Chapman and Hall/RCR Research Notes Math.,* Boca Raton, FL, 2001.

GONZALEZ, F. AND KAKEHI, T.

2003   Pfaffian systems and Radon transforms on affine Grassmann manifolds *Math. Ann.* **326** (2003), 237–273.

2004   Dual Radon transforms on affine Grassmann manifolds, *Trans. Amer.Math. Soc.* **356** (2004), 4161–4180.

2006   Invariant differential operators and the range of the matrix Radon transform, *J. Funct. Anal.* **241** (2006), 232–267.

GONZALEZ, F. AND QUINTO, E.T.

1994   Support theorems for Radon transforms on higher rank symmetric spaces, *Proc. Amer. Math. Soc.* **122** (1994), 1045–1052.

GOODEY, P. AND WEIL, W.

1991   Centrally symmetric convex bodies and the spherical Radon transform, preprint, 1991.

GRINBERG, E.

1985   On images of Radon transforms, *Duke Math. J.* **52** (1985), 939–972.

1986   Radon transforms on higher rank Grassmannians, *J. Differential Geom.* **24** (1986), 53–68.

1987   Euclidean Radon transforms: Ranges and restrictions, *Contemp. Math.* **63** (1987), 109–134.

1992   Aspects of flat Radon transform, *Contemp. Math.* **140** (1992), 73–85.

1994a   "Integration over minimal spheres in Lie groups and symmetric spaces of compact type," in: *Proc. Conf. 75 Years of Radon Transform,* Vienna, 1992, International Press, Hong Kong, 1994, 167–174.

1994b   "That kappa operator", in *Lectures in Appl. Math.* **30**, 1994, 93–104.

GRINBERG, E. AND QUINTO, E.T.

1999   "Morera theorems for spheres through a point in $\mathbf{C}^N$", in *Recent Developments in Complex Analysis and Commutative Algebra,* (Gilbert et al. eds) Kluwer Acad. Publ., 267–275.

2000   Morera theorems for complex manifolds, *J. Funct. Anal.* **178** (2000), 1–22.

GUILLEMIN, V.
1976   Radon transform on Zoll surfaces, *Adv. in Math.* **22** (1976), 85–99.
1985   The integral geometry of line complexes and a theorem of Gelfand-Graev, *Astérisque No. Hors Série* (1985), 135-149.
1987   Perspectives in integral geometry, *Contemp. Math.* **63** (1987), 135–150.

GUILLEMIN, V. AND POLLACK, A.
1974   *A Differential Topology*, Prentice-Hall, N.J., 1974.

GUILLEMIN, V. AND STERNBERG, S.
1977   *Geometric Asymptotics,* Math. Surveys, American Mathematical Society, Providence, RI, 1977.
1979   Some problems in integral geometry and some related problems in microlocal analysis, *Amer. J. Math.* **101** (1979), 915–955.

GURARIE, D.
1992   *Symmetries and Laplacians: Introduction to Harmonic Analysis and Applications,* North Holland, Amsterdam, 1992.

GÜNTHER, P.
1966   Sphärische Mittelwerte in kompakten harmonischen Riemannschen Mannigfaltigkeiten, *Math. Ann.* **165** (1966), 281–296.
1988   *Huygens' Principle and Hyperbolic Equations,* Academic Press, Boston, 1988.
1991   Huygens' Principle and Hadamard's Conjecture, *Math. Intelligencer* **13** (1991), 56–63.
1994 L^∞-decay estimations of the spherical mean value on symmetric spaces, *Ann. Global Anal. Geom.* **12** (1994), 219–236.

HALPERIN, I.
1962   The product of projection operators, *Acta Sci. Math. (Szeged)* **23** (1962), 96–99.

HAMAKER, C. AND SOLMON, D.C.
1978   The angles between the null spaces of X-rays, *J. Anal. Appl.* **62** (1978), 1–23.

HARINCK, P.
1998   Formule d'inversion des intégrales orbitales et formule de Plancherel, *J. Funct. Anal.* **153** (1998), 52–107.

HARISH-CHANDRA
1954   The Plancherel formula for comples semisimple Lie groups. *Trans. Amer. Math. Soc.* **76** (1954), 485–528.
1957   A formula for semisimple Lie groups, *Amer. J. Math.* **79** (1957), 733–760.
1958   Spherical functions on a semisimple Lie group I, *Amer. J. Math.* **80** (1958), 241–310.

HELGASON, S.
1956   Multipliers of Banach algebras, *Ann. of Math.* **64** (1956), 240–254.
1957   Topologies of group algebras and a theorem of Littlewood, *Trans. Amer. Math. Soc.* **86** (1957), 269–283.

1959	Differential Operators on homogeneous spaces, *Acta Math.* **102** (1959), 239–299.
1961	Some remarks on the exponential mapping for an affine connection, *Math. Scand.* **9** (1961), 129–146.
1962	*Differential Geometry and Symmetric Spaces*, Academic Press, New York, 1962.
1963	Duality and Radon transforms for symmetric spaces, *Amer. J. Math.* **85** (1963), 667–692.
1964	A duality in integral geometry: some generalizations of the Radon transform, *Bull. Amer. Math. Soc.* **70** (1964), 435–446.
1965a	The Radon transform on Euclidean spaces, compact two-point homogeneous spaces and Grassmann manifolds, *Acta Math.* **113** (1965), 153–180.
1965b	Radon-Fourier transforms on symmetric spaces and related group representation, *Bull. Amer. Math. Soc.* **71** (1965), 757–763.
1966a	"A duality in integral geometry on symmetric spaces," in: *Proc. U.S.-Japan Seminar in Differential Geometry,* Kyoto, 1965, Nippon Hyronsha, Tokyo, 1966, 37–56.
1966b	Totally geodesic spheres in compact symmetric spaces, *Math. Ann.* **165** (1966), 309–317.
1970	A duality for symmetric spaces with applications to group representations, *Adv. in Math.* **5** (1970), 1–154.
1972	"Harmonic analysis in the non-Euclidean disk," in: *Proc. International Conf. on Harmonic Analysis,* University of Maryland, College Park, MD, 1971, Lecture Notes in Math. No. 266, Springer-Verlag, New York, 1972.
1973	The surjectivity of invariant differential operators on symmetric spaces, *Ann. of Math.* **98** (1973), 451–480.
1978 [DS]	*Differential Geometry, Lie Groups and Symmetric Spaces,* Academic Press, New York, 1978. Amer. Math. Soc. 2001.
1980a	A duality for symmetric spaces with applications to group representations III: Tangent space analysis, *Adv. in Math.* **30** (1980), 297–323.
1980b	Support of Radon transforms, *Adv. in Math.* **38** (1980), 91–100.
1980c	*The Radon Transform,* Birkhäuser, Basel and Boston, 1980.
1980d	"The X-ray transform on a symmetric space," in: *Proc. Conf. on Differential Geometry and Global Analysis,* Berlin, 1979, Lecture Notes in Math. No. 838, Springer-Verlag, New York, 1980.
1981	*Topics in Harmonic Analysis on Homogeneous Spaces,* Birkhäuser, Basel and Boston, 1981.
1983a	"Ranges of Radon transforms," AMS Short Course on Computerized Tomography, January, 1982, in: *Proc. Symp. on Applied Mathematics,* American Mathematical Society, Providence, RI, 1983.

1983b   "The range of the Radon transform on symmetric spaces," in: *Proc. Conf. on Representation Theory of Reductive Lie Groups,* Park City, Utah, 1982, P. Trombi, ed., Birkhäuser, Basel and Boston, 1983, 145–151.

1983c   "Operational properties of the Radon transform with applications," in: *Proc. Conf. on Differential Geometry with Applications,* Nové Mesto, Czechoslovakia, 1983, 59–75.

1984 [GGA]   *Groups and Geometric Analysis: Integral Geometry, Invariant Differential Operators and Spherical Functions,* Academic Press, New York, 1984. Now published by American Mathematical Society, Providence, R.I., 2000.

1987   Some results on Radon transforms, Huygens' principle and X-ray transforms, *Contemp. Math.* Vol. **63** (1987).

1990   The totally geodesic Radon transform on constant curvature spaces, *Contemp. Math.* **113** (1990), 141–149.

1992   The flat horocycle transform for a symmetric space, *Adv. in Math.* **91** (1992), 232–251.

1994a   "Radon transforms for double fibrations: Examples and viewpoints," in: *Proc. Conf. 75 Years of Radon Transform,* Vienna, 1992, International Press, Hong Kong, 1994, 163–179.

1994b [GSS]   *Geometric Analysis on Symmetric Spaces,* Math. Surveys and
& 2008   Monographs No. 39, American Mathematical Society, Providence, RI, 1994. Second Edition, 2008.

2005   The Abel, Fourier and Radon transforms on symmetric spaces. *Indag. Math. NS.* **16** (2005), 531–551.

2006   "Non-Euclidean Analysis", in (Prékopa and Molnár Eds.) Non-Euclidean Geometries. *Proc. J. Bolyai Mem. Conf.* Budapest, July 2002. Springer, 2006, p. 367–384.

2007   The inversion of the X-ray transform on a compact symmmetric space, *J. Lie Theory* **17** (2007), 307–315.

HERGLOTZ, G.

1931   *Mechanik der Kontinua,* Lecture notes, University of Göttingen, Göttingen, Germany, 1931.

HERTLE, A.

1979   Zur Radon Transformation von Funktionen und Massen, Dissertation Univ. Erlangon–Nürenberg, Germany.

1982   A characterization of Fourier and Radon transforms on Euclidean spaces, *Trans. Amer. Math. Soc.* **273** (1982), 595–608.

1983   Continuity of the Radon transform and its inverse on Euclidean space, *Math. Z.* **184** (1983), 165–192.

1984a   On the range of the Radon transform and its dual, *Math. Ann.* **267** (1984), 91–99.

1984b   On the injectivity of the attenuated Radon transform, *Proc. Amer. Math. Soc.* **92** (1984) 201–205.

HILGERT, J.

1994   "Radon transform on half planes via group theory," in: *Noncompact Lie Groups and Some of Their Applications,* Kluwer Academic Publishers, Norwell, MA, 1994, 55–67.

HÖRMANDER, L.

1963   *Linear Partial Differential Operators,* Springer-Verlag, Berlin and New York, 1963.

1983   *The Analysis of Linear Partial Differential Operators* I, II, Springer-Verlag, Berlin and New York, 1983.

HOUNSFIELD, G.N.

1973   Computerized transverse axial scanning tomography, *British J. Radiology* **46** (1973), 1016–1022.

HU, M.-C.

1973   Determination of the conical distributions for rank one symmetric spaces, Thesis, MIT, Cambridge, MA, 1973.

1975   Conical distributions for rank one symmetric spaces, *Bull. Amer. Math. Soc.* **81** (1975), 98–100.

ISHIKAWA, S.

1997   The range characterization of the totally geodesic Radon transform on the real hyperbolic space, *Duke Math. J.* **90** (1997), 149–203.

2003   Symmetric subvarieties in compactifications and the Radon transform on Riemannian symmetric spaces of the noncompact type, *J. Funct. Anal.* **204** (2003), 50–100.

JENSEN, S.R.

2004   Sufficient conditions for the inversion formula for the k-plane Radon transform in $\mathbf{R}^n$, *Math. Scand.* **94** (2004), 207–226.

JOHN, F.

1934   Bestimmung einer Funktion aus ihren Integralen über gevisse Mannigfaltigkeiten, *Math. Ann.* **109** (1934), 488–520.

1935   Anhängigheit zwischen den Flächenintegralen einer stetigen Funktion, *Math. Ann.* **111** (1935), 541–559.

1938   The ultrahyperbolic differential equation with 4 independent variables, *Duke Math. J.* **4** (1938), 300–322.

1955   *Plane Waves and Spherical Means,* Wiley–Interscience, New York, 1955.

KAKEHI, T.

1992   Range characterization of Radon transforms on complex projective spaces, *J. Math. Kyoto Univ.* **32** (1992), 387–399.

1993   Range characterization of Radon transforms on $\mathbf{S}^n$ and $\mathbf{P}^n\mathbf{R}$, *J. Math. Kyoto Univ.* **33** (1993), 315–328.

1995   Range characterization of Radon transforms on quaternion projective spaces, Math. Ann. **301** (1995), 613–625.

1999   Integral geometry on Grassmann manifolds and calculus of invariant differential operators, *J. Funct. Anal.* 168 (1999), 1-45.

KAKEHI, T. AND TSUKAMOTO, C.

1993   Characterization of images of Radon transforms, *Adv. Stud. Pure Math.* **22** (1993), 101–116.

KATSEVICH, A.I.

1997   Range of the Radon transform on functions which do not decay fast at infinity, *SIAM J. Math. Anal.* **28** (1997), 852–866.

KLEIN, S. THORBERGSSON, G. AND VERHÓCZKI, L.

2009   On the Funk transform on compact symmetric spaces, *Publ. Math. Debrecen* **75** (2009), 485–493.

KOLK, J. AND VARADARAJAN, V.S.

1992   Lorentz invariant distributions supported on the forward light cone, *Compositio Math.* **81** (1992), 61–106.

KOORNWINDER, T.H.

1975   A new proof of the Paley-Wiener theorem for the Jacobi transform, *Ark. Mat.* **13** (1975), 145–149.

KORANYI, A.

1995   On a mean value property for hyperbolic spaces, *Contemp. Math., Amer. Math. Soc.*, **191** (1995), 107–116.

KUCHMENT, P.A. AND LVIN, S.Y.

1990   Paley-Wiener theorem for exponential Radon transform, *Acta Appl. Math.* **18** (1990), 251–260.

KUMAHARA, K. AND WAKAYAMA, M.

1993   On Radon transform for Minkowski space, *J. Fac. Gen. Ed. Tattori Univ.* **27** (1993), 139–157.

KURUSA, A.

1991a   A characterization of the Radon transform's range by a system of PDE's, *J. Math. Anal. Appl.* **161** (1991), 218–226.

1991b   The Radon transform on hyperbolic spaces, *Geom. Dedicata* **40** (1991), 325–339.

1991c   Translation-invariant Radon transforms, *Math. Balcanica* **5** (1991), 40–46.

1992   The invertibility of the Radon transform on abstract rotational manifolds of real type, *Math. Scand.* **70** (1992), 112-126.

1993   Support curves of invertible Radon transforms, *Arch. Math.* **61** (1993), 448-458.

1994   Support theorems for the totally geodesic Radon transform on constant curvature spaces, *Proc. Amer. Math. Soc.* **122** (1994), 429–435.

1997   The totally geodesic Radon transform on the Lorentz space of curvature −1, *Duke Math. J.* **86** (1997), 565–583.

2000   Orbital integrals on the Lorentz space of curvature −1, *Arch. Math.* **75** (2000), 132–146.

LAX, P. AND PHILLIPS, R.S.,

1967   *Scattering Theory,* Academic Press, New York, 1967.

1979   Translation representations for the solution of the non-Euclidean wave equation, *Comm. Pure Appl. Math.* **32** (1979), 617–667.

288   Bibliography

1981   A local Paley-Wiener theorem for the Radon transform in real
            hyperbolic spaces, *Math. Anal. and Applications* Part B, 483–487,
            L. Nachbin, Ed. Academic Press, 1981.

1982   A local Paley-Wiener theorem for the Radon transform of L^2
            functions in a non-Euclidean setting, *Comm. Pure Appl. Math.*
            35 (1982), 531–554.

LICHNEROWICZ, A. AND WALKER, A.G.

1945   Sur les espaces Riemanniens harmoniques de type hyperbolique
            normal, *C. R. Acad. Sci. Paris* **221** (1945), 394–396.

LISSIANOI, S. AND PONOMAREV, I.

1997   On the inversion of the geodesic Radon transform on the hyper-
            bolic plane, *Inverse Problems* **13** (1997), 1053–1062.

LUDWIG, D.

1966   The Radon transform on Euclidean space, *Comm. Pure Appl.
            Math.* **17** (1966), 49–81.

MADYCH, W.R. AND SOLMON, D.C.

1988   A range theorem for the Radon transform, *Proc. Amer. Math.
            Soc.* **104** (1988), 79–85.

MARKOE, A.

2006   *Analytic Tomography*, Encyclopedia of Mathematics and its Ap-
            pliations, Cambridge Univ. Press.

MATSUMOTO, H.

1971   Quelques remarques sur les espaces riemanniens isotropes, *C. R.
            Acad. Sci. Paris* **272** (1971), 316–319.

MELROSE, R.B.

1995   *Geometric Scattering Theory,* Cambridge University Press, Lon-
            don and New York, 1995.

MICHEL, L.

1972   Sur certains tenseurs symétriques des projectifs réels, *J. Math.
            Pures Appl.* **51** (1972), 275–293.

1973   Problèmes d'analyse geométrique liés a la conjecture de Blaschke,
            Bull. Soc. Math. France **101** (1973), 17–69.

1977   Un problème d'exactitude concernant les tenseurs symmetriques
            et les géodésiques. *C.R. Acad. Sci.* Sér. A, **284**, (1977), 183–186.

MINKOWSKI, H.

1911   Über die Körper kostanter Breite, Collected Works, II, pp. 277–
            279, Teubner, Leipzig, 1911.

NAGANO, T.

1959   Homogeneous sphere bundles and the isotropic Riemannian man-
            ifolds, *Nagoya Math. J.* **15** (1959), 29–55.

NATTERER, F.

1986   *The Mathematics of Computerized Tomography,* John Wiley, New
            York, 1986.

NIEVERGELT, Y.

1986   Elementary inversion of Radon's transform, *SIAM Rev.* **28**
            (1986), 79–84.

NOVIKOV, R.G.

2002   An inversion formula for the attenuated X-ray transformation, *Ark. Mat.* **40** (2002), 145–167.

ÓLAFSSON, G. AND E.T. QUINTO

2006   The Radon Transform, Inverse Problems and Tomography, *Proc. Symp. Appl. Math.* Am. Math. Soc. 2006.

ÓLAFSSON, G., ET AL. (Eds.)

2008   Radon Transforms, Geometry and Wavelets, *Contemp. Math.* Vol. 464, Amer. Math. Soc. 2008.

OLEVSKY, M.

1944   Some mean value theorems on spaces of constant curvature, *Dokl. Akad. Nauk. USSR* **45** (1944), 95–98.

ORLOFF, J.

1985   Limit formulas and Riesz potentials for orbital integrals on symmetric spaces, thesis, MIT, Cambridge, MA, 1985.

1987   "Orbital integrals on symmetric spaces," in: *Non-Commutative Harmonic Analysis and Lie Groups,* Lecture Notes in Math. No. 1243, Springer-Verlag, Berlin and New York, 1987, 198–219.

1990a   Invariant Radon transforms on a symmetric space, *Contemp. Math.* **113** (1990), 233–242.

1990b   Invariant Radon transforms on a symmetric space, *Trans. Amer. Math. Soc.* **318** (1990), 581–600.

ORTNER, N.

1980   Faltung hypersingularer Integraloperatoren, *Math. Ann.* **248** (1980), 19–46.

PALAMODOV, V.

1996   An inversion method for the attenuated X-ray transform, *Inverse Problems* **12** (1996), 717–729.

2004   *Reconstructive Integal Geometry. Birkhauser,* Boston, (2004).

PALAMODOV, V. AND DENISJUK, A.

1988   Inversion de la transformation de Radon d'apres des données incomplètes, *C. R. Acad. Sci. Paris Sér. I Math.* **307** (1988), 181–183.

PALEY, R. AND WIENER, N.

1934   *Fourier Transforms in the Complex Domain,* American Mathematical Society, Providence, RI, 1934.

PATI, V., SHAHSHAHANI, M. AND SITARAM, A.

1995   The spherical mean value operator for compact symmetric spaces, *Pacific J. Math.* **168** (1995), 335–343.

PETROV, E.F.

1977   A Paley-Wiener theorem for a Radon complex, *Izv. Vyssh. Uchebn. Zaved. Math.* **3** (1977), 66–77.

POISSON, S.D.

1820   *Nouveaux Mémoires de l'Académie des Sciences, Vol.* III, 1820.

PRUDNIKOV, A.P., BRYCHKOV, YU. A., AND MARICHEV, O.I.

1990   *Integrals and Series, Vol.* I–V, Gordon and Breach, New York, 1990.

QUINTO, E.T.

1978 On the locality and invertibility of the Radon transform, thesis, MIT, Cambridge, MA, 1978.

1980 The dependence of the generalized Radon transform on defining measures, *Trans. Amer. Math. Soc.* **257** (1980), 331–346.

1981 Topological restrictions on double fibrations and Radon transforms, *Proc. Amer. Math. Soc.* **81** (1981), 570–574.

1982 Null spaces and ranges for the classical and spherical Radon transforms, *J. Math. Ann. Appl.* **90** (1982), 405–420.

1983 The invertibility of rotation invariant Radon transforms, *J. Math. Anal. Appl.* **91** (1983), 510–521; erratum, *J. Math. Anal. Appl.* **94** (1983), 602–603.

1987 Injectivity of rotation invariant Radon transforms on complex hyperplanes in $\mathbf{C}^n$, *Contemp. Math.* **63** (1987), 245–260.

1992 A note on flat Radon transforms, *Contemp. Math.* **140** (1992), 115–121.

1993a Real analytic Radon transforms on rank one symmetric spaces, *Proc. Amer. Math. Soc.* **117** (1993), 179–186.

1993b Pompeiu transforms on geodsic spheres in real analytic manifolds, *Israel J. Math.* **84** (1993), 353–363.

1994a "Radon transforms satisfying the Bolker assumtion," in: *Proc. Conf. 75 Years of Radon Transform,* Vienna, 1992, International Press, Hong Kong, 1994, 231–244.

1994b "Radon Transform on Curves in the Plane", in: *Lectures in Appl. Math.* No. 30, American Mathematical Society, Providence, RI, 1994.

2006 Support theorems for the spherical Radon transform on manifolds, *Intl. Math. Research Notes*, 2006, 1–17, ID 67205.

2008 Helgason's support theorem and spherical Radon transforms, *Contemp. Math.*, 2008.

RADON, J.

1917 Über die Bestimmung von Funktionen durch ihre Integralwerte längs gewisserMannigfaltigkeiten, *Ber. Verh. Sächs. Akad. Wiss. Leipzig. Math. Nat. Kl.* **69** (1917), 262–277.

RAMM, A.G.

1995 Radon transform on distributions, *Proc. Japan Acad. Ser. A Math. Sci.* **71** (1995), 205–206.

RAMM, A. AND KATSEVICH, A.I.

1996 *The Radon transform and Local Tomography,* CRC Press, Boca Raton, FL, 1996.

RENARD, D.

1997 Formule d'inversion des intégrales orbitales tordues sur les groupes de Lie réductifs réels, *J. Funct. Anal.* **147** (1997), 164–236.

RICHTER, F.

1986a Differential Operatoren auf Euclidischen k-Ebenräumen und Radon Transformationen, Dissertation, Humboldt Universität, Berlin, 1986.

1986b   On the k-Dimensional Radon Transform of Rapidly Decreasing
             Functions, in Lecture Notes in Math. No. 1209, Springer-Verlag,
             Berlin and New York, 1986.
1990   On fundamental differential operators and the p-plane transform,
             Ann. Global Anal. Geom. **8** (1990), 61–75.

RIESZ, M.
1949   L'integrale de Riemann-Liouville et le problème de Cauchy, Acta
             Math. **81** (1949), 1–223.

ROSS, K.
2004   From Helgason's number to Khintchine's inequality, Gazette
             Austr. Math. Soc. **29** (2004), 1–4.

ROUVIÈRE, F.
1983   Sur la transformation d'Abel de groupes des Lie semisimples de
             rang un, Ann. Scuola Norm. Sup. Pisa **10** (1983), 263–290.
1994a   Transformations de Radon, Lecture notes, Université de Nice,
             Nice, France, 1994.
2001   Inverting Radon transforms: the group-theoretic approach, En-
             seign. Math. **47** (2001), 205–252.
2004   Geodesic Radon transforms on symmetric spaces, preprint 2004.
2006   Transformation aux rayons X sur un espace symétrique. C.R.
             Acad. Sci. Paris Ser. I **342** (2006), 1–6.
2008a   X-ray transform on Damek-Ricci spaces, preprint (2008).
2008b   On Radon transforms and the Kappa operator, preprint (2008).

RUBIN, B.
1998a   Inversion of k-plane transform via continuous wavelet transforms,
             J. Math. Anal. Appl. **220** (1998), 187–203.
1998b   Inversion of fractional integrals related to the spherical Radon
             transform, J. Funct. Anal. **157** (1998), 470–487.
1999   Inversion and characterization of the hemispherical transform, J.
             D'Analyse Math. **77** (1999), 105–128.
2002   Helgason–Marchand inversion formulas for Radon transforms,
             Proc. Amer. Math. Soc. **130** (2002), 3017–3023.
2004a   Radon transforms on affine Grassmannians, Trans. Amer. Math.
             Soc. **356** (2004), 5045–5070.
2004b   Reconstruction of functions from their integrals over k-planes,
             Israel. J. Math. **141** (2004), 93–117.
2008   Inversion formulas for the spherical mean in odd dimension and
             the Euler-Poisson Darboux equation, Inverse Problems **24** (2008)
             No. 2.

SAWA, J.
1985   The best constant in the Khintchine inequality for complex Stein-
             haus variables; the case of $p = 1$, Studia Math. **81** (1985).

SCHAEFER, H.H.
1971   Topological Vector Spaces, Springer, Berlin 1971.

SCHIMMING, R. AND SCHLICHTKRULL, H.
1994   Helmholtz operators on harmonic manifolds, *Acta Math.* **173**
            (1994), 235–258.

SCHWARTZ, L.
1966   *Théorie des Distributions,* 2nd Ed., Hermann, Paris, 1966.

SEKERIN, A.
1993   A theorem on the support for the Radon transform in a complex
            space, *Math. Notes* **54** (1993), 975–976.

SELBERG, A.
1963   "Discontinuous groups and harmonic analysis," in: *Proc. Inter-
            national Congress Math.,* Stockholm, 1962, 177–189, Almqvist &
            Wiksells, Uppsala, 1963.

SEMYANISTY, V.I.
1960   On some integral transforms in Euclidean space, *Soviet Math.
            Dokl.* **1** (1960), 1114–1117.

1961   Homogeneous functions and some problems of integral geome-
            try in spaces of constant curvature, *Soviet Math. Dokl.* **2** (1961),
            59–62.

SHAHSHAHANI, M. AND SITARAM, A.
1987   The Pompeiu problem in exterior domains in symmetric spaces,
            Contemp. Math. **63** (1987), 267–277.

SHEPP, L.A., ET AL.
1983   "AMS Short Courses on Computerized Tomography," Cincinnati,
            OH, January, 1982, in: *Proc. Sympos. Appl. Math.* 27, American
            Mathematical Society, Providence, RI, 1983.

SHEPP, L.A. AND KRUSKAL, J.B.
1978   Computerized tomography: The new medical X-ray technology,
            Amer. Math. Monthly **85** (1978), 420–438.

SMITH, K.T. AND SOLMON, D.C.
1975   Lower-dimensional integrability of L^2 functions, *J. Math. Anal.
            Appl.* **51** (1975), 539–549.

SMITH, K.T., SOLMON, D.C., AND WAGNER S.L.
1977   Practical and mathematical aspects of the problem of reconstruct-
            ing objects from radiographs, *Bull. Amer. Math. Soc.* **83** (1977),
            1227–1270. Addendum *ibid* **84** (1978), p. 691.

SOLMON, D.C.
1976   The X-ray transform, *J. Math. Anal. Appl.* **56** (1976), 61–83.

1987   Asymptotic formulas for the dual Radon transform, *Math. Z.* **195**
            (1987), 321–343.

STEIN, E.M.
1970   *Singular Integrals and Differentiability Properties of Functions,*
            Princeton Univ. Press, 1970.

STRICHARTZ, R.S.
1981 L^p Estimates for Radon transforms in Euclidean and non-
            Euclidean spaces, *Duke Math. J.* **48** (1981), 699–727.

1992   Radon inversion-variation on a theme, *Amer. Math. Monthly* **89** (1982), 377–384 and 420–425.

SUNADA, T.

1981   Spherical means and geodesic chains on Riemannian manifolds, *Trans. Amer. Math Soc.* **267** (1981), 483–501.

SYMEONIDIS, E.

1999   On the image of a generalized d-plane transform on $\mathbf{R}^n$, *J.Lie Theory* **9** (1999), 39–68.

SZABO, Z.I.

1991   A short topological proof for the symmetry of two-point homogeneous spaces, *Invent. Math.* **106** (1991), 61–64.

TAKIGUCHI, T. AND KANEKO, A.

1995   Radon transform of hyperfunctions and support theorem, *Hokkaido Math. J.* (1995), No. 1, 63–103.

TITCHMARSH, E.C.

1948   *Introduction to the Theory of Fourier Integrals* 2nd Ed. Oxford Uni. Press, 1948.

TITS, J.

1955   Sur certains classes d'espaces homogènes de groupes de Lie, *Acad. Roy. Belg. Cl. Sci. Mém. Collect.* **29** (1955), No. 3.

TRÈVES, F.

1963   Equations aux dérivées partielles inhomogènes a coefficients constants dépendent de parametres, *Ann. Inst. Fourier (Grenoble)* **13** (1963), 123–138.

VOLCHKOV, V.V.

1997   Theorems on injectivity of the Radon transform on spheres, *Dokl. Akad. Nauk* **354** (1997), 298–300.

2001   Spherical means on symmetric spaces, *Mat. Sb.* **192** (2001), 17–38.

2003   *Integral Geometry and Convolution Equations*, Kluwer, Dordrecht, 2003.

WANG, H.C.

1952   Two-point homogeneous spaces, *Ann. of Math.* **55** (1952), 177–191.

WEISS, B.

1967   Measures which vanish on half spaces, *Proc. Amer. Math. Soc.* **18** (1967), 123–126.

WHITTAKER, E.T. AND WATSON, G.N.

1927   *A Course of Modern Analysis*, Cambridge University Press, London, 1927.

WIEGERINCK, J.J.O.O.

1985   A support theorem for the Radon transform on $\mathbf{R}^n$, *Nederl. Akad. Wetensch. Proc. A* **88** (1985), 87–93.

WOLF, J.A.

1967   *Spaces of Constant Curvature*, McGraw–Hill, New York, 1967.

YOSIDA, K.

1960   *Lectures on Differential and Integral Equations*, Interscience, New York, 1960.

ZALCMAN, L.

1980   Offbeat integral geometry, *Amer. Math. Monthly* **87** (1980), 161–175.

1982   Uniqueness and nonuniqueness for the Radon transforms, *Bull. London Math. Soc.* **14** (1982), 241–245.

ZHANG, G.

2009   Radon transform on symmetric matrix domains, *Trans. Amer. Math. Soc.* **361** (2009), 351-369.

ZHOU, Y.

2001   Two radius support theorem for the sphere transform, *J. Math. Anal. Appl.* **254** (2001), 120-137.

ZHOU, Y. AND QUINTO, E.T.

2000   Two-radius support theorems for spherical Radon transforms on manifolds. *Contemp. Math.* **251** (2000), 501–508.

ZHU, F-L.

1996   Sur la transformation de Radon de $M_{2,n}(\mathbf{H})$, *Bull. Sci. Math.* **120** (1996), 99–128.

Notational Conventions

Algebra As usual, $\mathbf{R}$ and $\mathbf{C}$ denote the fields of real and complex numbers, respectively, and $\mathbf{Z}$ the ring of integers. Let

$$\mathbf{R}^+ = \{t \in \mathbf{R} : t \geq 0\}, \quad \mathbf{Z}^+ = \mathbf{Z} \cap \mathbf{R}^+.$$

If $\alpha \in \mathbf{C}$, $\operatorname{Re}\alpha$ denotes the real part of α, $\operatorname{Im}\alpha$ its imaginary part, $|\alpha|$ its modulus.

If G is a group, $A \subset G$ a subset and $g \in G$ an element, we put

$$A^g = \{gag^{-1} : a \in A\}, g^A = \{aga^{-1} : a \in A\}.$$

The group of real matrices leaving invariant the quadratic form

$$x_1^2 + \cdots + x_p^2 - x_{p+1}^2 - \cdots - x_{p+q}^2$$

is denoted by $\mathbf{O}(p, q)$. We put $\mathbf{O}(n) = \mathbf{O}(o, n) = \mathbf{O}(n, o)$, and write $\mathbf{U}(n)$ for the group of $n \times n$ unitary matrices. The group of isometries of Euclidean n-space $\mathbf{R}^n$ is denoted by $\mathbf{M}(n)$.

Geometry The $(n-1)$-dimensional unit sphere in $\mathbf{R}^n$ is denoted by $\mathbf{S}^{n-1}$, $\Omega_n = 2\pi^{n/2}/\Gamma(n/2)$ denotes its area. The n-dimensional manifold of hyperplanes in $\mathbf{R}^n$ is denoted by $\mathbf{P}^n$. If $0 < d < n$ the manifold of d-dimensional planes in $\mathbf{R}^n$ is denoted by $\mathbf{G}(d, n)$; we put $\mathbf{G}_{d,n} = \{\sigma \in \mathbf{G}(d, n) : o \in \sigma\}$. In a metric space, $B_r(x)$ denotes the open ball with center x and radius r; $S_r(x)$ denotes the corresponding sphere. For $\mathbf{P}^n$ we use the notation $\beta_A(0)$ for the set of hyperplanes $\xi \subset \mathbf{R}^n$ of distance $< A$ from 0, σ_A for the set of hyperplanes of distance $= A$. The hyperbolic n-space is denoted by $\mathbf{H}^n$ and the n-sphere by $\mathbf{S}^n$.

Analysis If X is a topological space, $C(X)$ (resp. $C_c(X)$) denotes the sphere of complex-valued continuous functions (resp. of compact support). If X is a manifold, we denote:

$$C^m(X) = \left\{ \begin{array}{l} \text{complex-valued } m\text{-times continuously} \\ \text{differentiable functions on } X \end{array} \right\}$$

$$\begin{aligned}
C^\infty(X) &= \mathcal{E}(X) = \cap_{m>0} C^m(X). \\
C_c^\infty(X) &= \mathcal{D}(X) = C_c(X) \cap C^\infty(X). \\
\mathcal{D}'(X) &= \{\text{distributions on } X\}. \\
\mathcal{E}'(X) &= \{\text{distributions on } X \text{ of compact support}\}. \\
\mathcal{D}_A(X) &= \{f \in \mathcal{D}(X) : \text{ support } f \subset A\}. \\
\mathcal{S}(\mathbf{R}^n) &= \{\text{rapidly decreasing functions on } \mathbf{R}^n\}. \\
\mathcal{S}'(\mathbf{R}^n) &= \{\text{tempered distributions on } \mathbf{R}^n\}.
\end{aligned}$$

The subspaces $\mathcal{D}_H$, $\mathcal{S}_H$, $\mathcal{S}^*$, $\mathcal{S}_o$ of $\mathcal{S}$ are defined in Ch. I, §§1–2.

S. Helgason, *Integral Geometry and Radon Transforms*,
DOI 10.1007/978-1-4419-6055-9, © Springer Science+Business Media, LLC 2010

While the functions considered are usually assumed to be complex-valued, we occasionally use the notation above for spaces of real-valued functions.

The Radon transform and its dual are denoted by $f \to \widehat{f}$, $\varphi \to \check{\varphi}$, the Fourier transform by $f \to \tilde{f}$ and the Hilbert transform by $\mathcal{H}$.

I^α, I_-^λ, I_o^λ and I_+^λ denote Riesz potentials and their generalizations. M^r the mean value operator and orbital integral, L the Laplacian on $\mathbf{R}^n$ and the Laplace-Beltrami operator on a pseudo-Riemannian manifold. The operators $\square$ and Λ operate on certain function spaces on $\mathbf{P}^n$; $\square$ is also used for the Laplace-Beltrami operator on a Lorentzian manifold, and Λ is also used for other differential operators.

Frequently Used Symbols

Ad:	adjoint representation of a Lie group, 259
ad:	adjoint representation of a Lie algebra, 259
$\mathfrak{a}$, $\mathfrak{a}_*$:	abelian subspaces, 272
$\mathfrak{a}'$:	regular set, 272
$\mathfrak{a}^+$:	Weyl chamber, 272
A_x:	antipodal set to x , 147
$A(r)$:	spherical area, 148
$B_r(p)$:	open ball with radius r, center p, 8
β_R:	ball in Ξ, 28
B:	Killing form, 260
$\mathbf{C}_n$:	special set, 241
$C(X)$:	space of continuous functions on X, 1
$\mathcal{D}(X)$:	$\mathcal{C}_c^\infty(X)$, 1
$\mathcal{D}'(X)$:	space of distributions on X, 70, 223
dg_k:	invariant measure on G/K, 69
$\mathcal{D}_K(X)$:	set of $f \in \mathcal{D}(X)$ with support in K, 221
$\mathbf{D}(G)$:	algebra of left-invariant differential operators on G, 256
$\mathbf{E}(X)$:	set of all differential operators on X, 253
$\mathcal{E}'(X)$:	space of compactly supported distributions on X, 70
$F(X)$:	space of rapidly decreasing functions, 179
$\mathcal{F}(X)$:	space of exponentially decreasing functions, 179
$f \to \dot{f}$:	mapping from G to G/K, 73
$f \to \widehat{f}$, $f \to \widehat{f}_p$:	Radon transforms, 77, 114
$\varphi \to \check{\varphi}$, $\varphi \to \check{\varphi}_p$:	dual transforms, 77, 114
φ_λ:	spherical function, 130
$\mathbf{G}(d,n)$:	manifold of d-planes in $\mathbf{R}^n$, 34
$\mathbf{G}_{d,n}$:	manifold of d-dimensional subspaces of $\mathbf{R}^n$, 34
$\mathbf{H}^n$:	hyperbolic space, 78, 118
$\mathcal{H}$:	Hilbert transform, 22, 57
I^α:	Riesz potential, 199, 236
Im :	imaginary part
$L^1(X)$:	space of integrable functions, 18
$L = L_X$:	Laplace-Beltrami operator, 3, 185
$L(g) = L_g$:	left translation by g, 254
Λ:	operator on $\mathbf{P}^n$, 22, 43
M_p:	tangent space to a manifold M at p, 253
M^r:	mean value operators, 8, 209
$\mathbf{M}(n)$:	group of isometries of $\mathbf{R}^n$, 3
$\mathbf{O}(n)$, $\mathbf{O}(p,q)$:	orthogonal groups, 9, 187
Ω_n:	area of $\mathbf{S}^{n-1}$, 9

S. Helgason, *Integral Geometry and Radon Transforms*,
DOI 10.1007/978-1-4419-6055-9, © Springer Science+Business Media, LLC 2010

$\mathbf{P}^n$:	set of hyperplanes in $\mathbf{R}^n$, 2
$\mathfrak{p}$:	part of Cartan decomposition, 268
r^α:	special distribution, 237
Re:	real part
Res:	residue, 237
$\mathbf{S}^n$:	n-sphere, 3
$S_r(p)$:	sphere of radius r, center p, 8
$\mathcal{S}(\mathbf{R}^n)$:	space of rapidly decreasing functions, 5
$\mathcal{S}^*(\mathbf{R}^n)$, $\mathcal{S}_0(\mathbf{R}^n)$:	subspaces of $\mathcal{S}(\mathbf{R}^n)$, 10
$\mathcal{S}'(\mathbf{R}^n)$:	space of tempered distributions, 223
sgn(s):	signum function, 28
$\mathcal{S}_H(\mathbf{P}^n)$:	subspace of $\mathcal{S}(\mathbf{P}^n)$, 5
$\tau(x)$:	translation $gH \to xgH$ on G/H, 64
θ:	Cartan involution, 268
$U(\mathfrak{g})$:	universal enveloping algebra, 257
Ξ:	dual space, 63
x_+^α:	special distribution, 236
$\mathbf{Z}(G)$:	center of $\mathbf{D}(G)$, 259
$\sim$:	Fourier transform, 4, 226
$\,\widehat{}\,$:	Radon transform, incidence, 2, 68
$\vee$:	Dual transform, incidence, 2, 68
$\langle\ ,\ \rangle$:	inner product, 106
$\natural$:	K-invariance, 175
$\square$:	operator, wave operator, 3, 196
$\|\ \|$:	norm, 4

Index

S. Helgason, *Integral Geometry and Radon Transforms*,
DOI 10.1007/978-1-4419-6055-9, © Springer Science+Business Media, LLC 2010

Printed in the United States
By Bookmasters